Electron Bombardment Induced Conductivity
and its applications

Electron Bombardment Induced Conductivity
and its applications

by

The late W. EHRENBERG
Formerly Emeritus Professor of Physics
Birkbeck College
University of London, U.K.

and

D. J. GIBBONS
Central Research Laboratories
THORN EMI plc
Hayes, Middlesex, U.K.

1981

ACADEMIC PRESS

A Subsidiary of Harcourt Brace Jovanovich, Publishers
London   New York   Toronto   Sydney   San Francisco

ACADEMIC PRESS INC. (LONDON) LTD
24/28 Oval Road,
London NW1

United States Edition published by
ACADEMIC PRESS INC.
111 Fifth Avenue,
New York, New York 10003

Copyright © 1981 by
ACADEMIC PRESS INC. (LONDON) LTD

All rights reserved
No part of this book may be reproduced in any form by photostat, microfilm,
or any other means, without written permission from the publishers

British Library Cataloguing in Publication Data

Ehrenberg, W.
 Electron bombardment induced conductivity and its applications.
 1. Electric conductivity   2. Electron beams
 I. Gibbons, D. J.
 537.6'2   QC610.4

ISBN 0-12-233350-0
LCCCN 81-66385

Printed by W & G Baird Ltd at The Greystone Press, Antrim

Preface

This is the first book to be written in which the subject matter is dedicated exclusively to electron bombardment induced conductivity. The treatment presented is suitable for the graduate or final-year undergraduate physicist, materials scientist or electronic engineer whose interests lie in the electronic properties of insulators or semiconductors or in the design and development of electron devices. Recent developments such as time-of-flight techniques and the scanning electron microscope are dealt with and, as this is the first publication of its kind, we have not only reviewed these but we have found it necessary to examine critically the foundations upon which the more recent studies and applications are now standing. However, the reader might then have been left with an almost intolerable burden of work if we had merely provided a long list of references, so we have attempted to incorporate the information supplied by the relevant authors as part of the overall story.

In the early years, the influence on the electrical conductivity of an insulator or semiconductor while bombarded by fast electrons (electron bombardment conductivity, or EBC) was treated in its own right in a very similar way to the closely related phenomena of photoconductivity and cathodoluminescence. The studies were largely phenomenological, and the greater part of the first few chapters treats EBC from this standpoint. However, the subject has now progressed to the stage where a somewhat more theoretical approach is necessary for interpreting the observations correctly. Such is the case where transient EBC methods are used for characterizing an insulating or semiconducting material. Some workers call these drift mobility techniques, and two chapters are devoted to this important application of EBC. But derivation of most of the mathematical formulae has been omitted here.

The equivalent phenomenon to the photovoltaic effect is known as the electron voltaic effect (EVE). Large non-permanent changes in the conductivity of semiconductor barriers have been applied in electron devices, where an electron bombarded p–n junction or Schottky barrier structure in a semiconductor target has been used to achieve virtually noise-free current amplification. Such electron bombarded semiconductor targets were first used extensively in the 1970s, and further work can be expected soon when radiation hardened semiconductor devices become more common.

An important new application has arisen recently because of the scanning electron microscope. In the electron beam induced current mode, an image of sub-surface structures (such as bulk defects, or dislocations) can be examined by collecting the charge carriers released internally by the beam. The corresponding current is used as a video signal. A brief description of this application is made in the ultimate chapter.

The book was completely planned and most of the earlier parts had been written when the senior author died. His untimely death has left the scientific community the poorer for the loss of a brilliant scientist. His interest in the subject extended for almost 40 years and it continued right up to a few days before he passed away. Among his particular attributes was an insistence on understanding the underlying physical reason for certain patterns of EBC behaviour, and his ability to overcome irritating practical difficulties which always confront the experimentalist. The book includes a number of such seemingly small details which have been found important from experience. The basic philosophy behind the approach adopted in the book, and which was agreed by the authors at the outset, has been maintained throughout.

Grateful thanks are expressed to Professor W. E. Spear who read most of the manuscript and to whom the authors are indebted for many corrections and suggestions. Thanks are also due to Dr J. Hirsch and Dr E. W. Williams who assisted in reading some portions of the manuscript. Finally, the untiring assistance of Mrs M. Johnson of the THORN EMI central library, and of my wife Jean, with the references has been much appreciated.

June 1981   D. J. Gibbons

Contents

Preface

Chapter 1. Physical principles of electron bombardment conductivity (EBC)
   I. Historical background, 1
   II. Energy levels in solids, 5
      A. Crystalline solids, 5
      B. Amorphous solids, 9
      C. Localized energy states, 11
   III. Conduction of electricity through insulators, 17
      A. Conduction of electricity through gases, 17
      B. Transport of electrons and holes through crystals, 20
      C. Transport of electrons and holes through amorphous solids and low mobility crystals, 24
      D. Drift experiments in gases or in insulating solids and liquids, 32

Chapter 2. Properties of electron beams and their interaction with matter
   I. Sources of electrons, 34
      A. Radioactive sources, 34
      B. Electron guns for EBC measurements, 34
      C. EHT supplies, 41
      D. Concentration of intensity with electron lenses, 41
      E. Transmission of control, 42
      F. The scanning electron microscope (SEM), 43
   II. Path length and energy loss of fast electrons, 43
      A. Cloud chamber tracks, 44
      B. Thomson–Whiddington–Bethe law of energy losses, 44
      C. The problem of penetration, 49
      D. Experimental measurements using phosphorescence glows, 54
      E. Scaling of variously defined ranges, 59
      F. Range–energy tables, 60
      G. Range measurements on primary electrons of energy below about 10 keV, 61
   III. Ionizing radiation and radiation effects, 65
      A. Production of electron–hole pairs, 65
      B. Luminescence, 67

　　　　C. Photoconduction, 67
　　　　D. Photovoltaic effect, 70
　　　　E. Radiation damage, 71
　　　　F. Cosmic rays, 75
　　　　G. Nuclear radiation, 77
　　　　H. Persistent polarization, 78

Chapter 3. Steady state EBC of thin insulating films
　　I. Preparation of films, 80
　　　　A. Evaporation, 80
　　　　B. Sputtering, 81
　　　　C. Glassy bubbles, 81
　　　　D. Deposition by glow discharge, 82
　　　　E. Anodizing and chemical deposition, 83
　　　　F. Other methods, 84
　　II. Specimens for measurements, 84
　　　　A. Solid electrodes, 86
　　　　B. Electron beam contacts, 86
　　III. Transfer of excitation, 89
　　　　A. X-rays, 89
　　　　B. Excitons, 90
　　　　C. Fluorescence, 91
　　IV. Conductivity induced by electrons, 92
　　　　A. Influence of bombarding voltage, 93
　　　　B. Gain, 103
　　　　C. Lateral induced conductivity, 114
　　　　D. The influence of contacts in normal EBC, 121

Chapter 4. The electron voltaic effect (EVE)
　　I. Semiconductor junctions, 123
　　　　A. Point contacts, 125
　　　　B. Schottky barriers, 125
　　　　C. p–n junctions, 128
　　II. Practical junctions, 129
　　　　A. Theory of the EVE, 129
　　　　B. Diffusion length of conduction electrons in silicon solar cells, 132
　　III. Particular semiconductor junctions, 133
　　　　A. Selenium, 133
　　　　B. Gallium arsenide, 135
　　　　C. Silicon, 135
　　　　D. EBC avalanche effect, 137

Chapter 5. Transient EBC time-of-flight measurements
- I. Time-of-flight technique, 141
  - A. Background of similar work prior to 1957, 142
  - B. Measurements on a crystal in its virgin state, 143
  - C. Types of transient EBC measurements, 144
  - D. Comparison with light-flash or α-particle excitation, 158
  - E. Insulators having a short carrier free lifetime, 160
  - F. Semiconductors having a short dielectric relaxation time, 161
  - G. Signal averaging, 164
- II. Transient EBC methods for measurements other than drift velocity, 165
  - A. Transient EBC pulse shape near $t = 0$, 165
  - B. Electric field profiles, 166
  - C. Carrier recombination, 166
  - D. Mean free carrier lifetime, 168
  - E. Amorphous semiconductors and the continuous time random walk, 170
  - F. Trap distributions, 173
  - G. Carrier release time from traps, 175
- III. Carrier velocity, 177
  - A. Mobility of either sign of carrier, 177
  - B. Trap-controlled mobility, 180
  - C. Trap density, 181
  - D. The motile trap model, 182
- IV. Apparatus, 183
  - A. Medium and low mobility insulating solids and liquids, 183
  - B. Solid gases, 186
  - C. High mobility solids, 188
  - D. Microwave time-of-flight technique, 192

Chapter 6. Specific materials properties determined by transient EBC techniques
- I. Group IV elemental semiconductors, 196
  - A. Silicon, 196
  - B. Germanium, 201
  - C. Diamond, 206
- II. Amorphous silicon, 208
- III. III–V intermetallic compound semiconductors, 222
  - A. Gallium arsenide, 223
  - B. Indium antimonide, 226
- IV. II–VI compounds, 228
  - A. Cadmium sulphide, 228

      B. Cadmium telluride, 235
      C. Other II–VI solids, 241
  V. Solid and liquid gases, 244
      A. Noble gases (neon, argon, krypton, and xenon), 244
      B. Molecular gases (nitrogen, oxygen, carbon monoxide, hydrogen, and methane), 250
 VI. Selenium and sulphur, 254
      A. Selenium, 254
      B. Sulphur, 256

Chapter 7. Applications of EBC and EVE in electron devices
   I. Devices using semiconductor junctions, 265
      A. Photosil EVE hybrid multiplier photocell, 265
      B. Digicon multichannel photocell, 269
      C. Intensified charge coupled devices (ICCD) and self-scanned arrays, 274
      D. Silicon intensifier target (SIT) television pickup tube and SIT scan converter, 279
      E. Evoscope fixed pattern generator, 284
      F. Electron bombarded semiconductor (EBS) microwave devices, 285
      G. Degradation phenomena, 291
  II. Miscellaneous applications of bombarded semiconductor targets, 294
      A. Barrier EVE and the scanning electron microscope, 294
      B. Nuclear radiation detectors, 305
 III. EBC of insulating films in electron devices, 311
      A. Ebicon (Ebitron, Uvicon) television pickup tube, 311
      B. Computer mass memory (Beamos, Ebam), 313
      C. Graphechon scan converter, 317

Appendix
   I. Two-layer dielectric, 321
  II. Secondary emission of insulators, 322
 III. Optimum scanning speed for constant charge imaging device, 323
 IV. Ramo's theorem, 325

References, 327

Index, 340

1
Physical principles of electron bombardment conductivity (EBC)

I. Historical background

The study of EBC is concerned with the conduction of electricity across gaps between electrodes. The gaps may be filled by a non-metallic solid and only a very small current can flow in the presence of an applied electric field if the electrodes are blocking. However, conductance can often be influenced by radiation, particularly electromagnetic rays, cathode rays, and nuclear particles. The effect of the first—photoconductivity—was discovered about 100 years ago and for many years it was considered to be the property of only a few substances. In Glazebrook's classic *Dictionary of Applied Physics* (1922) for example we read: "Arrhenius (1887) showed that by exposing silver halides to light an increase in electric conductivity was produced . . . various other substances have been studied by Coblentz and Emerson. The well-known sensitiveness of selenium to light, shown by a diminution in electrical resistance when thin films are illuminated, may be explained by supposing that the *incident light liberates slowly moving electrons* which remain within the substance and increase its conducting power". Actually the generation of an emf in a galvanic cell when one of its plates was illuminated by violet light was discovered by E. Becquerel (1839) in the form we now know as the photovoltaic effect. The photoconductive effect was discovered by Willoughby Smith (1873) who accidentally came across it in selenium. In 1916 it was discovered in cuprous oxide, and from then onwards a systematic search led to the discovery of many more photoconductive materials, in particular sulphides and coloured alkali halides.

Studies of the influence of X-rays and nuclear particles from sources such as uranium or radium were made towards the end of the 19th century. Within a year of Röntgen's (1895) discovery of X-rays, J. J. Thomson showed that the paraffin wax dielectric of a condenser became conducting under their influence. Curie showed that radium can cause insulating liquids to become conductive (1895), and H. Becquerel found that uranium can produce a similar effect in gases (for early history of these effects see

Becquerel, 1903). The first recorded account of measurements with cathode rays (as opposed to β-rays) was by Becker (1904) who bombarded paraffin wax with 25 kV electrons produced from a gas discharge.

Phosphors are excited with ultraviolet light and are also efficiently excited by cathode rays—this, of course, is utilized in TV receivers. Excitation of luminescence was probably the first property of cathode rays to be discovered. In 1859 T. Plücker of Bonn reported the "beautiful and mysterious green glow" produced by discharges in tubes exhausted by means of Geissler's mercurial pumps. Some years later Hittorf (1869) saw that this glow is caused by rays which proceed in straight lines and are intercepted by a solid placed in their path. Soon afterwards it was found that luminescence in some minerals is far more striking than that of glass.

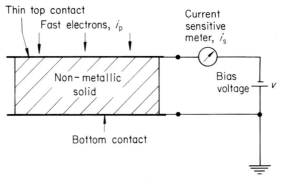

Figure 1.1. Basic circuit for measuring EBC. Here the conduction currents are through the thickness of the specimen and the beam of fast primary electrons is parallel to them; other orientations are possible. By convention, i_p is the primary (bombarding) current and i_s is the induced current.

It is almost obvious today that electrons should cause these effects (Fig. 1.1). For all we know of cathode rays, they ought to do everything of this nature that electromagnetic radiation does, only more so and less selectively. Before the advent of the laser, strong light carried 1 W cm^{-2} but electrons could be readily made to carry 1 kW cm^{-2}. However, what must astound us is that we had to wait until 1948 for the discovery of an effect corresponding to photoconductivity. True, a few papers had been published which reported a search for the effect. In photoconduction, the measure of the efficiency is the number, or fraction, of electronic charges carried round the circuit per absorbed photon. The corresponding measure for cathode rays, often called gain, g, is obtained by substituting the number of electrons impinging on the specimen for the number of photons. Since each electron can deposit in the specimen many thousand times the energy of a photon, we should expect gain values of this order (i.e. several thousand), but early reports by Becker, de Laer Krönig (1924), Lenz (1925), Distad (1938), and Bloembergen (1945) speak of gains as small as 10^{-5}–10^{-4} or even less. This is about eight orders of magnitude short of expectation; * so, what had gone wrong?

Renewed interest in studies of EBC came independently from two sources. In 1937 one of the present authors (W.E.) was developing high-voltage cathode ray tubes in the research laboratories of EMI Limited in England, and with a colleague found that the normally insulating phosphor screens became appreciably conducting when exposed to fast electrons. They developed an experimental high-voltage cathode ray tube with a transparent conducting faceplate (see Ehrenberg, 1940). The same principle was employed to stabilize the screens of their infrared image converter tubes (1942–5) whose glass faceplate was coated with a fine-pitch sputtered platinum grid.

A similar aim was pursued in the research laboratories of RCA at Princeton by Epstein and Pensak, but their quest was for a conducting metallic film to be deposited on the phosphor rather than on the faceplate. After experimenting with numerous metals and different techniques, they eventually came up with the method (1946) that is still used. A modern television cathode ray tube with its aluminized screen is probably the most familiar device today that makes use of the EBC effect. The first numerical results by Pensak for the steady-state EBC of thin films of an insulator were presented orally on 1st May 1948; the gain of a silica film was more than 100. The first results by Ehrenberg, by then assistant director of research at Birkbeck College, London, were demonstrated by a research student of his, Frederick Ansbacher. On 1st July 1948, at the official opening of their new biomolecular research laboratory, he demonstrated his apparatus which at that time was measuring an EBC gain of about 30 on thin films of As_2S_3.

The second source of impetus towards a revival of interest came about as a result of a thesis, then unpublished, by van Heerden (1945). Using melt-grown crystals of AgCl and improved electronic pulse amplifiers of the type used by Kosmata and Huber (1941), he found considerable response from single α-particles. This encouraged a group at the Bell laboratories in the U.S.A. to study a variety of crystals to find those most suitable for nuclear particle counters. Investigations under α-particle excitation showed that a number of crystals responded, but some diamond chips yielded a particularly strong output; the especially pronounced sensitivity of some diamonds had previously been reported by Stetter (1941). One such diamond was subjected to electron pulse excitation by McKay in a special sealed-off

---

* de Laer Krönig's results form an exception; he exposed a selenium photoconductive cell of conventional design to low-energy cathode rays of up to 100 eV energy and found an effect comparable to that produced by visible light.

cathode ray tube. On 31st January 1948, at the Columbia University meeting of the American Physical Society, McKay reported pulse gains of up to 160. His paper inspired Rittner of the Philips research laboratories at Irvington U.S.A., at that time unaware of Krönig's 1924 results, to measure the EBC of a selenium photoconductive cell under bombardment by 2 keV electrons; the gain was 123.

Ehrenberg, Pensak, McKay, and Rittner were some of the first people to suggest applying the EBC effect to electron devices. In December 1948 McKay published his first results for diamond, and in February 1949 Pensak reported his for silica. In early 1949 Ansbacher and Ehrenberg published initial results for the steady-state EBC of thin films of As_2S_3 and Al_2O_3 and reported gains of over 1000.

These groups independently came to the conclusion that at least part of the failure of the earlier investigators was due to their underestimating the magnitude of trapped space charges which can occur in dielectrics. McKay minimized space charge development in his diamonds under electron bombardment by applying this in short pulses and by using an alternating field. Ansbacher and Ehrenberg let cathode rays fall on thin films of dielectrics and obtained high values of gain under steady bombardment if the energy of the electrons exceeded a threshold. The thin-film threshold bombarding voltage was interpreted as the minimum for the exciting electrons to penetrate the whole thickness of the specimen and thus permanently release any trapped charges. Pensak performed similar experiments and came to the same conclusion. McKay eventually reported gains of up to about 1000, Pensak and Rittner of about 100, and Ansbacher and Ehrenberg, who studied the effect over a wide range of temperature, obtained gains of up to 500 for the primary effect and 40 000 for the secondary effect.

The experiments made it obvious that electron beams are a potent means of liberating carriers in insulators and semiconductors, and thus of studying their electronic properties independently of the restrictions that optical excitation imposes, such as high selectivity and exponential absorption. The wide range of their power has already been mentioned. The high values of EBC gain immediately suggested applications in which the insulator or semiconducting target could be made the current amplifying structure of an electron device. Another important feature is the relative ease with which controlled pulses of electrons down to nanosecond length and shorter can be produced. Since the rate of ionization caused by cathode rays is almost independent of the properties of the material traversed, it might have been expected that all materials would respond to electron bombardment. This, however, was not found to be the case, and the first successful materials were undoubtedly selected by trial and error, as well as consideration of their mechanical and physical suitability.

So the seed was sown for studies of the curious behaviour of dielectrics and semiconductors exposed to fast electrons, the application of the EBC effect in electron devices, and the use of electron beams as a tool in the service of the science of non-metallic materials. Today, on the foundations laid during studies spanning the period from about 1948 to 1970, the scanning electron microscope in the charge collection mode and the transient EBC drift mobility techniques have become standard methods for determining materials properties (see Holt *et al.,* 1974; Spear, 1969; Ottaviani *et al.*, 1975; Canali *et al.*, 1976; Leamy *et al.*, 1978).

II. Energy levels in solids

The cohesive forces that bind the atoms, ions, or molecules from which a solid are built cause the energy levels associated with the isolated components to be modified. If the solid is a perfect crystal—a convenient theoretical concept but something that never actually exists—these modifications are well defined. Crystal defects can often be described in a fairly precise manner, and thus the perfect crystal with defects can be used as a model for which several patterns of behaviour can be predicted and put to experimental tests.

In about 1959 amorphous solids came under study; liquid semiconductors posed the same theoretical problem. The concepts that had been applied so successfully to crystals had to be severely modified, but all the time there was the constraint that any new concept introduced to take into account the random nature of the packing had to allow for the fact that many of the macroscopic properties such as colour, dielectric constant, density, and work-function were not vastly different from their crystalline counterparts. Many well known EBC materials such as amorphous selenium and As_2Se_3 are glassy, so their importance to an understanding of EBC phenomena is paramount. An understanding of the crystalline solid is important for junction phenomena. Both are important for interpretation of carrier transport properties, which are measured extensively by transient EBC methods.

The question of the available energy levels in crystalline solids and in amorphous solids will now be considered, and their respective modes of charge carrier transport in Section III.

A. Crystalline solids

As a solid is built up from its constituent atoms or molecules, interactions between them cause their atomic (or molecular) energy levels to be bunched into bands. The energy band structure of a solid is one of the factors that influence the transport of electrons. The quantum mechanical analysis of the available states in a solid also allows for the existence of holes. At one time, the motion of holes was looked upon as the motion of electrons in the opposite direction, but this could not explain why the behaviour of holes is an exact mirror image of that of electrons. Today we look upon holes as precisely equivalent to electrons in questions of electrical conduction in condensed phases within the framework of the quantum theory.

Briefly the argument is this. One electron in a periodic structure is described in terms of a wavefunction, ψ. This satisfies the Schrödinger equation, $H\psi = E\psi$, where H is the Hamiltonian operator describing the kinetic energy and the interaction between the electron and its surroundings, and E is the energy of the electron. The solutions that satisfy the Schrödinger equation must be periodic with the lattice. This means that ψ must be of a form having antinodes at every lattice point—a way of reiterating the equivalence of all points from which the periodic structure is built. These are known as the Bloch solutions of the wave equation.

A more accurate statement of the quantum mechanical description of a hole is that the "velocity" of a hole is the same as that of a missing state; the wavepacket constructed from the determinant containing all wavefunctions in a band (except the missing state) will be carried through space at the same rate as its neighbours on either side. Similarly, it can be shown that the acceleration of a hole by an electric field is the same as that of a positive charge having the same effective mass as a particle in the range normally associated with electrons.

Assuming the band structure of solids, if all the allowed energy states in the highest occupied band are filled, and there is an energy gap of several electron-volts between the highest energy state in the filled band and the lowest energy state in the band above it, the material is an insulator. For example, in the insulating crystal NaCl, the partly occupied 3p energy levels in the free Cl^- ions are broadened into an energy band whose occupancy is completed by the 3s electrons of the Na^+ ions. Although individual electrons can pass from ion to ion, no net state of motion can take place and the crystal fails to conduct, merely because there are few free charge carriers.

In a real solid, the existence of lattice irregularities gives rise to localized energy levels lying between the filled band and the next empty band. These irregularities might arise by deformation of the lattice, the existence of impurity atoms, or a stoicheiometric excess of one component in a compound lattice. In Fig. 1.2, F is the filled band (known as the valence band) and C is the next higher unoccupied band (known as the conduction band) in a typical insulating solid.

A substitutional impurity, such as a pentavalent atom substituted for a tetravalent carbon atom in the lattice of a diamond crystal, gives rise to a

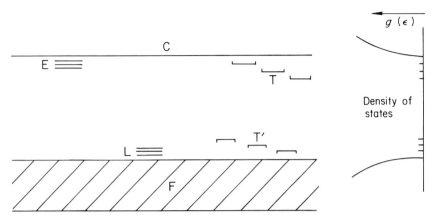

Figure 1.2. Band structure in a crystalline insulator or semiconductor. There is a well defined energy gap between the full valence band F and the empty conduction band C. Local impurities, such as a pentavalent or a trivalent atom in a substitutional position in the diamond lattice, can give rise to donor levels E near C, or acceptor levels L near F, respectively. Lattice defects or neutral impurities can act as traps, shown as T and T', lying within the band gap. The density of states distribution is shown on the right.

localized energy level or levels near C; these are shown at E on the diagram. A substitutional trivalent atom similarly gives rise to a vacant energy level or levels L in the neighbourhood of F.

A typical semiconductor, such as silicon or gallium arsenide, is a covalent structure based on the diamond-type crystal lattice. The band gap of a semiconductor is smaller than that of an insulator (e.g. in the range about 1·5–0·2 eV), and the localized levels E or L arising from certain kinds of substitutional pentavalent or trivalent impurity are thermally ionized at room temperature. In the former case we have an n-type semiconductor (carriers predominantly negative) and in the latter a p-type semiconductor (carriers predominantly positive).

Other types of lattice impurity centres can give rise either to the one type of localized energy level or to the other in the same crystal, i.e. occupied levels near F and unoccupied levels near C, depending on the degree of coordination of the impurity atom.

Thermal excitation or infrared excitation can raise an electron from E to C, leaving behind an ionized impurity centre. Under the influence of an electric field this electron can now move through the body of the crystal and cause electrical conduction. Similarly, an electron may be excited from F to L, leaving a vacant energy level in F. This now permits the passage of a positively charged hole in the opposite direction. Once in the conduction band, an electron might fall back to an unoccupied level E which thus sometimes acts as an electron trap. Similarly, the levels L might act as a hole trap when occupied by an electron; in this case, the trapping of a hole can be regarded as a downward transition of an electron to F.

Lattice defects such as irregularities in the normally periodic structure of the lattice or substitutional or interstitial neutral impurities can also produce localized levels between F and C by distorting the internal crystal fields. These levels act as traps. Levels due to neutral impurities or non-stoicheiometric structures occur less frequently in a carefully prepared crystal, but the density of point defects increases exponentially with the absolute temperature at which it was grown. These levels are shown at T and T' in Fig. 1.2.

Excitation of the lattice by irradiation with light, or by bombardment with electrons, nuclear particles, or X-rays, may cause the transition of an electron from F to C (Fig. 1.2).

Once in C, an electron may return either to F or to L or be trapped by one of the levels T. Electrons that are not trapped may pass through the body of the crystal under the influence of an applied electric field until they arrive at one face; here they may escape from the crystal completely if contact is made to that face. Electrons that are trapped may be released subsequently by an upwards transition to C by thermal excitation or by the absorption of infrared radiation. Alternatively, the electron might make a downwards transition to L or F depending on the depth of the trap, and the cross-section for capture.

The law governing the transition of an electron from one level to another is that the total momentum shall be the same before and after the transition. The downward transition of an electron from C or E to L or F is often accompanied by the emission of a quantum of radiation. In order to conserve momentum, the electron may now emit or absorb a phonon (a quantum of lattice vibration) or a number of phonons.

If an electron can undergo a transition between two energy bands without a change in momentum, these are known as "direct gap" transitions. Of course, in all circumstances the total momentum (electron + lattice) is conserved, but in many materials non-direct or indirect transitions dominate the observed optical excitations. Kurtin *et al.* (1969) drew attention to the importance of the difference in electronegativity of the constituent atoms from which the lattice is composed (see also Mead and McGill, 1976). It was found empirically that, if the difference exceeds about 0·6, the single-particle crystal momentum either is not an important selection rule ("non-direct") or is conserved only with phonon participation ("indirect"). If the electronegativity difference is less than this, most transitions are direct. Figure 1.3 shows how the relative contribution of non-direct transitions depends on electronegativity difference.

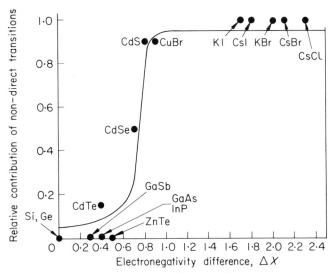

Figure 1.3. Relative contribution of non-direct to direct transitions in a solid as a function of electronegativity difference between its constituent atoms ΔX (from Pauling, 1960). (Kurtin *et al.*, 1969)

If a localized state such as a recombination centre is involved, a transition from a band to the localized state is more likely than a band–band transition. Thus transitions between E or L and F or C are more probable than those directly between C and F.

B. Amorphous solids

The atoms or molecules condensed into a glassy, liquid, or amorphous semiconductor or insulator are packed in a more or less random manner about a mean separation distance. In place of the precise unit cell describing a crystal, we now use a radial distribution function $4\pi r^2 \rho dr$. This is the average number of pairs of atoms separated by a distance lying between r and $(r + dr)$. It is normally assumed that the atomic density is uniform over a given radius. Figure 1.4 shows the difference between the radial distribution functions for crystalline and amorphous silicon; in this example it is very prominent how the density of next-but-two nearest neighbours is a peak in the crystalline whereas it is a minimum in the amorphous form. The main result of this randomized structure on the available energy states is to give rise to a finite density in the band gap, "tail states" at the band edges [Fig. 1.5(a)], and also to localized regions of higher density which arise from the presence of units in metastable lattice positions [Fig. 1.5(b)]. As shown later, localized states can appear and disappear if a charge carrier moves slowly

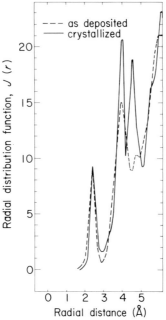

Figure 1.4. The radial distribution function for an amorphous silicon film 100 Å thick, before and after recrystallization. (Moss and Graczyk, 1970)

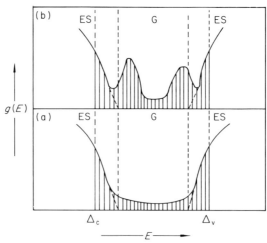

Figure 1.5. Schematic diagrams of the density of states distribution in an amorphous solid: (a) a "perfect" amorphous semiconductor or a glass; (b) an amorphous semiconductor with defect states. ES = extended states; G = gap states; Δ_c and Δ_v mark the limits of the extended states, the gap between them is called the mobility gap, and the localized states between them is shaded. (Owen and Spear, 1976)

through the disordered lattice. A specific example is also given later for amorphous silicon (a-Si) which has two (permanent) local regions of higher density (see Fig. 1.11). The number of regions of higher density of states may be as high as three for amorphous selenium and four for amorphous As_2Se_3 (Owen and Spear, 1976).

A review of electrical and optical properties of amorphous semiconductors was published by Brodsky (1979), and a more theoretical treatise on the subject in the book by Mott and Davis in the same year.

C. Localized energy states

1. Traps

Localized energy states which can capture pseudo-free carriers and from which they can only be released subsequently by thermal transition into an appropriate band have already been mentioned and are shown at T and T' in Fig. 1.2. Such centres are electrically neutral and can arise by the presence of impurities, misplaced atoms or molecules, mechanical strain, or unionized acceptor or donor impurities. Other methods of emptying traps are absorption of infrared radiation or recombination of a carrier of the opposite sign from a band.

Knock-on misplaced atoms can be produced by irradiation of the solid by neutrons, β-rays, or γ-rays, but there is a threshold energy below which bulk damage cannot occur directly. Section III.E of Chapter 2 will discuss the types of radiation damage that can occur. Konopleva et al. (1966), Sevchenko et al. (1967), and Milevskii and Garnyk (1979) have cited a number of electron or hole traps in single-crystal silicon mainly caused by vacancy-interstitial pairs due to energetic irradiation.

A trap is a localized state which may hold a charge carrier temporarily and from which its release thermally into the appropriate band is more likely than a transition in the opposite direction and so causing recombination. Since the transport of carriers may be regarded as dominated by random thermal motion, the rate of trapping can be simply expressed as the product of the carrier thermal velocity, the trap cross-section, and the trap density:

$$-(1/n)dn/dt = u_{th}\sigma N_t. \tag{1.1}$$

where n is the free electron density in the conduction band, u_{th} is their thermal velocity, σ is the trapping cross-section, and N_t is the trap density. This equation applies whether the electrons are drifting under the influence of an applied electric field or not because, at all drift velocities below about 10^6 cm s^{-1}, the steady drift is only a small perturbation on the steady-state thermal distribution. Similar considerations apply to holes.

Two features concerning the numerical value of σ are worth mentioning at this point. (i) The value of σ may change once a carrier is trapped, provided that enough time elapses to allow the lattice to relax around the trapping centre to take into account the introduction of an extra charge. This time is of the order of a period of lattice vibration or, say, about 10^{-12} s. (ii) The value of σ can be changed by a strong electric field, and field-assisted detrapping is sometimes important at fields above about 10^4–10^5 V cm^{-1} (Poole–Frenkel effect). The secondary EBC effect and secondary photoconductivity can both be explained in terms of one sign of carrier being trapped in centres having a high capture cross-section, and thereafter a low recombination cross-section for carriers of the opposite sign. Carriers trapped in centres of this kind (class II centres) thus assist in entry of carriers from the electrodes, and a secondary "gain" arises through this effect. It is easily shown that the secondary gain in EBC or photoconductivity is numerically given by

$$g' = t_1/T \tag{1.2}$$

where t_1 is the recombination lifetime of the trapped carrier and T is the transit time of the carrier of the opposite sign. Secondary gain must not be confused with EBC gain which can easily exceed unity even in the absence of the entry of carriers from the electrodes.

Rose (1951) showed how σ is related to the thermal release rate of trapped charges. All considerations mentioned in the preceding paragraph are not included, so this analysis is not universally applicable, but it can be applied in many circumstances. The reasoning is as follows. In thermal equilibrium, the rate of trapping into a particular group of centres must equal the rate of release. Also, under steady-state conditions, at a temperature θ the distribution of carriers between the conduction band and traps at a distance ΔE below it is given by the Maxwell–Boltzmann distribution, $n/n_t = (N_c/N_t)\exp(-\Delta E/k\theta)$, where N_c and N_t are the effective density of states in the conduction band and the density of the traps respectively. By detailed balance, it then leads directly to a rate of thermal release given by

$$-(1/n)\mathrm{d}n/\mathrm{d}t = N_c u_{\mathrm{th}}\sigma\exp(-\Delta E/k\theta) \tag{1.3}$$

where the pre-exponential factor has the dimensions of frequency and is often termed "attempt-to-escape frequency". It is regarded as the product of the highest lattice frequency multiplied by a factor less than or equal to unity. A theoretical value was calculated by Kubo (1952).

If we rearrange equation (1.3) and write f for the factor $N_c u_{\mathrm{th}}\sigma$, we obtain an expression for the depth of trap that will empty thermally as a function of temperature, using the mean emptying time t'' as a parameter. Thus

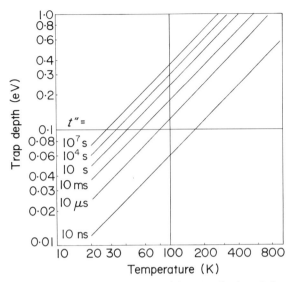

Figure 1.6. Relationship between the trap depth ΔE (eV) and the temperature θ (K) for which the thermal release time has a given value t''. This is calculated from $\Delta E = k\theta \ln(t''f)$, where a value of $f = 10^{11}$ s^{-1} is assumed.

$\Delta E = k\theta \ln(t''f)$. This expression is plotted in Fig. 1.6 using a value of 10^{11} for f. The sulphide phosphors (e.g. ZnS) have an unusually small value for f, this being in the region of 10^8 s^{-1}.

2. Recombination centres

Again localized impurities in an otherwise perfect lattice can give rise to energy states lying in the gap, but in the case of recombination centres they are distinguished by encouraging the simultaneous trapping of a hole and that of an electron. This kind of recombination centre is thus a two-level centre and the subsequent recombination process is sometimes accompanied by emission of energy in the form of light (Prenner–Williams model of luminescence, 1956). Momentum is conserved by emission of a phonon or a number of phonons.

Light emission is not always associated with recombination. Transition metals in Si are well known impurities causing a reduction of mean free carrier lifetime, and Ni and Co are well known impurities causing carrier recombination without light emission in the sulphide phosphors or a reduction in carrier lifetime in GaAs or Ga(As,P) (Partin *et al.*, 1979). Such impurities are intentionally added to a semiconductor or a phosphor to modify its properties in a controlled way.

14   ELECTRON BOMBARDMENT INDUCED CONDUCTIVITY

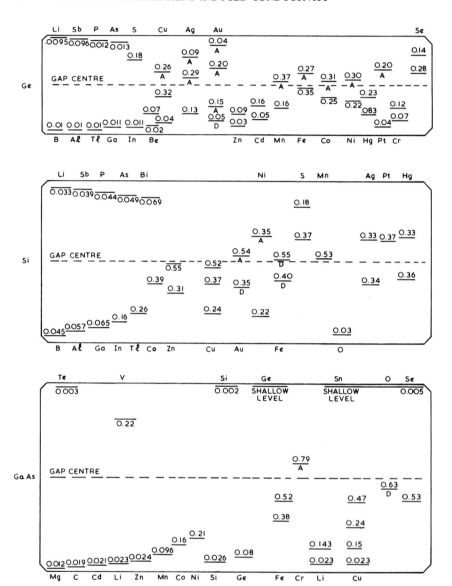

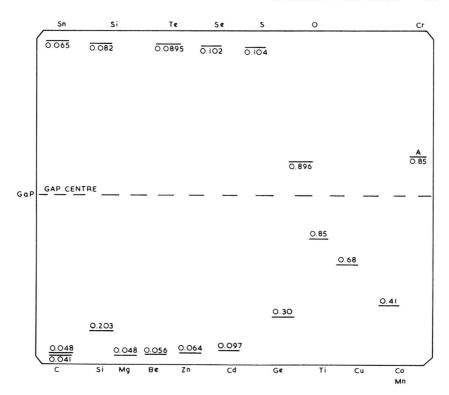

Table 1.1. Impurity energy levels for various elements in Ge, Si, GaAs, and GaP. The levels below the gap centres are measured from the top of the valence band and are acceptor levels unless indicated by D for donor levels. The levels above the gap centres are measured from the bottom of the conduction band level and are donor levels unless indicated by A for acceptor levels. The band gaps at 300 K are 0·67, 1·13, 1·42, and 2·24 eV for Ge, Si, GaAs, and GaP respectively. (Partin *et al.*, 1979; Milnes, 1973; Sze and Irvin, 1968)

3. Luminescence centres

A phosphor is usually prepared in a highly purified form and controlled amounts of impurities are added to provide the right colour and persistence of light emission when an electron and a hole recombine at one of these centres. The luminescence might also arise through a self-activation process which is usually the result of native defects. Usually an "activator" and "co-activator" are added in such a way that charge neutrality in the normally stoicheiometric base material is maintained.

If we now refer to the energy level scheme shown in Fig. 1.2, luminescence can be produced by the downwards transition of a trapped electron in one of the levels T to a vacancy in F (a free hole); this is known as the Lambe–Klick model (1955). Alternatively, the Schön–Klasens model (Schön, 1951; Klasens, 1953) allows recombination of a free electron in C with a trapped hole in T'. The third model, that of Prenner and Williams, has been described above in Section 2. Transitions involving a localized level have a greater probability than transitions involving an extended band, and for this reason the direct transition of an electron from C to F is swamped by those in which localized states are an intermediary, except of course in very perfect intrinsic semiconductors.

4. Donors and acceptors

A brief description of a trivalent substitutional impurity atom in a diamond-type lattice and a pentavalent substitutional impurity in a similar lattice has already been given in Section A. If these impurities form shallow centres they are thermally ionized at room temperature and the solid becomes a p-type semiconductor (carriers predominantly positive) or an n-type semiconductor (carriers predominantly negative) respectively. The ionization energies of the useful impurities of this type are typically less than about 0·07 eV.

Table 1.1 gives impurity energy levels for various elements in Si, Ge, GaAs, and GaP. Table 1.2 gives the band gap in a number of cubic semiconductors.

Table 1.2. Band gap of some cubic semiconductors at room temperature.

Si 1·13 eV	AlP 2·4 eV	GaP 2·24 eV	InP 1·26 eV	ZnS 3·6 eV
Ge 0·67	AlAs 2·16	GaAs 1·42	InAs 0·35	ZnSe 2·7
	AlSb 1·6	GaSb 0·67	InSb 0·17	ZnTe 2·2
				CdTe 1·5

III. Conduction of electricity through insulators

A. Conduction of electricity through gases

The conduction of electricity through gases provides a good model for induced electronic conduction in dielectrics. Apart from obvious differences the similarities are striking. In both cases the critical region forms a gap between electrodes, which normally does not allow an electric current to pass. In both cases the gap can be made conductive by irradiation. Hence it can be inferred that the results obtained for gases can be adapted to dielectrics in the first approximation. However, experiments confirm that crystals contain traps, so a more or less permanent space charge can build up in an insulator which opposes the applied field if the traps are deep. The presence of traps is one of the features by which solids differ from gases, and this alters the situation inasmuch as the space charges in an ionization chamber are due to ions near the electrodes, whereas in a solid they are due to carriers in traps.

Let us now consider what happens in a gas-filled ionization chamber when the space between two parallel plate electrodes is uniformly and steadily irradiated by weakly absorbed ionizing radiation such as X-rays (Fig. 1.7). The left electrode at $x = -l/2$ is kept at a potential $+P/2$ and the right electrode at $+l/2$ at a potential $-P/2$. Then, in any layer dx the density of negative carriers increases because they are being created at the rate βdx per unit cross-section, and because carriers enter across the plane x at the rate nu_n; here n is the density of negative carriers and u_n their velocity. The

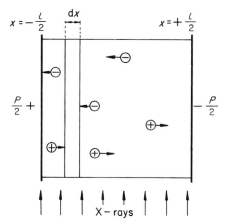

Figure 1.7. Thomson's model of an ionization chamber. The gas is ionized uniformly by weakly absorbed X-rays, and the electrodes do not allow the entry of charge carriers.

density decreases because $(n + dn)(u_n + du_n)$ depart across the plane $(x + dx)$, and on account of bimolecular recombination between the negative and positive charge carriers at the rate of αnp. Collisions of carriers of opposite sign will occur at a rate proportional to the density of either sign of carrier; in a certain fraction of cases this will lead to an exchange of charges and to a reversion of the particles to neutral states. The constant of proportionality α is obtained experimentally from the variation of current with field.

Hence, when a stationary state has been established,

$$\beta - \alpha np - d(nu_n)/dx = 0 \tag{1.4}$$

and for positive carriers,

$$\beta - \alpha np - d(pu_p)/dx = 0 \tag{1.5}$$

so that

$$pu_p - nu_n = \text{const} \quad \text{if } u_n > 0 \text{ and } u_p < 0 \tag{1.6}$$

In Thomson's model of a gas-filled ionization chamber, carriers do not enter from the electrodes, and hence only one sign of carrier is present near each. Thus, no steady-state current can flow if only one sign of carrier is mobile. The current density due to the positive carriers is $-qpu_p$ and that due to negative carriers qnu_n, where q is the charge of the electron; the total current density is constant over the excited insulator.

When the field is high, the carriers are swept out rapidly, and their density remains low, so the recombination term can be neglected. Equations (1.4) and (1.5) can be integrated:

$$pu_p = \beta x + a_1 \quad \text{and} \quad nu_n = \beta x + a_2 \tag{1.7}$$

where

$$a_1 = -\beta l/2 \quad \text{and} \quad a_2 = +\beta l/2 \tag{1.8}$$

so the current is

$$I = -qpu_p + qnu_n = -q\beta(x - l/2) + q\beta(x + l/2) = q\beta l \tag{1.9}$$

The current increases with the separation of the electrodes, simply because more ion pairs are created in a larger volume; the current is saturated, i.e. independent of the applied voltage.

In order to get an idea of the potential (V) distribution in the ionization chamber we simplify the problem by assuming an equal mobility (μ) of both carriers so that

$$-pu_p = \mu p(dV/dx) \tag{1.10}$$

and

$$nu_n = \mu n(dV/dx) \tag{1.11}$$

With equations (1.7) and (1.8) we obtain

$$pu_p + nu_n = (n - p)\mu(dV/dx) = 2\beta x \quad (1.12)$$

Where n and p are not equal, a space charge exists of density

$$\sigma = q(n - p)$$

so, by Poisson's equation,

$$\frac{d^2V}{dx^2} = q(p - n) = -\frac{2q\beta x}{\mu dV/dx} \quad (1.13)$$

$$\frac{d}{dx}\left(\frac{dV}{dx}\right)^2 = 2\frac{dV}{dx}\frac{d^2V}{dx^2} = -\frac{4\beta xq}{\mu} \quad (1.14)$$

$$\frac{dV}{dx} = \sqrt{\left(\text{const} - \frac{2q\beta}{\mu}x^2\right)} \quad (1.15)$$

This shows that for small values of β, i.e. a low degree of ionization, the field is almost constant and hence given by P/l. As the power of the X-rays increases, the field is distorted by the second term under the square-root, so it is stronger than P/l near the electrodes and weaker in the middle. This result is of general validity.

Confirmation of this behaviour when applied to a solid was achieved theoretically by Roberts (1967) whose results are summarized in Fig. 1.8. The graphical representation shows the electrostatic potential inside the solid in the cases (a)–(c) of the metal making an ohmic contact, (d) when the Fermi level of the dielectric coincides with the Fermi level of the metal and no band bending occurs, and (e) when the metal makes a rectifying contact to the solid. The same considerations apply to a uniformly ionized insulator.

If P is very small, the charge carrier velocities are small and we can neglect the third term in equation (1.4). Then

$$n \approx p = (\beta/\alpha)^{1/2} \quad (1.16)$$

There is no space charge, and

$$dV/dx = P/l \quad (1.17)$$

$$I \approx -(Pq/l)(\beta/\alpha)^{1/2}(\mu_n + \mu_p) \quad (1.18)$$

so Ohm's law holds. The current is shared between the positive and the negative carriers in proportion to their mobilities. Space charge however might occur in the immediate neighbourhood of the electrodes where only attracted carriers are available.

It should be mentioned that the coefficient of recombination (α) is not independent of the rate of ionization under all conditions. Langevin (according to Moulin, 1908) drew attention to the particular nature of the ionization

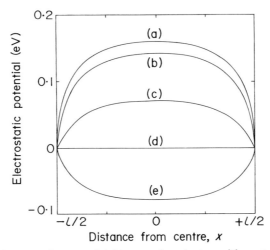

Figure 1.8. Electrostatic potential in a semiconductor with metallic electrodes. Curves (a), (b), and (c) correspond to the condition where the metal makes an injecting contact and the semiconductor has different donor impurity concentrations; the curvature of the electrostatic potential is due to an accumulation layer near the metal. The line (d) corresponds to the case where the work function of the semiconductor coincides with that of the metal, and no band bending occurs. Curve (e) represents the situation where the metal makes a rectifying contact and the band bending is due to a depletion layer. (Roberts, 1967)

by α-particles, viz. the high linear density of ion pairs along the track (columnar ionization) which must greatly increase the rate of recombination beyond that given by the average rate of pair production.

B. Transport of electrons and holes through crystals

The transport of electricity through matter is variously described as the movement of wavepackets or as the motion of particle-like ions or electrons. On the one hand, the wavepackets and associated undulatory phenomena interact with phonons and electric and magnetic fields. On the other hand, the particles are accelerated by the local electric and magnetic fields between collisions with atoms, molecules, phonons, or other obstacles. It can be shown that a phonon is equivalent, as far as carrier scattering is involved, to a particle of momentum $\hbar\omega$ where ω is its angular frequency. Diffusion of the particles becomes relevant if their distribution is non-uniform and the electric field low—the latter condition rarely met in problems of conductivity induced by an external agency such as electron bombardment. This (conventional) theory of carrier mobility was substantially complete by the 1950s.

If the problem of electron and hole mobility in a solid is simplified to its ultimate limit, the motion of carriers is described as their drift in a viscous medium although, of course, there is no associated net movement of atoms or molecules. The faster electrons have a greater probability of making a collision, so collisions may be represented by a resistive force proportional to the velocity. This immediately leads to the mobility $\mu = qt_0/m$ where m is the mass of an electron and t_0 is the mean time between scattering events. In modern terms, m in this expression must be replaced by the effective mass, m^*. Conversely, mobility can always be defined through the current density, since this is equal to $+q\mu_p p$ or $-q\mu_n n$.

Quite generally it must be realized that in any theory the quantity primarily derived is the electric current. If Ohm's law is satisfied, the concept of mobility can then be introduced. At one time it seemed as if this concept, which can be attributed to Drude (1900) and to Lorentz (1905) and therefore dates from pre-wave-mechanical times, would no longer serve a useful purpose. However, it has survived remarkably well. If carriers are electrons, as in most metals, the current entails no chemical changes, but if they are electrically charged atoms they transport matter and thus cause chemical changes locally. Hence, as a matter of principle, it should be very easy to decide whether in a non-metallic current-carrying medium the carriers are electrons or ions. As a matter of fact, the decision was at times difficult when the currents were low or of short duration. At a current of 10^{-10} A, for instance, less than 10^{-3} μg is deposited per hour. It is now recognized that such low currents are usually electronic, unless there is good reason to consider them ionic; this applies in particular to current induced by any external ionizing radiation.

The scattering processes that may be involved in a solid under different circumstances are lattice scattering (either acoustic or optical phonons), ionized impurity scattering, neutral impurity scattering, and "heated" electron scattering. These processes were well reviewed by Nag (1972). Some of them also apply to dielectric liquids.

1. Lattice scattering

Seitz (1948) and Bardeen and Shockley (1950) showed that, if the scattering in a non-polar solid is isotropic and it is possible to define a mean free path (λ) which is independent of the velocity (u) of the carriers, then $t_0 = \lambda/u$. By finding averages over a Maxwellian distribution this leads to $\mu = (4/3)q\lambda/(2\pi mk\theta)^{1/2}$, and combining this with the fact that $\lambda \propto 1/\theta$ gives $\mu = \text{const.}\theta^{-3/2}$. A similar result was obtained from deformation potential theory. This assumes that the band gap is distorted by the passing carrier, and the mobility $\mu = qt_0/m^*$ is expressed, in terms of the longitudinal elastic constant

c_{11} and the shift in band edge per unit dilatation Δ_1, as $\mu =$ const. $c_{11}/\Delta_1^2(k\theta)^{3/2}$, which again leads to a temperature dependence of the form $\mu =$ const.$\theta^{-3/2}$ if the deformation potential Δ_1 is independent of temperature.

The influence of the shape of the conduction band on mobility has been discussed by Dumke (1956) and by Herring (1955). If this is of the many-valley type both shear waves and intervalley transitions play an important part in determining the mobility. In the simple case of acoustic mode scattering considered so far, only longitudinal modes are involved. If both shear wave and longitudinal modes are involved, Dumke's treatment still gives $\mu \propto \theta^{-3/2}$. If a degeneracy occurs near the band edge, as in the case of the valence band in silicon and germanium, the situation may be complicated by the presence of more than one type of carrier of the same sign (in this case holes). By using independent scattering parameters for each type, Ehrenreich and Overhauser (1956) were able to explain the observed temperature dependence of the mobility of holes for germanium in the range 100–400 K which is $\mu_h \propto \theta^{-2.3}$.

Optical mode scattering may give rise to a temperature dependence of mobility stronger than $\theta^{-3/2}$, and explicit variation of m^* with temperature could also cause this law to differ. An energy dependent effective mass has been used by Spear and Le Comber (1969) to explain hot electron effects in solid noble gases.

2. Ionized impurity scattering

The original treatment by Conwell and Weiskopf (1950) was subsequently modified by Brooks (1955). Blatt (1957) examined the approximations made and confirmed a mobility proportional to $\theta^{+3/2}$. This analysis makes use of the Rutherford formula for scattering of an electron by an ion through a given angle ϕ. The relaxation time is then given by $t_0 = (a_i/u)(1 - \langle \cos \phi \rangle)^{-1}$ where $\langle \cos \phi \rangle$ is the average of the cosine of the scattering angle, a_i is the mean separation of the ionized impurities, and u is the velocity of an electron.

3. Neutral impurity scattering

Neutral impurity scattering may be important in the case of an unionized donor or acceptor atom in a semiconductor, as well as in the case of an impurity in an insulator. Mathematically this is the same as the scattering of slow electrons by a hydrogen atom (Pearson and Bardeen, 1949). Using this approach, Dexter and Seitz (1952) calculated the mobility when associated with this type of scattering alone. They found $\mu = m^*q^2/20N\varepsilon\hbar^3$, where N is the density of neutral impurities and ε is the dielectric constant. Neutral impurity scattering is inversely proportional to ε, whereas ionized impurity scattering is inversely proportional to ε^2. Thus neutral impurity scattering is more pronounced in materials of high dielectric constant.

4. Scattering by dislocations

Scattering by dislocations arises through the dilatation caused to the lattice. Dexter and Seitz (1952) showed that the probability for scattering is proportional to the number of dislocation lines per unit area, and directly proportional to temperature. Thus, if N_s is the number of dislocations per unit area, $\mu \propto (1/N_s)\theta^{-1}$.

5. High field mobility

In a high mobility solid or liquid the carriers may no longer remain in thermal equilibrium with the lattice at high values of applied field. In these circumstances, although the random velocity distribution of the electrons may still be Maxwellian ("hot" electrons) the concept of proportionality of the carrier drift velocity to applied field then breaks down. It is possible to describe the carrier motion in terms of a field-dependent mobility but it is more usual to describe it in terms of the drift velocity.

By *assuming* that the electron velocity distribution remained Maxwellian, Shockley (1951) developed his theory of hot electrons in terms of an effective electron temperature θ^*. If the electric field ($\mathscr{E}$) is high, the power supplied to electrons from it may be more rapid than the rate at which they lose energy to phonons. As a result, the average electron energy will rise until electrons can supply energy to the phonons fast enough to create a steady state. This condition leads to a drift velocity $u = 1\cdot 23 c\mu_0 \mathscr{E}^{1/2}$ where $\mu_0 = qt_0/m^*$ is the low field mobility as governed by a relaxation time t_0 for scattering by acoustic modes and c is the velocity of sound. Shockley also predicted a saturation of the drift velocity in the high field limit. It was originally assumed that the carriers emitted an optical phonon as soon as they reached the phonon energy, $\hbar\omega_0$. This yielded a value for the saturation drift velocity of $u_s = (8\hbar\omega_0/3\pi m^*)^{1/2}$. Subsequent calculations have shown that this expression can only give a rough guide. A more general transport theory, which takes into account the additional influence of the momentum transfer through the introduction of a structure factor, has been considered by Cohen and Lekner (1967). Their analysis shows that the structure factor is of importance in the case of electronic conduction in liquids or when multi-phonon collision processes are included.

Jacoboni and Reggiani (1979) have reviewed hot-electron properties of cubic semiconductors.

C. Transport of electrons and holes through amorphous solids and low mobility crystals

Since about 1960 considerable progress has been made towards understanding electron and hole motion in amorphous solids and liquids. Conventional band theory of electrical conduction as summarized in Section B can be applied, provided that the bandwidth is not too small or the lattice too disordered. In either of these cases, however, the band structure is at least partly destroyed and normal band motion cannot occur. This can be simply demonstrated by Heisenberg's uncertainty principle.

For the band approach to be a good one, the uncertainty in the energy of a carrier due to scattering by lattice vibrations should not exceed the bandwidth, W. Thus

$$W \geqslant \hbar/t_0 \tag{1.19}$$

where t_0 is the scattering relaxation time. This enables a lower limit to be set on mobility for the band theory of conduction to apply.

Since $\mu = qt_0/m^*$, inequality (1.19) leads to

$$\mu \leqslant q\hbar/m^*W \tag{1.20}$$

In terms of the Bloch tight binding scheme applied to a simple cubic lattice, the carrier bandwidth is simply $W = 12J$, where J is the exchange energy between an atom and its nearest neighbours.

On the same model, the carrier effective mass at the band edge is given by

$$1/m^* = \pm(2/\hbar^2)Ja^2$$

where a is the lattice constant. We can write $1/m^*W = a^2/6\hbar^2$ and by substitution in inequality (1.20) obtain

$$\mu \geqslant qa^2/6 \tag{1.21}$$

For a representative liquid or solid in which, say, $a = 3$ Å we obtain $\mu = 0\cdot2$ cm^2 V^{-1} s^{-1} as a lower limit for the band approach to apply.

An alternative formulation takes into account the requirement that an electron in a band must have a mean (thermal) kinetic energy that is larger than the uncertainty in energy resulting from the scattering process. Thus, if t_0 is the mean time between scattering events,

$$(3/2)k\theta > \hbar/t_0 = q\hbar/m^*\mu \tag{1.22}$$

If μ is expressed in cm^2 V^{-1} s^{-1} equation (1.22) leads to

$$\mu \geqslant [30/(m^*/m)](300/\theta) \tag{1.23}$$

Inequality (1.23) is only valid when the bandwidth is not too small and thus the effective mass m^* is not too large. In terms of the tight binding scheme, the maximum velocity in a band is given by

$$u_m = Wa/6\hbar \approx Ws/k\Theta \tag{1.24}$$

where the lattice spacing (a) is expressed in terms of the sound velocity (s) and the Debye temperature (Θ).

Whenever $u_m < s$, i.e. $W \leq k\Theta$, absorption and emission of single acoustic phonons becomes impossible and the normal theory of acoustic phonon scattering of carriers within a band can no longer be applied. This limitation implies that, for normal band motion to prevail,

$$m^*/m \leq \hbar/sam \tag{1.25}$$

If we take inequalities (1.22) and (1.25) in conjunction, this leads to

$$\mu > 10^4 sam/\hbar\theta \tag{1.26}$$

where μ is expressed in units of cm^2 V^{-1} s^{-1} and the velocity of sound is in units of cm s^{-1}.

With a typical value of 3×10^5 cm s^{-1} for s, and 3 Å for a, we get room temperature mobilities in the region of 0·1 cm^2 V^{-1} s^{-1} as a lower limit for which the band approach is applicable. These two estimates thus agree within a factor of 2.

Transport of electrons and holes in a liquid occurs by processes that are probably—though not necessarily—similar to those applicable to a low mobility solid. Some liquids (for example some of the liquid noble gases) transport excess electrons as freely as in conventional wide-band semiconductors such as silicon or germanium. However, in a molecular liquid we should normally expect one or other of the transport processes that would apply to an amorphous or narrow-band solid to be more likely, because in many ways an amorphous solid is like a supercooled liquid. These mechanisms will now be discussed.

1. The concept of electron or hole transport

As discussed in Section B, motion of an excess electron or hole in a conventional solid can be worked out on the basis of the carrier undergoing random thermal excursions within a fairly wide energy band, and superimposed on this is a slight velocity perturbation in the direction of an applied electric field. The drifting carrier is occasionally scattered by thermal vibrations of the constituent atoms or molecules, and such phonon interactions can be treated in the strict quantum mechanical sense as small perturbations to the precise periodic nature of the crystal. However, in a low mobility solid or liquid the carrier may be travelling so slowly through the lattice that this can adjust itself to the presence of the carrier. The problem has to be treated with electron–phonon coupling as the prime interaction, dealing with the periodicity of the lattice as a perturbation. How does this mode of "electron–phonon interaction" differ from that already discussed in the Bardeen–Shockley deformation potential model for band transport governed by lattice scattering? This is illustrated by a simple example of molecular events which occur when the carrier pauses on a molecular site for longer than about 10^{-12} s.

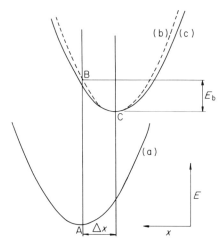

Figure 1.9. Diagram showing the formation of a small polaron in a deformable lattice. Curve (a) represents the potential energy of a diatomic molecule as a function of the separation between its two nuclei, and (b) is that of its ion. When an electron or hole becomes localized on the molecule for more than about 10^{-12} s, the potential minimum moves a distance Δx, from A to C, and the molecule expands or contracts by the same amount.

Figure 1.9 shows the potential energy of a diatomic molecule and its negative ion as a function of the internuclear spacing (x). In the ground state, the atoms undergo small amplitude oscillations of frequency ω_0 about point A. The force constant (K) in the range of interest is assumed to be independent of x, and in terms of the reduced mass (M) of the system,

$$K = M\omega_0^2 \tag{1.27}$$

Curve (a) is the potential energy curve for the unionized molecule, while (b) is the corresponding curve for the molecular ion. It is seen that the minimum at C is displaced relative to A by an amount Δx. This corresponds to the molecule contracting or expanding when it is ionized; of course, in either the neutral or the ionized state the nuclei will be separated on the average by the value of x corresponding to A or C. On ionization the internuclear separation will follow the path from B to C on curve (b), and the energy of the entity consisting of the molecule and its associated electron will be reduced. If K is unchanged, the configurational diagram of the ion can be represented by (c) which has the same shape as (a). The combined entity—the distorted molecule and the electron—is known as a small polaron. In the circumstances just postulated, the polaron binding energy is given by

$$E_b = \tfrac{1}{2} K(\Delta x)^2 \tag{1.28}$$

The way in which a polaron can move from site to site is by a process known as "hopping". From the site-to-site transition rate (P) the drift mobility can be expressed

$$\mu = (qa^2/k\theta)P \tag{1.29}$$

A site transition occurs when the energy of a neighbouring neutral molecule coincides with that of the molecular ion. This might occur, for example, by thermal fluctuations. P [=(probability of coincidence) × (probability of transfer when a coincidence occurs)] is given by

$$P = (\omega_0/2\pi)\exp(-E_b/2k\theta) \times p \tag{1.30}$$

The value of p determines whether the polaron proceeds with or without phonon assistance. The limiting cases of non-adiabatic and adiabatic hopping will be discussed in Sections 3 and 4 respectively. Spear (1974) has reviewed these theories.

2. Small polarons

The small polaron is a particularly important entity when conduction mechanisms in a low mobility solid or liquid are considered because it is concerned with the transport of localized electrons or holes. In Section B it was shown how an excess electron situated in a perfect crystalline solid interacts with atoms or molecules situated many tens—or even hundreds—of lattice spacings away, as well as those nearby. This is another way of expressing the fact that the Bloch solutions of the Schrödinger equation describing the wavefunction of an excess electron in a non-deformable periodic lattice lead to solutions extending over many lattice spacings. However, if the lattice is deformable, or the energy of deformation is large compared with the bandwidth, the spatial extent of the interaction might be severely limited—perhaps to only a few atoms. This is known as the small polaron approximation, and the combined entity consisting of the carrier with its local region of lattice distortion moves in a particular fashion from site to site. The wavefunction of a polaron is now the product of a purely electronic part and a purely vibrational part, and site-to-site transitions are governed by the product of the wavefunction overlaps of both. Thus dispersion of vibrational frequencies is necessary for a polaron to move.

3. Non-adiabatic hopping

If we now return to Section 1, it is shown there that the transition probability of a localized carrier from site to site is given by equation (1.30). In the non-adiabatic regime the carrier adjusts itself slowly to the motion of the lattice and it misses many coincidence events before making a transition. In this case $p \ll 1$, and

$$p = (2\pi/\hbar\omega_0)(\pi/2E_b k\theta)^{1/2} J^2 \tag{1.31}$$

where the exchange energy resulting from the purely electronic wavefunction overlaps between a molecule and its neighbours is denoted by J. The non-adiabatic case seems to apply to most molecular solids, and by combining equations (1.29) and (1.31) we get

$$\mu = (qa^2/k\theta\hbar)(\pi/2E_b k\theta)^{1/2} J^2 \exp(-E_b/2k\theta) \tag{1.32}$$

If we use $\gamma = E_b/\hbar\omega$ as a dimensionless parameter describing the strength of the electron–lattice interaction, equation (1.30) can be plotted. Figure 1.10 shows values of P according to Holstein's analysis, using γ as a parameter and $E_b/k\theta$ as an independent variable.

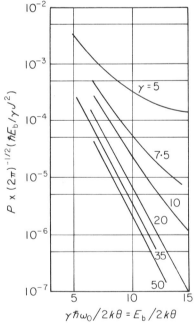

Figure 1.10. Carrier transition probability (P) to an adjacent site as a function of temperature, using the full expression for non-adiabatic hopping in a linear array of diatomic molecules derived by Holstein (1959).

4. Adiabatic hopping

If the carrier adjusts rapidly to the thermal motions of the lattice, when a coincidence does occur the carrier is very likely to make a transition to a neighbouring site. We can now put $p = 1$ in equation (1.30), and therefore

$$\mu = (qa^2\omega_0/2\pi k\theta)\exp(-E_b/2k\theta) \tag{1.33}$$

It may be noted that this equation predicts that there is a maximum of mobility as the temperature is varied. Thus, in the adiabatic hopping regime, the mobility at first rises with increasing temperature, reaches a maximum at $\theta = E_b/2k$, and then falls again with increasing temperature.

5. Electronic bubble state or cluster formation

It has been shown that an essential feature of electronic transport mechanisms in a low mobility solid or liquid is the formation first of a localized electronic state and then of hopping or tunnelling between such states. The description of the small polaron formation showed how the carrier can be self-trapped in a potential well produced when the lattice becomes distorted owing to its presence. However, other localization mechanisms can occur, such as the formation of the R_2^+ molecular ion in liquid Ar, Kr, and Xe, or the formation of an electronic bubble or cluster as in liquid He, Ne, or H_2. The latter mechanism is explained as follows.

The ground-state energy (V_0) of a quasi-free electron in a liquid is the sum of two terms, a short-range repulsive interaction (T) and a long-range attractive interaction (U_p) which is mainly due to polarization. V_0 represents a balance between these, thus $V_0 = T - U_p$. If V_0 is negative, an energetically stable quasi-free electron state is formed. However, a positive value of V_0 implies a repulsive force between the excess electron (or hole) and the surrounding atoms or molecules. If this repulsive force is strong, the surrounding atoms will be pushed away from the electron, so leaving the electron in a void; this is an electronic bubble. The basic stability criterion for a bubble state to give rise to a localized electronic state below V_0 is given by

$$V_0 > E_e + 4\pi r^2 \Gamma + (4/3)\pi r^3 p \tag{1.34}$$

Here E_e is the ground state electronic energy, r is the radius of the void, p is the pressure, and Γ is the surface tension. This condition has been approximated by Springett et al. (1968) by

$$(2\pi/m)\hbar^2(\Gamma/V_0^2) < 0.047 \tag{1.35}$$

Inequality (1.35) has been successfully applied to liquid He and Ne.

30 ELECTRON BOMBARDMENT INDUCED CONDUCTIVITY

A similar situation applies in the case of cluster formation. The residual force is predominantly attractive and the surrounding atoms or molecules then form a small localized region of higher density (Atkins, 1959).

6. *Ionic transport in an insulating liquid and Walden's rule*

Ions move slowly compared with quasi-free electrons and holes. To a good approximation one can calculate their mobility by equating the viscous drag given by Stokes's law to the electrostatic force on the ion. This yields

$$\mu = q/a\eta r \tag{1.36}$$

where η is the viscosity and a is a constant between 4π and 6π.

Equation (1.36) also leads to Walden's rule which states that the product $\pi\eta$ for ions should be constant and independent of temperature.

7. *Amorphous solids*

The Bloch solutions of the wave equation assume that the lattice is strictly periodic. If the lattice is random, the atomic or molecular units from which it is built are only disposed about a certain average separation distance, and the normal electronic band structure is thus modified. The main features of the band structure of an amorphous solid have been shown in Fig. 1.5. It is seen that the lower edge of the conduction band and the upper edge of the valence band are no longer sharply defined, and the band edges extend well into the gap where the density of states is no longer zero. In addition to this continuum there are localized regions where there is a higher density of states within the gap; these correspond to atoms lying in metastable positions where the surroundings are favourable but energetically not positions of a true minimum (see Fig. 1.11).

Localization in amorphous solids does not depend on the lattice being deformable, as it does in the case of a small polaron. The localization mechanism will be discussed in Section 9 (Anderson localization).

Near the band edges the density of states is low and the carrier mean-free-path between scattering events is of the same order as the lattice spacing (a). In this case, $\mu = (1/6)(qa^2/k\theta)\nu_{el}$ where ν_{el} is the site jump frequency of the order of an atomic frequency (10^{15} s^{-1}). Mobilities of about 5 cm^2 V^{-1} s^{-1} are expected.

8. *Random phase model*

The concept of an uncorrelated site jump as just described does not require the assistance of phonons, and the wavefunctions describing the localized carriers are built up as a linear combination of atomic wavefunctions without the usual condition for a phase relationship as in the Bloch solutions.

9. Anderson localization

Anderson (1958) investigated a model of an amorphous semiconductor composed of a number of potential wells each having a slightly different energy and the ensemble covering a range U_0. He analysed the model using the tight binding approximation, and took into account the disorder parameter U_0. If W was the carrier bandwidth, and if the ratio U_0/W exceeded a

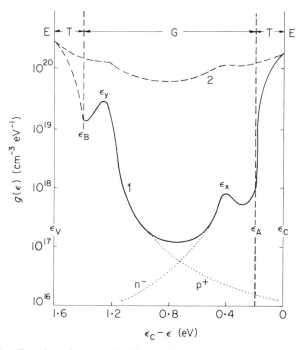

Figure 1.11. Density of states distribution in a-Si specimens: curve 1, glow discharge specimens; curve 2, evaporated specimens. (Le Comber and Spear, 1979)

certain critical value (of the order of ½), diffusion of a carrier to an adjacent site was impossible at a temperature $\theta = 0$. This localization mechanism, which is a critical feature of carrier motion in certain random lattices, is marked on a typical density of states diagram (Fig. 1.11) at energy ε_C. The corresponding energy for holes is marked on the same figure at ε_V and the gap between them is known as the mobility gap. Above 0 K, between ε_C and ε_V, carriers can diffuse by phonon assisted hopping, and

$$\mu(\varepsilon) = \nu_{\text{ph}}(qR^2/6k\theta)\exp(-2\alpha R)\exp(-W/k\theta) \quad (1.37)$$

where R is the mean distance of a "hop".

10. Variable range hopping

At low temperatures the occupancy at ε_C or ε_V is generally small, and the dominant current paths are then via the localized state distribution lying within the mobility gap. A mechanism known as variable range hopping is then likely to occur (Mott, 1969). In this case the transitions determining the carrier motion are not to the nearest neighbour but to sites further away and for which W is small. The well known relationship between electrical conductivity and temperature is $\sigma = \exp(-c/\theta^{1/4})$ where c is a constant. The implications to "mobility" will be discussed in Chapter 5.

11. Mean free path of a slowly moving electron

In the comments that follow we refer to the mean distance travelled by a pseudo-free carrier in the direction of the applied field before being trapped or recombining (schubweg). If the carrier is trapped by small polaron formation, by a localized defect state, or as a result of disorder, recombination is improbable while the carrier is localized, especially if the carrier of the opposite sign is not in an extended state. The trapped state can thus be regarded as a "protected" condition, and such processes as those that give rise to a more or less permanent loss cannot occur without a transition into the appropriate band or by direct tunnelling to the impurity site or recombination centre. This clearly can take much longer than in conventional band motion where the carriers spend much of their time moving about randomly in a band at a velocity greater than the speed of sound with a mean energy of about $(3/2)k\theta$.

Thus, in the low mobility solids and liquids of the kind discussed here, it is quite common to find fundamental drift mobilities as low as 2×10^{-3} cm^2 V^{-1} s^{-1} but mean free lifetimes as long as 0·1 s, provided that the specimen is pure. Thus mean free paths as high as several mm or more are quite possible.

12. Representative amorphous solids

Very few materials can be regarded as truly amorphous; many so described are really polycrystalline. They are relevant in the present context if they are supercooled liquids. The glassy solids and glow discharge a-Si and a-Ge are good examples, and the transport properties of a number of them will be discussed in Chapter 6.

D. Drift experiments in gases or in insulating solids and liquids

Drift velocity measurements in ionized gases by a time-of-flight technique were first made by Tyndall and Grindley (1926). Their experimental arrangement involved the use of mesh electrodes as a kind of electronic shutter to allow short bursts of electrons or positive or negative ions to drift in the presence of an applied electric field from one plate to the other in a gas-filled cell. A galvanometer included in the circuit allowed the charge carrier transit time to be measured, and thence their mobilities.

A modern method for determining the drift mobilities of charge carriers in gases has been described by Saelee *et al.* (1977).

Schmidt *et al.* (1974) used a technique for determining excess charge carrier mobilities in insulating liquids. Their cell was similar to that used by Tyndall and Grindley but ionization was produced by a pulse of X-rays. Since the mobility of electrons in a liquid is nearly always greater than that of positive ions, it is not difficult to separate their contributions to the induced current pulse when its shape is examined on an oscilloscope.

The transient EBC time-of-flight technique was first used to measure the drift velocity of generated electrons in a single crystal of AgCl in 1949 but it was not until 1957 that Spear developed it. This has now become the most widely accepted method for measuring the drift velocity of either sign of carrier in insulators and in some semiconductors and liquids. Chapter 5 will be devoted entirely to this important application of EBC.

The transient technique must not be confused with the Haynes–Shockley (1951) method for determining the drift velocity of carriers in semiconductors. The latter applies to semiconductors for which the dielectric relaxation time is short compared with the transit time and, after injecting excess majority carriers in the form of a short pulse, minority carriers enter the specimen from the remote electrode. Strict charge neutrality is thus maintained within the specimen to conform to the mass-action equilibria and the drifting charge packet is screened. The transit time is always equal to the transit time of the minority carriers, and it is not possible to use this technique for measuring the drift velocity of both signs of carrier on the same specimen. However, it is possible to do so on high resistivity insulators or semiconductors by transient EBC methods or related transient charge techniques using photoexcitation. These will be described in detail in Chapters 5 and 6.

2
Properties of electron beams and their interaction with matter

I. Sources of electrons

A. Radioactive sources

Nuclear sources of electrons are now easily bought (e.g. from The Radiochemical Centre, Amersham, England, or one of their overseas agents) and they are useful if thick specimens are to be uniformly ionized. Their drawbacks are a limited range of power, and a slow response to on–off operations since these must involve the mechanical movement of either the source or a shutter. The most common monochromatic sources are strontium-90 (+yttrium-90) which has a half-life of 28·5 years and a maximum β energy of 2·27 MeV, nickel-63 with a half-life of 100 years and a maximum β energy of 66 keV, promethium-147 with a half-life of 2·62 years and a maximum β energy of 230 keV, and thallium-24 with a half-life of 3·78 years and a maximum β energy of 760 keV. Above about 300 keV energy the range of electrons is proportional to the voltage. The energy deposited does not follow a simple exponential law.

An arrangement that has been used for the examination of silica specimens is shown in Fig. 2.1. The strength of the irradiation is controlled by the depth of an activated silver disc in a lead castle, the on–off control being a sliding shutter. With a 50 mCi disc source, an electron current of 10^{-12} A impinges on a specimen area of 0·078 cm², at the closest approach of 1·5 cm between specimen and source.

B. Electron guns for EBC measurements

The art of producing beams of electrons, and of controlling them, is highly developed. Demountable electron guns may employ a hairpin filament of tungsten or tantalum 0·1 mm thick as emitter, similar to those used in electron microscopes. Figures 2.2 and 2.3 are diagrams of a typical gun and of a particular focus coil design.

2. ELECTRON BEAMS AND THEIR INTERACTION WITH MATTER   35

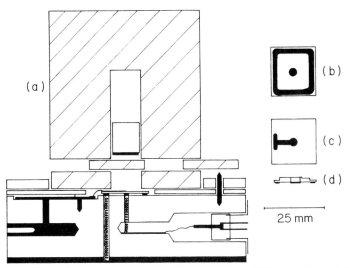

Figure 2.1. Experimental arrangement for measuring EBC using β-particles from a radioactive source. (a) Testing base with source in position and shutter closed; (b) top electrode; (c) bottom electrode; (d) miniature ionization chamber. The specimen rests on a block of Perspex 10 cm square and 2·5 cm high. Connections to the guard ring and the bottom electrode are made by two very light helical springs, the former going to a grounded metal plate below the base and the latter to a coaxial connector. The top electrode is contacted by a small clip. The radioactive source is attached to a Perspex block inside a cylindrical lead castle located by a Perspex ring. The ionization chamber is essentially a Perspex washer with electrodes corresponding to those of the specimen.

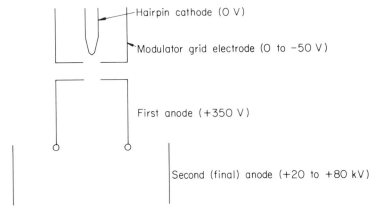

Figure 2.2. Diagram of a tetrode electron gun; the field in the neighbourhood of the filament comes from the first anode so that the electron current is independent of the potential difference between the filament and the second anode which determines the energy of the electrons.

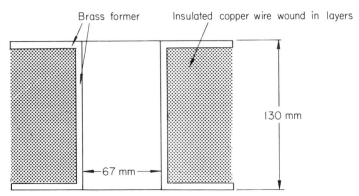

Figure 2.3. Design and dimensions of a typical focus coil.

The main accelerating voltage is applied between the first and the second or final anode which, together with the rest of the apparatus, is normally at earth potential. The negative EHT is then connected to the filament which is 0–50 V positive with respect to the grid. The cut-off voltage depends on the grid aperture (1–3 mm) and the position of the filament (flush with the aperture or a few mm inside). The first anode, normally at a few hundred volts positive with respect to the grid, screens the filament from the final anode, so that the current is independent of the EHT voltage and, in particular, of its fluctuations. The filament or the filament and the whole gun should be independently adjustable from outside so that the desired grid-base can be set and the beam can be made to emerge axially with respect to the anode. This may be conveniently effected while the gun is working by using a sliding O-ring seal and set screws which can be adjusted by a screwdriver made from a long rod of Tufnol. Alternatively, a vacuum-tight seal which is still capable of some flexibility can be made from phosphor-bronze "bellows"; again adjustments can be made by set-screws and a long insulated screwdriver (a Tufnol rod 500 mm long is quite adequate for 50 kV).

Diagrams of some typical electron guns are shown in Figs 2.4–2.6. Figure 2.7 shows a vacuum chamber and specimen holder using a commercial electron microscope gun (AEI triode EM-3). The glass chamber is made from a standard cathode ray tube having a thickened rim where the window is normally affixed. The glass–metal demountable joint uses a strengthened glass cone, a conical fibre collar, and an O-ring. The beam position can be adjusted by the mechanical deflecting arrangements provided with the gun mount by the manufacturers, or alternatively by external magnetic deflection coils. As in most electron microscopes, the heater power is derived here from an rf oscillator operating at about 5 kHz. This is operated at earth potential, and coupling to the gun is through a small transformer

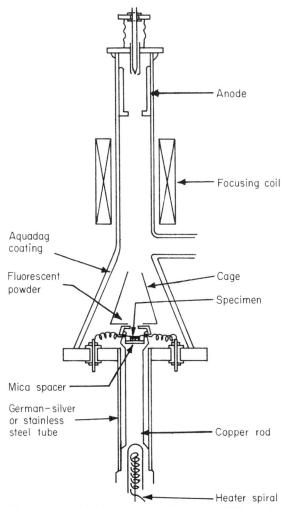

Figure 2.4. Tube used by Ansbacher and Ehrenberg (1951); this makes use of a demountable cathode ray tube—a glass flask with an extended neck. A tungsten hairpin filament terminates with a long stalk attached to flexible metal bellows by which it can be adjusted inside the grid cylinder. The specimen is placed inside a shallow opening in a thick copper rod which acts as a thermal conductor; the other end of the copper rod can be immersed in liquid nitrogen or electrically heated.

built to withstand the full EHT potential. The whole emission control unit including the first anode voltage supply and the heater power source is at negative EHT potential and it may be enclosed so as to prevent corona discharge (a cardboard box is often sufficient). The filament is usually supplied by a 6 V lead–acid accumulator via a transistorized current

38 ELECTRON BOMBARDMENT INDUCED CONDUCTIVITY

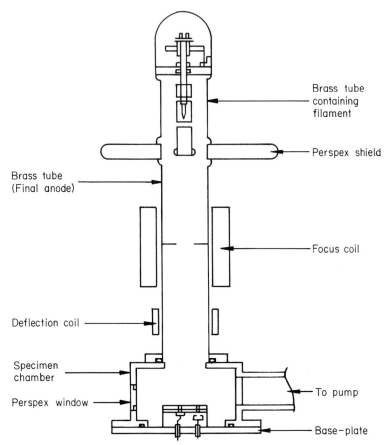

Figure 2.5. Typical electron gun for EBC measurements; such a gun can be made readily using techniques familiar to most laboratory model shops and no special equipment is needed. The Perspex window is held in place by atmospheric pressure and seating on an O-ring in the cylindrical base section of the final anode; the latter makes a vacuum-tight seal with the base-plate by another O-ring.

stabilizing circuit. Although dry cells for the first anode and modulator were used for many years, a transistorized dc converter is now usually employed, again using a 6 V lead–acid accumulator. Gun supplies of this nature (up to an EHT of 20 kV) can even be provided using an oil-immersed 50 Hz isolating transformer. There are a number of ways for accommodating the necessary equipment associated with the "high voltage end" of the gun. For laboratory purposes it is usually convenient to support the filament control, the filament on–off switch, the filament current meter, the grid bias and anode voltage supplies and switches etc. along with batteries and other

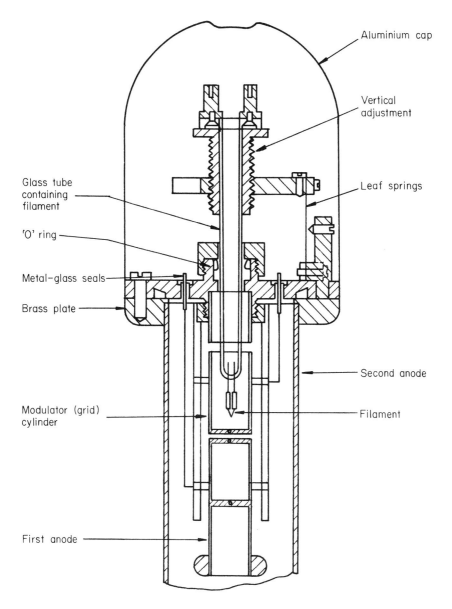

Figure 2.6. Electron gun design shown in more detail. The filament sits at the end of a long glass tube sealed with an O-ring through which it can slide and be tilted; a set of three leaf-springs and a screw thread allow the filament to be adjusted. (Hidden, 1960)

40 ELECTRON BOMBARDMENT INDUCED CONDUCTIVITY

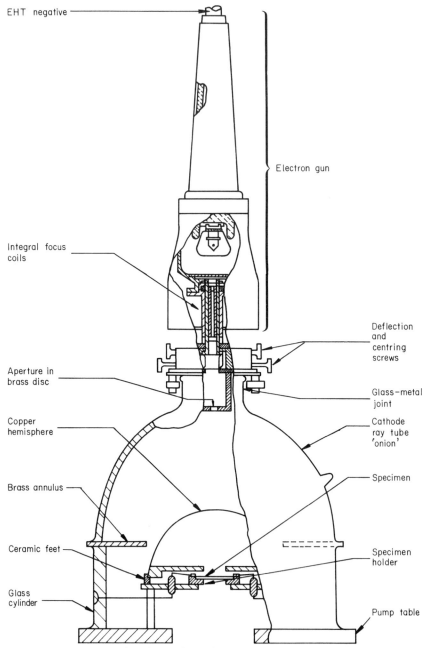

Figure 2.7. Vacuum chamber and specimen holder using an electron microscope gun. (Varol, 1978)

essential apparatus on a large strong wooden platform supported well above the "earthy" end of the gun on stout polythene rods (e.g. 300 mm long and 35 mm in diameter). The controls are switched by long Tufnol or Perspex rods (e.g. 600 mm long and 6 mm in diameter).

C. EHT supplies

The usual voltages of interest lie in the range about 5–50 kV although in a few cases voltages as low as 1 kV or as high as 100 kV have been used. A number of units are now available commercially which serve the purpose well. Often a supply of 1–2 mA and of variable voltage within the range of interest is quite adequate. Up to about 1958, when such supplies were not readily available, X-ray type sets with rectifiers and high voltage condensers were used, or sets using high frequency oscillators and Cockcroft–Walton type voltage multipliers. The EHT voltage should be adjustable continuously from a low value up to its maximum, and be indicated with reasonable accuracy. A reliable way of measuring the EHT voltage is carefully to make up a series chain of a number of 10 megohm high stability resistors on a Perspex tower so as to draw perhaps 50 μA at maximum voltage. Both the resistor set and the meter can then be readily calibrated and checked.

D. Concentration of intensity with electron lenses

The electrons emerge from the gun as a divergent bundle and are focused by a magnetic lens on to a plane near the specimen. It is useful to make this plane a sheet of non-magnetic metal separating the specimen region from the gun region in order to prevent stray electrons from reaching the specimen or the leads connecting it with the measuring devices. The sheet should be coated with some phosphor (e.g. willemite) so that the appearance of the beam is readily seen and adjusted. Before focusing (i.e. energizing the magnetic lens) the luminous patch should be round, fuzzy edged (if there is no stop in the tube), and reasonably well centred. As the focusing coil is energized it should shrink about its original centre. Centring is achieved by aligning or slightly tilting the coil, which must be given some play about the tube. It is advisable to wind the coil carefully in layers, and not to make it too small (perhaps 10 cm long), in order to avoid overheating. For instance, a coil suitable for a low voltage transistorized power supply unit has 20 layers each of 90 turns of Lewmex T18 wire insulated with 0·1 mm glass cloth. This provides a coil of resistance 9 ohms. It can be operated up to 3000 amp-turns without excessive heating, with 29 watts at 16 V (see Fig. 2.3). Design data are readily found in texts on electron optics (e.g. Klemperer and Barnett, 1971). The beam is brought into the desired position by the use of either permanent magnets or a pair of deflecting coils attached to the tube between the focusing coil and the specimen holder. The coils are toroidal with diameter somewhat larger than the tube diameter, and shaped to fit the tube (see Klemperer and Barnett, pages 404–5).

Some experimenters provide the phosphor-coated sheet as mentioned with two equal holes, one leading to a Faraday cage by which the electron current is measured, the other above the specimen. In order to work with a well defined current density, the spot on the screen is defocused to a few times the size of the apertures, and its centre portion is selected for the tests. Currents of between 10^{-5} and 10^{-13} A are readily available, with electron energies between 1 and 100 keV.

E. Transmission of control

For slow operations it is usually adequate to operate the switches and other controls by means of long insulating rods. Rotary switches are more convenient than toggle types. A meter is usually placed in series with the filament supply so that the current can be set to a predetermined value; this can be seen from a safe distance when the EHT is on.

When the switching has to be performed quickly, or if the EHT is very high, very good isolation can be achieved by using a lamp and a small self-contained photoactivated switch circuit. The latter can be operated from a 12 V dry battery. A suitable circuit using a relay switch is shown in Fig. 2.8.

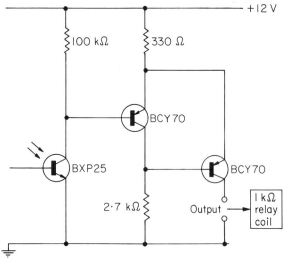

Figure 2.8. Phototransistor circuit for reception of optical control signals which operate a relay at negative EHT potential.

F. The scanning electron microscope (SEM)

It is becoming more common today to find a scanning electron microscope in a university or in industry, and since the apparatus is already equipped to provide constant magnitude focused electron beams in the energy range about 5–50 keV it is often convenient to make EBC or EVE (electron voltaic effect) measurements using the apparatus available. This is likely to be the case if a short-term programme of measurements or a spot-check is to be made. Such methods have been used to measure the EBC avalanche effect on a silicon photodiode (Gibbons, 1975), and the steady-state EBC of SiO_2 and Si_3N_4 (Hezel, 1979).

If the specimen is small, as might be the case if a device is made using photolithographic masking techniques, it is essential to make quite sure that the primary electron beam is striking the correct area on the top electrode. This may be done quite simply by using the normal SEM secondary emission image; this is performed to align the scan and position the beam for each value of the bombarding voltage (V_p) used. For an actual EBC or EVE measurement, the scans are then reduced to a small amplitude and the specimen is measured in the usual EBC steady-state current circuit using a sensitive current meter (see Fig. 7.28) but with a meter in place of the amplifier. In some cases, because the normal methods for earthing the different component parts of the SEM involve fairly complex paths through the equipment, it is necessary to employ a battery-driven self-contained meter; this avoids "earth loops" which otherwise upset accurate current measurements in the range 10^{-11}–10^{-7} A. One of the connections to the current-sensitive meter is then joined to the SEM positive potential via a single connection (probably to the body of the instrument). The primary current (i_p) is measured in the usual way with a small Faraday cup fixed on the SEM stage beside the specimen. Again the normal secondary emission image of the cup can be used to make sure that all of the beam enters it. The cup connnection is brought out via an independent pin on the multiwire socket provided with the instrument.

Rapid switching is not convenient in a standard SEM, and measurements of this kind are not suitable for determining the time response of the EBC effect.

II. Path length and energy loss of fast electrons

In the range of energies here of interest, electrons interact with matter mainly by producing secondary knock-on electrons of much lower energy, i.e. by ionizing atoms in the outer orbits and passing some kinetic energy on to the electrons released, by elastic scattering by atomic nuclei, inelastic scattering by atomic electrons, and the production of bremsstrahlung (X-rays). Another possible process is through the creation of excitons by which the released atomic electron remains loosely bound to the core. Normally an electron is gradually slowed down, all the time changing its direction until its range is exhausted and it has become so slow that it attaches itself to an atom, a molecule, or an electrode. The production of X-rays in which the electron loses a large fraction of its energy is a rare event. It can, however, be of importance because of its reverse process, the photoelectric absorption of X-rays, by which a bound electron is set free. Since X-rays travel much further than cathode rays, the double conversion electron to X-ray to electron can make the effective range of electrons appear much extended. Also, exposure to X-rays is a means of producing electron tracks uniformly in thick specimens. Normally only a very small fraction of the electron energy, perhaps 10^{-3}–10^{-4}, is transformed into X-rays.

A. Cloud chamber tracks

The passage of electrons through air at 1 atmosphere pressure is illustrated by a cloud-chamber photograph (Fig. 2.9). High energy β-rays go in straight lines (if they are not deflected by the presence of a magnetic field). The relative proportion of energy loss is small and scattering is slight. But at low energies the electron suffers erratic deflections and its trajectory curls up as shown in Fig. 2.10. Corresponding tracks in denser matter would be similar but reduced in size in inverse proportion to the densities of the materials. The average dissipation of energy of electrons entering a medium as a function of depth below the surface is therefore a combined effect of energy loss along the track and its changes of direction. As a result, the dissipation of energy poses a very difficult theoretical problem which, in spite of many efforts by physicists like Lenard, Bothe, Fermi, and others (see e.g. Birkhoff, 1958), evaded solution until Lewis (1950) opened up a correct approach which was nevertheless useless in pre-computer days. This calculation became the basis of extended work and computations by Spencer (1959).

B. Thomson–Whiddington–Bethe law of energy losses

Let us look at the energy losses of a fast electron along its path. This was first studied by Whiddington who based his ideas on concepts initiated by Thomson (1906). Their idea was that a fast electron of kinetic energy $E = \frac{1}{2}mu^2 = qV$ passes a lightly bound electron within the stopping medium. For a time Δt it exerts on it a mean force $\bar{F}$, and imparts on it an energy $\Delta E = \frac{1}{2}mv^2$ where v is the velocity of the lightly bound electron. The time of

2. ELECTRON BEAMS AND THEIR INTERACTION WITH MATTER   45

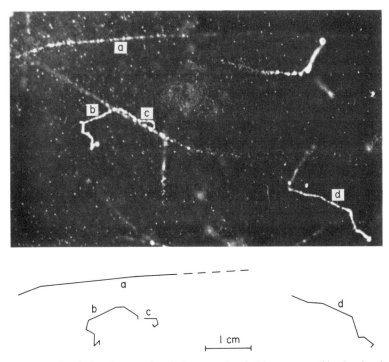

Figure 2.9. Cloud chamber tracks of electrons in air (the source of ionization here is random radioactivity). Tracings of selected tracks from the photograph are underneath. (a) Track of an electron having an energy above about 100 keV, for which the number of deflections per unit path is small and little energy is lost; this electron left the cloud chamber through its wall before coming to rest. (b)–(d) Tracks of electrons with energies in the range below about 30 keV; the deflections become more frequent as the electrons slow down until eventually they make no further progress. The behaviour of electrons in condensed matter would be similar but with the tracks reduced in inverse proportion to the density of the material traversed. (Photograph courtesy of D. J. H. Wort)

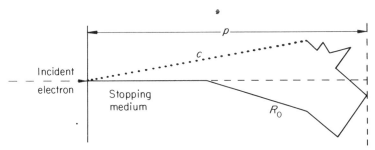

Figure 2.10. Electron track showing true range or integrated path length R_0 (continuous line), chord range c (dotted line), and projected range p.

interaction is inversely proportional to u and thus $\Delta t \propto u^{-1} \propto E^{-1/2}$. Whence

$$\Delta E = \tfrac{1}{2} m(\bar{F}\Delta t/m)^2 \propto E^{-1} \tag{2.1}$$

The kinetic energy lost by the passing electron per unit path length (R) will be equal to the energy lost per collision multiplied by the number of collisions (n). Thus

$$-\mathrm{d}E/\mathrm{d}R = -n\Delta E \propto E^{-1} \tag{2.2}$$

or

$$E^2 \propto (R + \text{const}) \tag{2.3}$$

and the path travelled by a fast electron starting with an energy qV_0 and ending with an energy qV_R comes to

$$R = (V_0^2 - V_R^2)/a \tag{2.4}$$

It has zero energy after travelling a path $R_0 = V_0^2/a$, where a is a constant and R_0 is called the integrated path length or residual range. As might have been expected, the constant a was found to be proportional to the density ρ of the matter traversed. Hence, by making use of the relation $E = qV$ and putting $a = b\rho$, we get

$$V_R^2 = V_0^2 - b\rho R \tag{2.5}$$

This is one form of the "Thomson–Whiddington law" where the subscripts have the same meaning as those in equation (2.4).

For a long time this formula (which predicts $R_0 \propto V_0^2$) was accepted for fast electrons. An improved formula, taking into account the binding energy of the atomic electron was given by Bohr (1915). The wave-mechanical theory of the passage of fast electrons through matter was developed by Bethe (1930, 1933). According to Bethe, the path traversed (in cm) per unit energy lost comes to

$$-(\mathrm{d}R/\mathrm{d}E)_{\mathrm{av}} = 65 \cdot 12(A/\rho Z)(\beta^2/S) \tag{2.6}$$

where E is in units of keV, and

$$S = \ln E + \ln \beta^2 - \ln(1-\beta^2) + 21\cdot 05 - 0\cdot 3069\beta^2 - 1\cdot 386\sqrt{(1-\beta^2)} - 2\ln J \tag{2.7}$$

The absorbing material is characterized by its density ρ (g cm^{-3}), atomic number Z, atomic weight A, and the mean excitation energy of its electrons J (eV). The symbol β represents the ratio of the instantaneous electron velocity to the velocity of light. The path length

$$R_0 = -\int_0^{E_0} (\mathrm{d}R/\mathrm{d}E)_{\mathrm{av}} \mathrm{d}E \tag{2.8}$$

will be given in μm if the instantaneous energy E and the initial energy E_0 are measured in keV. For a complex molecule (formula U_pY_q), effective values of Z/A and J may be calculated following the formulae due to Lane and Zaffarano (1954):

$$(Z/A)_{\text{eff}} = (pZ_U + qZ_Y)/(pA_U + qA_Y) \quad (2.9)$$

$$\ln J_{\text{eff}} = (pZ_U \ln J_U + qZ_Y \ln J_Y)/(pZ_U + qZ_Y) \quad (2.10)$$

Bloch (1933) suggested

$$\ln J = \ln Z + 2 \cdot 6 \quad (2.11)$$

A better approximation can be obtained by the use of the values for J derived from α-particle stopping experiments published by Bakker and Segrè (1951) as evaluated by Bethe and Ashkin (1953); these are shown in Table 2.1. It is seen that a good approximation to $\ln J$ can be taken as a

Table 2.1 Mean excitation energy J(eV) for elements of atomic number Z. (Bethe and Ashkin, 1953)

Z	1	3	4	6	13	26	29	47	50	74	82	92
$\ln J$	2·74	3·52	4·10	4·33	5·08	5·48	5·61	6·04	6·13	6·52	6·55	6·69

function of Z only, so missing values can be interpolated. R_0 depends on $\ln J$, which increases only for low values of Z, and on A/Z which is approximately constant; but mainly, it is inversely proportional to ρ. It was often postulated by experimentalists, such as Holliday and Sternglass (1959) and Kanter (1961), that the range–energy relation depends mainly on density. This should not be taken too seriously; Fig. 2.11 summarizes published data for ranges in mg cm^{-2} vs. electron energy. For low energies in particular, the ranges are seen to vary by as much as a factor of 7.

Now, it appears (King, 1960) that to a good approximation a variant of the Thomson–Whiddington relation, used at least since 1952 (Katz and Penfold),

$$R_0 = BE_0^n \quad (2.12)$$

is equivalent to equations (2.7) and (2.8). King in particular has confirmed this equivalence by computing R_0 for a large number of parameters for values of E_0 between about 10 and 100 keV. We can therefore determine B and n analytically by using equations (2.8) and (2.12) to obtain

$$dR/dE = nBE^{n-1} = (dR/dE)_{\text{av}}E \quad (2.13)$$

for two values of E. With $E = 10$ and 100 keV respectively, a simple calculation then leads to

$$n = 1 \cdot 898 + \log_{10}(18 \cdot 75 - 2\ln J) - \log_{10}(23 \cdot 56 - 2\ln J) \quad (2.14)$$

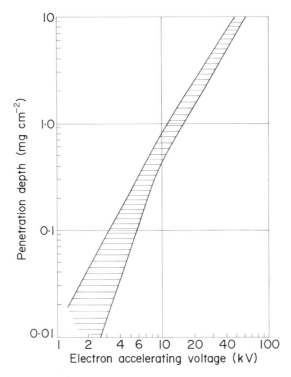

Figure 2.11. Approximate penetrating power of fast electrons into solids when P is assumed to be a function of density only; measured values lie within the shaded part of the figure. Especially at low energies, the ranges so deduced can be seen to vary by as much as a factor of 7.

and

$$\log_{10}B = 1\cdot 3939 - n - \log_{10}(nZ\rho/A) - \log_{10}(18\cdot 75 - 2\ln J) \quad (2.15)$$

More rapidly, by reference to a table of range–energy such as Table 2.5, we obtain, in terms of the range R_{10} of 10 keV electrons and R_{100} of 100 keV electrons,

$$n = \log_{10}(R_{100}/R_{10}) \quad (2.14a)$$

and

$$B = R_{10}^2/R_{100} \quad (2.15a)$$

It must be emphasized that R is the length of the devious path traversed by the electron, not the depth to which the electron on the average penetrates into the dielectric.

The "integrated path length" or "residual range", R_0, being the mean of the sum of a large number of elements, is subject to considerable straggling.

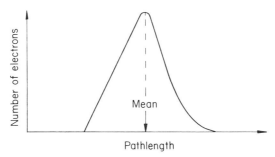

Figure 2.12. Distribution of path lengths of 19·6 keV electrons in oxygen gas, showing the "straggling" of ranges. (King, 1960)

An illustration of this is shown in Fig. 2.12 which has been given by King (1960) using the results of Williams (1932) for 19·6 keV electrons in oxygen gas.

The elastic scattering of particles by atomic nuclei was first calculated by Rutherford (1911). He found, for the differential cross-section (dϕ) per atom for scattering an electron through an angle between θ and dθ into a solid angle,

$$d\phi = 2\pi b db = \frac{\pi Z^2 q^4}{2m_0^2 u^4} \frac{\sin\theta d\theta}{\sin^4(\theta/2)}, \qquad (2.16)$$

The shielding effect of the atomic electrons, neglected by Rutherford, was taken into account by Bethe (1930), Bullard and Massey (1930), and Molière (1947). Mott (1929, 1932) calculated cross-sections for the relativistic Dirac electron. This "Mott cross-section" was corrected for shielding by Doggett and Spencer (1956). The contribution to scattering made by atomic electrons was given by Fano (1953).

Ruthemann (1948) discovered that electrons passing through very thin films suffer discrete energy losses of between 3 and 300 eV which are characteristic of the material of the films and independent of the energy of the cathode rays. These discrete losses are attributed to plasma oscillations. But this does not mean that they are additional to the energy losses discussed above. In condensed matter, electronic processes differ from those in gases owing to the mutual coupling of the outer shells. Because of this coupling, individual energy losses may appear as plasma oscillations rather than as excitations or ionizations.

C. The problem of penetration

The theories discussed so far refer to the events that occur along the path of the electron. But in studies of induced conductivity a plane surface is usually exposed to a beam of cathode rays falling normally on it, and the effect will depend on the energy dissipated as a function of depth below the surface, or the total energy dissipated and the depth to which the cathode rays penetrate. This would equal the residual range if the electrons go straight, but they do not do this.

Hence the simple idea of a gradual loss of energy according to the Thomson–Whiddington law or its successors does not correspond to the experimental conditions. Early investigators such as Lenard tried to account for the behaviour of cathode rays by introducing notions such as exponential absorption and detour-factors in addition to energy loss. Later a considerable number of attempts were made to combine the theory of energy loss of single particles with that of scattering or with experimental results (empirical formulae) concerning the fraction of electrons transmitted through films, but it was not until Lewis (1950) and Spencer (1955) succeeded in setting up and solving a general differential equation for the flux of electrons that reliable theoretical data for the distribution of energy losses in solids came in sight.

Consider a unit spherical probe, at a distance z from the source plane, i.e. the plane at which the electrons travel with energy E_0 and with a residual range R_0. The flux through the probe can be written as

$$F(R,\theta,z) \, dR \, d\Omega \, d\theta$$

This is the number of electrons making angles between θ and $(\theta + d\theta)$ with the normal to the source plane, having residual ranges between R and $(R + dR)$, crossing the probe a distance z from it. This flux changes with z and, after introducing the scattering cross-section and integrating over the angular variable, a differential equation for F with the variables R and z is obtained. This, after a considerable amount of manipulation, is solved in terms of moments of the distribution, from which finally the dissipation of energy is obtained as a function of depth below the reference plane.

No assumptions are made that restrict the deflection or path of the electrons. As a final result Spencer (1959) computed extensive tables of the spatial distribution of energy dissipation by electrons from 25 or 50 keV upwards, in a number of representative substances, i.e. carbon, aluminium, copper, tin, lead, air, and polystyrene, using previous results due to Nelms (1956) and Doggett and Spencer (1956). Interpolation procedures are given.

There is one difference between the experimental situation and that described by Spencer's theory. In the experiment the zero plane separates medium and vacuum, whereas in his theory the medium extends over all space. Thus in the experiment the energy dissipated in the medium is E_0 less the energy E_1 lost by back-scattering, while the theory applies to the case where a small portion E_2 of the back-scattered energy is regained by the medium and dissipated near the zero plane. This must be allowed for separately but it is normally a small and sometimes a negligible correction. An extract from Spencer's tables is given by way of illustration (Table 2.2). The residual range r_0 is given in g cm^{-2}, so the range in cm is $R_0 = r_0/\rho$ where ρ is the density of the material; $x = s/R_0$ measures depth below the surface in units of R_0. On average the energy lost per cm by an electron at s cm below the reference plane is

$$dE/ds = (dE/dr)_{E_0} K(x)\rho \text{ MeV cm}^{-1} \quad (2.17)$$

$$x = s\rho/r_0 = s/R_0 \quad (2.18)$$

Figure 2.13 shows $K(x)$ as a function of x for aluminium; $E_0 = 50$ keV.

There are some sources of doubt concerning Spencer's results. One concerns straggling (Fig. 2.12) neglected in Spencer's calculations. Berger (1963) has, however, repeated some of Spencer's work by a Monte Carlo method which allows for straggling of the residual range. He has also shown how solid–vacuum interfaces can be introduced in the calculation and how details of the back-scattering and transmission process can be calculated, giving specimen computations. Table 2.5 gives values of R_0 for 10 kV and 100 kV electrons in a number of substances. The results are not significantly different from Spencer's. Secondly, Spencer's computations stop at $E_0 = 25$ keV. One question is, down to what energy can the data be extrapolated? Experimental evidence will be given below which suggests that an extrapolation down to at least 10 keV is justified. This extension down to 10 kV was undertaken by Berger and Seltzer (1964). Finally, it should be pointed out that in dielectrics, at high electron currents, very large charges can be set up which can alter the distribution of energy dissipation. However, for low currents, or low doses, no evidence has appeared against the validity of Spencer's or Berger's tables.

Tables like 2.2 give $(dE/dr)_{E_0}$, r_0, and $K(x)$ as a function of x for the elements Cu, Al, Sn, C, and Pb, and for air and polystyrene, for initial energies between 25 and 200 keV. $(dE/dr)_{E_0}$ is the stopping power at the initial energy E_0. This is the energy dissipated g^{-1} cm^{-2} along the actual path. $x = 1$ denotes the depth that the electron would have reached if it had remained permanently travelling normal to the surface. It is seen, however, from Fig. 2.13 that $K(x)$ vanishes at $x = 0.875$. This is, of course, because no electron will pass through the medium without being deflected. It is also seen that $K(x)$ decreases with x very rapidly once it is down to 0.1. Let x_R be the corresponding value of x in the neighbourhood of $K(x_R \approx 0)$. One would therefore expect the corresponding value of s, $s_R = x_R R_0$, to be near to the observed range of the electron. Moreover, for all values of E_0, $K(0.77) = 0.1$,

Table 2.2. Energy dissipation function $K(x)$ and input data for aluminium. (Spencer, 1959)

$E_0 = 0.025$ MeV
Stopping power
$(dE/dr)_{E_0} = 8.50$ MeV(cm^2 g^{-1})
Residual range
$r_0 = 0.00167$ g cm^{-2}

x	$K(x)$	x	$K(x)$
0.000	2.009	0.500	1.939
0.025	2.224	0.525	1.743
0.050	2.419	0.550	1.545
0.075	2.593	0.575	1.347
0.100	2.743	0.600	1.151
0.125	2.870	0.625	0.959
0.150	2.973	0.650	0.774
0.175	3.054	0.675	0.601
0.200	3.111	0.700	0.443
0.225	3.145	0.725	0.305
0.250	3.153	0.750	0.193
0.275	3.135	0.775	0.108
0.300	3.090	0.800	0.051
0.325	3.018	0.825	0.019
0.350	2.920	0.850	0.005
0.375	2.797	0.875	0.001
0.400	2.653	0.900	0.000
0.425	2.492	0.925	0.000
0.450	2.316	0.950	0.000
0.475	2.131	0.975	0.000

$E_0 = 0.05$ MeV
Stopping power
$(dE/dr)_{E_0} = 5.14$ MeV(cm^2 g^{-1})
Residual range
$r_0 = 0.00561$ g cm^{-2}

x	$K(x)$	x	$K(x)$
0.000	1.972	0.500	1.919
0.025	2.180	0.525	1.710
0.050	2.372	0.550	1.501
0.075	2.546	0.575	1.293
0.100	2.703	0.600	1.089
0.125	2.842	0.625	0.892
0.150	2.960	0.650	0.706
0.175	3.055	0.675	0.534
0.200	3.124	0.700	0.382
0.225	3.167	0.725	0.254
0.250	3.181	0.750	0.153
0.275	3.168	0.775	0.081
0.300	3.125	0.800	0.035
0.325	3.053	0.825	0.012
0.350	2.952	0.850	0.003
0.375	2.826	0.875	0.000
0.400	2.675	0.900	0.000
0.425	2.505	0.925	0.000
0.450	2.319	0.950	0.000
0.475	2.122	0.975	0.000

$E_0 = 0.10$ MeV
Stopping power
$(dE/dr)_{E_0} = 3.24$ MeV(cm^2 g^{-1})
Residual range
$r_0 = 0.0183$ g cm^{-2}

x	$K(x)$	x	$K(x)$
0.000	1.916	0.500	1.857
0.025	2.129	0.525	1.660
0.050	2.324	0.550	1.462
0.075	2.501	0.575	1.263
0.100	2.659	0.600	1.067
0.125	2.797	0.625	0.877
0.150	2.913	0.650	0.696
0.175	3.004	0.675	0.528
0.200	3.068	0.700	0.379
0.225	3.105	0.725	0.252
0.250	3.112	0.750	0.152
0.275	3.091	0.775	0.081
0.300	3.040	0.800	0.035
0.325	2.961	0.825	0.012
0.350	2.856	0.850	0.003
0.375	2.726	0.875	0.000
0.400	2.577	0.900	0.000
0.425	2.412	0.925	0.000
0.450	2.235	0.950	0.000
0.475	2.049	0.975	0.000

$E_0 = 0.20$ MeV
Stopping power
$(dE/dr)_{E_0} = 2.22$ MeV(cm^2 g^{-2})
Residual range
$r_0 = 0.0570$ g cm^{-2}

x	$K(x)$	x	$K(x)$
0.000	1.822	0.500	1.750
0.025	2.022	0.525	1.566
0.050	2.205	0.550	1.380
0.075	2.372	0.575	1.194
0.100	2.521	0.600	1.009
0.125	2.653	0.625	0.829
0.150	2.763	0.650	0.658
0.175	2.850	0.675	0.499
0.200	2.910	0.700	0.358
0.225	2.943	0.725	0.238
0.250	2.948	0.750	0.143
0.275	2.924	0.775	0.075
0.300	2.872	0.800	0.033
0.325	2.793	0.825	0.011
0.350	2.690	0.850	0.002
0.375	2.566	0.875	0.000
0.400	2.425	0.900	0.000
0.425	2.270	0.925	0.000
0.450	2.104	0.950	0.000
0.475	1.930	0.975	0.000

2. ELECTRON BEAMS AND THEIR INTERACTION WITH MATTER

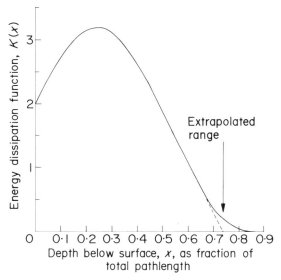

Figure 2.13. Calculated energy loss distribution from a plane unidirectional source of 50 keV electrons in aluminium; the horizontal scale is in units of the integrated path length R_0. (Data from Spencer, 1959)

so $0 \cdot 77 R_0$ would be the "Spencer range" s_R. Similar rules apply to the other materials listed in Spencer's tables (see Table 2.3); very roughly,

$$x_R = 0 \cdot 9 - 0 \cdot 05 \sqrt{Z} \qquad (2.19)$$

and finally,

$$s_R = x_R B E^n \; \mu m \qquad (2.20)$$

In other words, for each material the penetration depth is a range defined by the dissipation of energy proportional to the residual range.

Table 2.3. Ratio of penetration depth to integrated path length.

Material	x_R
Polystyrene	0·75
6 Carbon	0·75
13 Aluminium	0·77
29 Copper	0·6–0·63
50 Tin	0·5–0·52
82 Lead	0·4

D. Experimental measurements using phosphorescence glows

A range is observed either as the depth to which electrons impinging normally on a platelet penetrate into it, or by the maximum thickness of platelet that allows the electrons to pass. This point of vanishing transmission depends critically on the conditions of observation; it is called the "ultimate" range. A "practical" or "extrapolated" range is defined by the intercept with the abscissa of an extrapolated straight portion of the curve of transmitted number vs. initial electron energy. As an alternative to measuring the *number* of the transmitted electrons, their *energy* can be measured by the ionization caused in an ionization chamber or by the glow caused in a phosphor. Luminescence also permits an examination of the energy dissipated in the platelet itself under test. Dissipation of energy is of course the quantity responsible for EBC.

The proportionality of energy loss to luminescence produced was used by Ehrenberg and Franks (1953) and Ehrenberg and King (1963) to determine energy losses and ranges under conditions equivalent to those established in EBC experiments. Figure 2.14 shows the principle of the method employed.

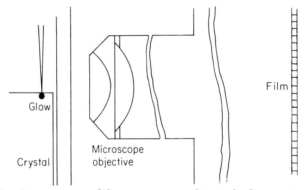

Figure 2.14. Arrangement of the apparatus used to study the penetration of electrons into phosphors by observing the glow. (Ehrenberg and King, 1963)

A narrow beam of electrons was focused so as to impinge normally upon a flat horizontal face of a phosphor crystal, and the resultant luminous glow was observed through the adjacent vertical face. The microscope was fitted with a specially corrected objective lens (40×, N.A. 0·65) and a film carrier in the focal plane of the eyepiece lens. Seven successive exposures could be made on each film. The glow was always accompanied by its mirror image in the horizontal face of the specimen.

The phosphors used in these tests were polystyrene, calcium tungstate, cadmium tungstate, potassium iodide, rubidium iodide, and caesium iodide. Both of the tungstates are birefringent and thus two penetration figures

2. ELECTRON BEAMS AND THEIR INTERACTION WITH MATTER   55

could be seen through the microscope. When these overlapped, one of them was suppressed by the insertion into the microscope of a correctly orientated Polaroid filter. Exposures were made at selected electron energies in the range 10–80 keV, and the dose was kept at a low level to avoid radiation damage (formation of F-centres).

Some enlargements of photographs of typical glows are shown in Fig. 2.15, though only a little of the true range of luminosities can be reproduced. Relative brightnesses within a glow were found by taking six photographs on one film, the exposure times being in the ratio 1:2:4:8:16:32 (usually a seventh exposure was made as a check of one of these). The seven developed images were enlarged by a high contrast two-stage process. Then well defined contours of equal brightness were obtained as the boundaries of the images, and the relative brightness of these contours was taken to vary inversely as their initial exposure time. These contours thus covered a

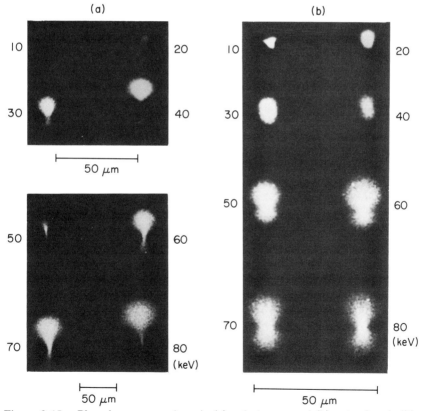

Figure 2.15. Phosphorescence glows in (a) polystyrene and (b) potassium iodide. (Ehrenberg and King, 1963)

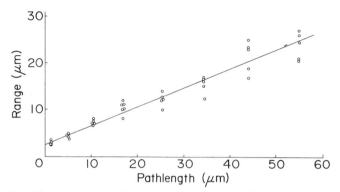

Figure 2.16. Measured practical ranges in potassium iodide as a function of calculated integrated path lengths.

brightness range of 2^5 which could be extended by combining two film-strips. The evaluation of the enlargements was carried out in two stages, the first concerned with the range of the electrons and the second with the dissipation of their energy.

The ranges so measured plotted against the residual ranges were found to lie on straight lines. This confirmed the principle of scaling. An example is shown in Fig. 2.16. The finite intercept is probably instrumental. The results are given in the last two columns of Table 2.4. This table gives the derived values of the parameters $(Z/A)_{\text{eff}}$ and J_{eff} which could be used in Spencer's

Table 2.4. Penetration of fast electrons into phosphors. (Ehrenberg and King, 1963)

	$\rho(\text{g cm}^{-3})$	$(Z/A)_{\text{eff}}$	$(\log J)_{\text{eff}}$	B	n
Polystyrene	1·05	0·534	4·120	0·041	1·77
KI	3·13	0·434	6·044	0·039	1·65
RbI	3·55	0·424	6·158	0·037	1·65
CsI	4·51	0·416	6·312	0·031	1·64
CaWO$_4$	6·12	0·438	5·898	0·019	1·65
CdWO$_4$	7·90	0·427	6·058	0·015	1·66

	p	u	$a(\mu\text{m})$
Polystyrene		0·79 ± 0·01	3·4 ± 0·6
KI	0·40 ± 0·01	0·52 ± 0·02	4·5 ± 0·5
RbI	0·31 ± 0·02	0·51 ± 0·01	3·3 ± 0·3
CsI	0·47 ± 0·02	0·59 ± 0·02	2·5 ± 0·4
CaWO$_4$	0·39 ± 0·01	0·54 ± 0·01	1·5 ± 0·2
CdWO$_4$	0·38 ± 0·03	0·37 ± 0·024	3·0 ± 0·2

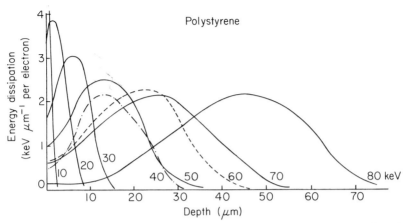

Figure 2.17. Dissipation of the energy of fast electrons in polystyrene as a function of depth below the surface; the parameter is the energy of the incident electrons. (Ehrenberg and King, 1963)

formulae for range–energy. From the data, the integrated path length is given by $R = BE^n$, the "practical" range (penetration depth) is $P = pR$, and the "ultimate" range is $U = uR + a$, where E is in units of keV and ranges are in μm.

Curves showing the dissipation of energy as a function of depth are in Figs 2.17 and 2.18, each curve being normalized by equating the energy represented by the area under the curve to the incident beam energy. The energy dissipation has a maximum rate at a depth that increases with increasing beam energy. In order to allow for back-scattering, the experi-

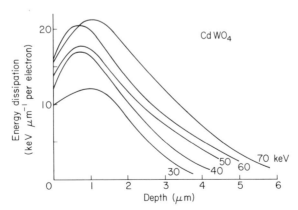

Figure 2.18. Decrease in energy of fast electrons in cadmium tungstate with depth below the surface, for several values of the incident energy. (Ehrenberg and King, 1963)

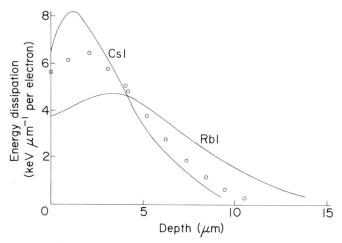

Figure 2.19. Energy dissipation for 50 keV electrons in rubidium iodide and caesium iodide. The points (○) indicate Spencer's values for a material (tin) of atomic number 50 and density 4 g cm^{-3} at the same electron energy.

mental curves at 50 keV for RbI ($Z = 45$) and CsI ($Z = 54$) were reduced by 25% before they were reproduced in Fig. 2.19 together with Spencer's values for tin ($Z = 50$). CdWO$_4$ has an average atomic number 26 and should not differ greatly from Cu for which it is 29. The correction for back-scattered energy is estimated at 15%, and the experimental curve was similarly reduced in Fig. 2.20. It is seen that agreement between theory and experiment is very satisfactory.

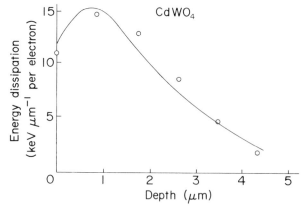

Figure 2.20. Energy dissipation for 50 keV electrons in cadmium tungstate; the experimental curve is taken from Fig. 2.18 but reduced by 15%. The points (○) show Spencer's values for a material (copper) of atomic number 29 and density 7·9 g cm^{-3}.

According to Sternglass (1955) and Holliday and Sternglass (1959) the energy lost by back-scattering is a fraction of the incident energy given approximately by $(0{\cdot}45 + 2 \times 10^{-3}Z)\eta$, where Z is the atomic number and η the probability of back-scattering. This has been reviewed by Archard (1961). A diagram consolidating all results suggests $\eta = 10^{-2}(Z - Z^2/200)$.

As mentioned above, prior to the work by Lewis and Spencer a considerable number of attempts were made to provide empirical or semi-empirical formulae for ranges and energy dissipation in individual materials and for the parameters that control them. They are no longer relevant or justified.

E. Scaling of variously defined ranges

It is not obvious that the variously defined ranges should be closely related to each other, but it has been found that they are related by constant factors of the order of unity. This property of "scaling" was first suggested to account for the constancy of the back-scattering coefficient with respect to electron energy. Practical ranges measured in aluminium by Schonland (1923, 1925), Lane and Zaffarano (1954), Young (1956), and Holliday and Sternglass (1959) are close to the values of the corresponding integrated path lengths multiplied by a factor of 0·72, which is in good agreement with the value derived from Spencer's calculations. Cosslett and Thomas (1965) confirmed scaling for films of Al, Cu, Ag, and Au at incident energies between 10 and 30 keV. In a higher energy range (160–960 keV) Seliger (1955) has shown that the shapes of transmission curves for aluminium, silver, and lead are independent of energy if the foil thickness is plotted in units of the integrated path length. Fleeman (1954) obtained similar evidence of scaling for 500–1500 keV electrons in beryllium, aluminium, copper, cadmium, and gold, and Makhov (1960) for silicon, germanium, Al_2O_3, and bismuth.

Scaling in this sense implies a balancing of energy loss and scattering (Seliger, 1955). It is easily seen that scaling would hold if both the range of relaxation for scattering and the integrated path length were proportional to the square of the energy of the electrons, the former according to Rutherford's formula and the latter by the Thomson–Whiddington law. An advanced theory of scaling was put forward by Blanchard (1954) who showed that scaling can be related to more realistic approximations for energy loss and scattering. As shown above, the property of scaling applies equally to a range defined by energy dissipation.

The ratio between the practical range, the ultimate range, and the total integrated path length as a function of Z for a number of elements is shown in Fig. 2.21.

60 ELECTRON BOMBARDMENT INDUCED CONDUCTIVITY

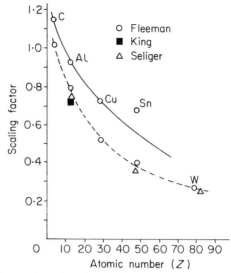

Figure 2.21. Scaling ratios of variously defined ranges; the scaling ratio for elements can be taken as a function of atomic number (Z) only. The solid curve gives the factor by which the total integrated path length (R_0) must be multiplied to give the ultimate range, and the dashed curve gives the appropriate factor for the projected range.

F. Range–energy tables

The original range problem as formulated by Bethe made the basic assumptions that the electron velocity was small compared with the velocity of light but the energy of the penetrating electrons was large compared with the ionization energy of the material being traversed. Moreover, Bethe assumed that the electrons belonging to the atoms of the stopping medium could be represented by hydrogen-like wavefunctions. Lewis and Spencer took this further along the lines already described and took into account the relativistic scattering cross-section.

Berger and Seltzer (1964) then used a Monte Carlo method of computation to allow for straggling of the residual ranges, and they also allowed for back-scattering by introducing a vacuum–solid interface at the point of entry of the incident electrons. As with Spencer, the energy loss was treated in the continuous slowing down approximation. The values of R_0 (Table 2.5) are not significantly different from those already given. From these figures for 10 and 100 keV electrons a good approximate range formula of the form $R_0 = BE_0^n$ can be derived by methods described above [equations (2.14a) and (2.15a)]. The penetration depth, practical range, and ultimate range are obtained using the appropriate scaling ratios from Fig. 2.21 or Tables 2.3

Table 2.5. Total integrated path length R_0 of fast electrons (initial energy 10 or 100 keV) in various stopping media according to the continuous slowing down approximation.* (Berger and Seltzer, 1964)

	$10^4 R_0$ (10 keV)	$10^2 R_0$ (100 keV)		$10^4 R_0$ (10 keV)	$10^2 R_0$ (100 keV)
Hydrogen	1·07 g cm^{-2}	0·663 g cm^{-2}	Water	2·44	1·40
Helium	2·47	1·46	Carbon dioxide	2·88	1·62
Lithium	2·80	1·67	Silver chloride	5·01	2·34
Beryllium	2·99	1·73	Silver bromide	5·38	2·46
Carbon	2·82	1·60	Sodium iodide	5·49	2·51
Nitrogen	2·87	1·62	Lithium iodide	5·82	2·60
Oxygen	2·90	1·63	Methane	2·00	1·18
Neon	3·23	1·75	Ethylene		
Magnesium	3·40	1·81	(Ethene)	2·28	1·33
Aluminium	3·52	1·86	Polyethylene		
*Silicon	*3·6	*1·9	(Polythene)	2·44	1·41
Argon	4·05	2·08	Xylene	2·44	1·41
Iron	4·26	2·11	Toluene	2·47	1·42
Copper	4·56	2·21	Acetylene		
Krypton	5·19	2·43	(Ethyne)	2·50	1·44
Silver	5·66	2·52	Polystyrene	2·50	1·44
Tin	6·00	2·64	Stilbene	2·54	1·46
Xenon	6·35	2·74	Lucite (Perspex)	2·51	1·44
Tungsten	7·60	2·98	Anthracene	2·58	1·48
Gold	7·94	3·03	Air	2·89	1·63
Lead	8·25	3·10	Photographic		
Uranium	9·24	3·25	emulsion	4·61	2·22

*The values for silicon were obtained by interpolation, using a plot of R against Z.

and 2.4. The range in cm is obtained by dividing the range given in units of g cm^{-2} by the density of the stopping medium (in whatever state) in units of g cm^{-3}.

G. Range measurements on primary electrons of energy below about 10 keV

We shall now discuss some particularly relevant experimental results for EBC. Vyatskin and Makhov (1958) carried out carefully planned experiments on the transmission of electrons having between 2 and 20 keV energy through thin films of Ge, Cu, Si, and Bi. The films between 50 and 180 nm thick were prepared by evaporation on smooth rock-salt surfaces, and subsequent removal of the salt base by dissolving it in distilled water. For each film the proportion of transmitted electrons was measured as a function of the electron energy, allowing for reflected electrons and secondary emission. Makhov (1960) re-analysed these results, dissatisfied with the accuracy

Table 2.6. Range–energy parameters. (Makhov, 1960)

Material	E_0 (keV)	C (μg cm^{-2})	n
Al	2·5–27	3·8	1·68
Si	2–20	3·4	1·65
Cu	2–20	5·8	1·53
Ge	2–18	6·5	1·47
Bi	3–20	4·2	1·44
Al$_2$O$_3$	1–9	4·6	1·65

with which the point of zero transmission could be determined. He suggested that a more reliable measure of electron penetration is the material thickness (S) at which the flux of electrons is reduced to 1/e of the incident flux, and found—taking into account the results obtained by Lane and Zaffarano (1954), Young (1956), and Butkevich and Butslov (1958)—that, within perhaps 10%, S satisfied relations

$$S = CE_0^n$$

with values for C and n given in Table 2.6. Makhov also suggested that the current through the film depends on its thickness y (measured in units of S) as

$$N \propto e^{-y^2} \quad \text{for Al, Si, Cu, Ge, Al}_2\text{O}_3$$

$$N \propto e^{-y} \quad \text{for Bi}$$

The practical range then is about 1·5S. Makhov's parameters check well with the Bethe and Spencer values. Similar measurements by Vyatskin and Khramov (1975) refined these results and showed that the exponent of y is higher for light substances.

Feldman (1960), following Young (1956) and Koller and Alden (1951), used the luminescence excited in fluorescent films to measure the range of electrons of energy between 1 and 10 keV both in the phosphors and in metallic films deposited on the phosphor films. In the first case the light output of a fluorescent film ceases to increase with increasing electron energy when the range S_0 of the electrons equals the thickness of the film; in the second case the phosphor commences to glow when the range of the electrons in the metallic deposit exceeds its thickness. The quantity observed in these experiments is the energy dissipated by the electrons (which also causes EBC), and thus the difficulty of producing and handling self-supporting films is circumvented. Feldman has summarized his results in the form $S_0 = BE_0^n$ (E_0 in keV, S_0 in 10^{-9} m) with the parameters shown in Table 2.7. His measured range–energy points and curves are reproduced as Fig.

2. ELECTRON BEAMS AND THEIR INTERACTION WITH MATTER

Table 2.7. Range measurements at 10 keV. (Feldman, 1960)

Solid	n	B	Range (μm)	Solid	n	B	Range (μm)
Mg	1·7	42	2·46	Pb	2·2	3·2	0·49
Al	1·9	18·3	1·75	CaF$_2$	2·9	3·8	2·74
Ni	1·8	9·2	0·57	ZnS	2·4	6·3	1·51
Ag	1·7	9·7	0·51	MgSiO$_3$	2·7	4·0	1·75
Sn	1·9	10·5	0·92	Zn$_2$SiO$_4$	3·0	1·2	1·37
Au	2·9	0·27	0·24	CaWO$_4$	2·7	1·7	0·81

2.22. Feldman's measurements, and his parameters, *if restricted to electron energies between 1 and 10 keV*, appear to be reliable, allowing for a reasonable margin of error, but the parameters *cannot be credited* as valid over a range not covered by his observations. In fact, his points can also be reproduced by parameters very different from those given in the table. Moreover, it must be realized that at low electron energies the power law need not hold, and that appreciable percentage errors can arise owing to space charges in the phosphors or inaccuracies in the correction for the "dead" layer which,

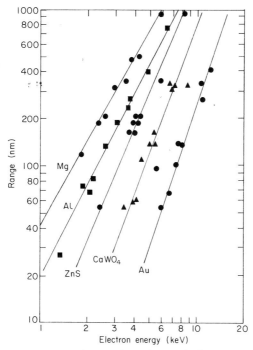

Figure 2.22. Some representative measured practical range–energy curves for electrons below 10 keV in energy (Feldman, 1960)

incorporated in the parameters, lead to gross errors at higher electron energies. For example, $CaWO_4$ is also listed in Table 2.4, and the parameters given there ($n = 1.65, B = 0.01$) lead to values acceptable with respect to the points marked on Fig. 2.16 whereas Feldman's parameters ($n = 2.7$, $B = 1.7$) lead to the unacceptably large range of $68\,\mu m$ for 50 keV electrons.

Measurement of transmission (practical ranges) through films of As_2S_3—a material much studied in EBC experiments—was reported by Spear (1955) and Bowlt and Ehrenberg (1969). Figure 2.23 includes these values together with the range–energy relation calculated. The agreement is quite satisfactory. The best fit to the experimental points is obtained from a relation $d = 0.013E^{1.67}$ (d in μm, E in keV), $Z_{av} = 24$, $(Z/A)_{av} = 0.488$, $\rho = 3.43\,g\,cm^{-3}$, $\ln J_{eff} = 5.51$.

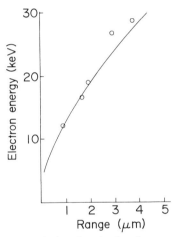

Figure 2.23. Range–energy relation for As_2S_3; the range is defined as the maximum thickness of a film at which transmission of electrons occurs.

Measurements of ranges not involving a lower boundary have been carried out on cloud tracks in air and argon by Tsein et al. (1947). Sacton (1956) determined ranges in nuclear emulsions. The scaling property was again confirmed. Herzog et al. (1972) measured photoresist exposure by high voltage electron beams.

Stinchfield (1940) and others have suggested calculating the dissipation of energy of electrons on the basis of a combination of the Thomson–Whiddington law (as confirmed by Terrill, 1923) for the most probable energy of electrons leaving a foil in the direction of the primary beam and having lost less than half of the original energy, and of Lenard's "law" $N = N_0 e^{-\gamma x}$ for the decrease of the electron current N with x. The fraction of the original energy remaining in the beam after the electrons have passed,

x, is taken as NV/N_0V_0, so $W/W_0 = 1 - NV/N_0V_0$. In Lenard's law, Stinchfield further follows Terrill (1924) by taking $\gamma = k\rho(u/2 + u_0/2)^4$, i.e. setting γ proportional to the density of the material and inversely proportional to the fourth power of the mean of the initial and the most probable final velocity. Terrill's γ refers to all the electrons leaving a foil with the exception of back-scattered and slow secondary electrons. W/W_0 then becomes a function of x/R_0,

$$W/W_0 = (1 - x/R_0)^{c/b+1/2}$$

where R_0 is the Thomson–Whiddington range of incident electrons and c/b is a constant lying between 1·5 and 2·5 derived from experimental data. A number of authors have been able to match certain portions of the observed EBC vs. bombarding voltage curves to this or to similar formulae.

III. Ionizing radiation and radiation effects

A. Production of electron–hole pairs

Exempting rare processes, since electrons lose their energy in small doses it may be guessed, and is consistent with theory (Bethe, 1933), that production of ion pairs, as well as the excitation of luminescence in phosphors, is proportional to the loss of energy and independent of the energy of the electrons. A collision leading to an ionization will normally give the secondary electron a kinetic energy up to a few hundred eV. If it is high enough, the secondary electron can ionize a further atom, and so on. No measurement can distinguish between the first and the second generation of ions.

Table 2.8. Energy required to produce an ion pair in various gases.

Gas	air	N_2	O_2	CO_2	C_3H_8	SO_2	Ar	Ne
W (eV)	33	34	30	32	24	30	27	29

The total number of ion pairs produced can be measured directly in gases, and a large number of studies have been reviewed by Valentine and Curran (1958). The measurements confirm that, almost independently of the velocity of the electrons, the production of an ion pair requires a definite amount of energy W. Representative values are given in Table 2.8. The determination is, in principle, simple. A beam of electrons of known energy is shot into an ionization chamber, and the saturation current is measured. This current divided by q is the number of pairs created per second. All values are small multiples of the ionization energy but the exact correlation is not clear.

Whereas in gases the primary result of an electron impact is the splitting of an atom or molecule into a positive ion and an electron, in solids it is the creation of the electron–hole pair. In gases it is usually possible to collect all the ions and electrons produced, but in solids some recombination of electrons and holes can occur so rapidly that it becomes impossible to observe the total number of pairs created.

Following Klein (1968), we call the average amount of energy given up by the incident radiation in the primary process of generating a single electron–hole pair the radiation ionization energy (E^*). In Chapter 3 we shall distinguish this from the mobilization energy which is the energy expended per *pair playing a role* in the subsequent conduction. The radiation ionization energy has two well confirmed relations: (a) it is essentially independent of the nature and energy of the ionizing particle, provided only that this energy is a multiple of E^*; (b) there is a strong correlation between E^* and the band gap energy. Obviously E^* should be at least as large as this, but normally the electrons and holes will also be given considerable kinetic energy which, in some cases, might exceed the band gap energy. More often it will be lost in exciting phonons (i.e. warming the crystal). In view of considerations such as these, Klein (1968), following Shockley (1961a,b), suggested that E^* has three contributory sources: the intrinsic band gap energy E_g, the mean kinetic energy imparted to the carriers, and optical phonon losses. The optical phonon loss is the product of the number (r) of

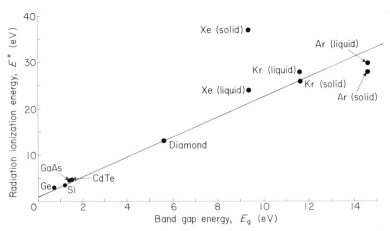

Figure 2.24. Radiation ionization energy (γ-rays, fast electrons, α-particles) consumed from the incident radiation to generate an electron–hole pair, as a function of band gap. Within experimental uncertainty, the data fit a semi-empirical law reflecting constancy of the phonon loss term $r\hbar\omega_0$. (Klein, 1968) Data for Si, Ge, CdTe, and GaAs are from Kozlov *et al.* (1975), for noble gases from Spear and Le Comber (1976), and for diamond from Nava *et al.* (1979a).

phonons involved and their average energy ($\hbar\omega$). The average residual kinetic energy of the carriers was originally estimated as $1\cdot 8E_g$, but later experimental data suggest a value of $1\cdot 1E_g$ to be more appropriate, so $E^* = 2\cdot 1E_g + 1\cdot 3$. Figure 2.24 shows the experimental data on which this is based.

B. Luminescence

The luminescence of solids under the influence of radiation is closely associated with the phenomenon of EBC. It has become a matter of intense technical development with the advent of fluorescent lamps and television. In manufacturing the phosphors it is necessary to prepare the required compound in a highly purified form and then to introduce particular impurities as activators which determine the colour of fluorescence. The light emission follows the irradiation within perhaps a microsecond if the activator is excited directly. If the delay is much longer the process is distinguished by being called phosphorescence. Then the primary process is the production of pairs, as in photoconductivity. The mobile partner wanders about until it acts on an activator or is temporarily trapped. The deeper the traps, and the lower the temperature, the longer it will take for the carrier to be released and the light emission to occur.

The investigation of crystalline phosphors has greatly contributed to the recognition of the importance of traps. When phosphors are irradiated at low temperature and the luminescence has died down, it reappears in bursts as the phosphors are heated up. The trapped carriers are rapidly released as the thermal energy becomes sufficiently high to overcome the forces tying carriers to particular kinds of traps. In this way it has been shown that curves of luminescence against time, and hence against temperature, give information about the depths and availability of traps. Such "glow curves" were first made the object of extensive studies by Randall and Wilkins (1945) and Garlick and Gibson (1949). In detail, the interpretation of glow curves is far from simple (but see Milnes, 1973, page 226).

Apart from giving rise to phosphorescence, as the solid is heated the free charge carriers released in this way can render the normally insulating solid conducting. Thus "conductivity glow curves" can also be used for measuring traps (see Milnes, Ch 9).

C. Photoconduction

Basic photoconductive phenomena have been described with these concepts of the transport of electrons and holes and of traps also discussed in Chapter 1, but to complete the picture it is necessary to discuss the nature of a metal–nonmetal contact. This question was examined in some detail by

Mead (1966) and has been reviewed intensively by Rhoderick (1978). Insulators and semiconductors may be broadly classified in two groups, those where the potential barrier adjacent to a metal contact is determined by the surface states of the nonmetal and those where it is determined by the electron affinity. In either case the barrier is set by the work function of the metal or its ionization potential. In general it can be said that, if an insulator barrier is work function controlled, a high work function metal is more injecting for holes and a low work function metal is more injecting for electrons. A simple rule-of-thumb for semiconductors whose contact barrier is surface state controlled is that any contact metal that could otherwise substitutionally enter the lattice of the semiconductor as a p-type impurity would be injecting for holes, and if it could enter as an n-type impurity it would be injecting for electrons. Criteria deciding whether a given solid falls into one or other of these classes are the same as those governing "direct" and "indirect" transitions (Kurtin et al., 1969; Mead and McGill, 1976). This is a convenient simplification of the problem, because the detailed nature of the contact barrier depends critically on the way in which the surface of the semiconductor is prepared, and a cleaved surface might have quite different properties from an etched one.

These notions concerning photoconductive phenomena can be carried over to EBC provided that account is taken of the differences in the distribution of energy deposited by the exciting radiation. These latter differences are sometimes substantial.

The primary process in photoconductivity is the absorption of a photon, by which an atom is ionized, and an electron or a hole (or both) becomes mobile. Normally this process requires the energy of the photon to be near the ionization energy of the atom. The atom can be a constituent of the matrix of the specimen (idiochromatic, intrinsic photoconductors) or it can be an impurity (allochromatic, extrinsic photoconductors). The same specimen can have both types of photoconductivity. Transport by holes can be initiated only in intrinsic photoconduction, because an ionized impurity atom has no neighbour from which it can recover its lost electron without involving a change in energy.

As a consequence of thermodynamical equilibrium during their growth, all crystals have imperfections such as vacancies (positions in which an atom is missing, Schottky defects) and interstitial atoms (atoms in positions not normally occupied, Frenkel defects), besides incidental defects due to their history or due to contamination. Defects as described are often "frozen in". Near any such defects, an electron (or hole) will have an energy higher or lower than that in the perfect matrix. In the case when it is lower, the defect can trap the carrier which then loses the excess energy by emission of a photon or a phonon. The alkali halides tend to have halogen vacancies or excess metal, and this state of affairs can be encouraged by exposing the crystals to the metal vapours; they then become ideal allochromatic photoconductors. The crystals are electrically neutral if the halogen vacancies are occupied by electrons. These sites are then called F-centres with characteristic absorption spectra which give the crystal a characteristic colour (KCl crystals with F-centres are beautifully mauve).

Gudden and Pohl (1923) were the first to distinguish between an instantaneous effect, which they called primary photocurrent and which is proportional to the light intensity (at given spectral distribution), and field. The measured photocurrent equals

$$\beta q(w_+ + w_-)/L \tag{2.21}$$

where β, q, and L are respectively the rate at which pairs are produced, the charge of the electron, and the separation of the electrodes, and w_+ and w_- are the mean distances travelled by the positive and negative carriers before they become immobile. Blocking contacts are assumed. At low fields, w_+ and w_- are proportional to the field but obviously neither can exceed L. In the limiting case $w_+ + w_- = L$ when the current becomes βq, i.e. one electronic charge passes the circuit for each quantum absorbed. The primary current saturates at this value.

In detail, the path of the mobile carrier is a random walk, biased by the applied field, and the traps are distributed at random; the values of w_+ and w_- are subject to statistical fluctuation. The resulting mean values have been calculated by Hecht (1932); they depend of course on the position of their origin relative to the electrodes. Hecht's formula has found wide application, and in Chapter 5 it will be applied to two particular cases, viz. when all carriers originate in one plane near the surface, and when the pairs are created uniformly throughout the specimen.

The primary current is normally only observed immediately after the start of the illumination for two reasons. (1) As carriers are settling in traps a space charge is established which alters the field in the specimen. (2) While the traps are being filled the number and properties of the carriers change and w_+ and w_- with them, but traps also release carriers thermally if the depth is not too large compared with $k\theta$ or under the influence of the radiation. When the bias is withdrawn a reverse current is observed because the field due to the trapped carriers is opposite to that originally applied.

In addition to the primary photocurrent, a secondary photocurrent is often observed; this need not be simultaneous with the illumination—it rises slowly and it continues after the illumination ceases. It is more prominent at high applied fields, at high temperatures, and at high intensities of illumination; the spectral response in this case depends on the field. The current can be greater than βq by many orders of magnitude, and the number of elementary charges transported through the circuit can be a high multiple of the number of photons absorbed. Such effects are impossible if both contacts are blocking as in Thomson's model of an ionization chamber. On the other hand, if carriers can enter freely from the electrodes, the current would not need promotion by illumination. It is therefore assumed that primary photocarriers initially approaching an electrode are trapped in its vicinity, so the field at the electrode is increased and a kind of field-enhanced emission takes place.

A similar process can take place in EBC. An example was cited by Gibbons (1974b) where a secondary EBC gain of 10^{10} or higher was observed with a single crystal of CdS using indium electrodes; the response time was many seconds.

The most recent and comprehensive text on photoconductivity and allied phenomena is the book edited by Mort and Pai (1976).

D. Photovoltaic effect

A semiconductor p–n junction, or Schottky barrier, can be used as a source of emf if the region near the junction or within the depletion region of the semiconductor is excited by ionizing radiation such as light. The widest part of the depletion region will be within the more lightly doped side of the semiconductor or, in the case of a Schottky barrier, it will be wider for high resistivity or intrinsic semiconductors. An example is the silicon solar cell (Prince, 1955), where boron is diffused into n-type silicon to a depth typically of 2 μm. This forms a junction between the diffused region (which is now heavily p-type) and the n-type starting material. Electron–hole pairs generated by incident light in the n-region are separated, and holes diffuse or drift under the influence of the junction field into the p-type side of the junction. Here the excess holes in the p-region represent a non-equilibrium population which can only be sustained by growth of the junction field or dynamically by the flow of a current through the device.

The equivalent circuit of a photovoltaic cell under irradiation is a constant current source, i_L, in parallel with the diode. If i_S is the reverse bias saturation current of an ideal junction, the current–voltage characteristic of the device is

$$i = i_S(e^{qV/k\theta} - 1) - i_L \qquad (2.22)$$

and the open-circuit photovoltage is given by

$$V_{oc} = (k\theta/q)\ln(i_L/i_S + 1) \qquad (2.23)$$

Photovoltaic cells such as these are used in spacecraft power supplies, and widespread research efforts are now being made to develop less expensive versions for use as electric power sources in remote locations on earth (Carlson and Wronski, 1979; Carlson, 1980).

The related electron voltaic effect (EVE) is the equivalent phenomenon occurring when a junction is bombarded by electrons; it will be considered in detail in Chapter 4 where the sign convention for the direction of i is the opposite of that implied in equation (2.22). Applications of the EVE will be discussed at greater length in Chapter 7.

E. Radiation damage

1. Damage above threshold

At the usual electron accelerating voltages encountered in EBC, the bombarding electrons do not have sufficient momentum to dislodge the atoms from which the insulator or semiconductor lattice is built. This type of radiation damage only occurs if the bombarding particles are in the energy range above about 100 keV. The main results of radiation damage in semiconductors subjected to bombardment with photons or high energy particles in this range are the formation of Frenkel defects (dislodged atoms) and a concurrent reduction in bulk carrier lifetime and an increase in "dark" conductance. Good reviews of our knowledge of radiation damage and defect formation have been published by Townsend *et al.* (1976) and by Saidoh and Townsend (1975). Sub-threshold damage such as decomposition might also be important.

In an insulator or semiconductor there is a definite relationship between this threshold energy for defect formation by ionizing radiation and the displacement energy of the atoms. The lattice structure is also important. As an example, the semiconductors of the Group IV elements and the intermetallic compounds of the type III–V are of the same structure as diamond or zinc blende. These are relatively open, and there are gaps which may be filled by interstitial atoms without substantial distortion to the surroundings. In general, it is found that the displacement energies of intermetallic compound semiconductors are below those of the elements of Group IV, and in the III–V compounds the displacement energy of the constituent of Group III is smaller than that of the element of Group V. Table 2.9 shows the displacement energy in a number of semiconductors.

2. Damage below threshold

The most extensive studies of sub-threshold radiation damage have been undertaken with the alkali halides mainly because this particular class of solids can be studied very conveniently using optical absorption data.

Table 2.9. Displacement threshold energy of semiconductors. (Urli and Corbett, 1977; Bäuerlein, 1962)

Substance	Displaced atom	Threshold energy (keV)	Displacement energy (eV)	Activation energy (eV)
Diamond	C	150–350	50–80	
Si	Si	250–300	25	2·98
Ge	Ge	360–400	14	
InP	In	274		
	P	111		
GaAs	Ga	233		
	As	256		
InAs	In	274		
	As	231		
GaSb	Ga			3·5
	Sb			3·45
InSb	In			1·82
	Sb			1·94
CdS	S	115	8·7	
CdSe		250		
CdTe		340	at 77K	
ZnTe		110–300		
ZnSe		236		
ZnS		185		
ZnO		310–900		
BeO		400		

Similar reasoning can be used to explain sub-threshold defect formation in other classes of insulating or semiconducting compounds. The mechanism, first put forward by Pooley (1966), depends on the formation of excitons which undergo a radiationless transition to their ground state and give up their energy to a pair of halogen atoms. Thus the generation of an exciton is sufficient to cause a lattice relaxation by trapping a hole on the halogen ions. Townsend describes the necessary steps as follows:

(i) Energy absorption $\rightarrow$ excitons $(e^- + h^+)$

(ii) $X^- + X^- + h^+ + e^- \rightarrow (X^- + X^0) + e^-$

$\rightarrow (X_2^-)^* + e^-$

Here X represents the halogen atom and the * symbol is used to signify an excited state. If this entity now returns to its ground state, the kinetic energy may be sufficient to separate the halogen from its parent cation and initiate a replacement collision sequence along an energetically favourable lattice direction. This leaves a vacancy at the origin and an interstitial centre further along the chain. In a single crystal the favoured direction is along the $\langle 110 \rangle$ line.

The various types of colour centre in the alkali halides are listed in Table 2.10; for convenience the older notation is given in brackets.

Zinc oxide in particular is a compound semiconductor that decomposes markedly under the influence of electron bombardment. This effect was investigated by Heiland (1952) who distinguished the increase in conductance of thin films under the influence of electron bombardment due to a loss of oxygen by the lattice from the normal EBC effect by the relative speeds of the two. He called the loss of oxygen the slow process, and the EBC effect the fast process. Similar effects in ZnO under ultraviolet irradiation were investigated by Mollwo and Stöckman (1948). Lead oxide also loses oxygen through decomposition in a similar way.

Careful measurements by Norris (1972) showed that many anomalous results suggesting that radiation damage can occur in bulk silicon and germanium films at energies far below threshold can be attributed to experimental difficulties associated with surface effects. Spear (1958) showed that irradiation of germanium samples by electrons having energies below the threshold caused an increase in surface conductance in p-type and a decrease in n-type specimens. These effects were interpreted in terms of electronic transitions in the surface region. The results of Norris for silicon

Table 2.10. Point defect structures in alkali halides. (Townsend *et al.*, 1976)

F^+ (α)	Halogen ion vacancy
F	An electron trapped by a halogen ion vacancy
F^- (F')	Two electrons trapped by a halogen ion vacancy
F_A	An F centre modified by an alkali impurity at one of the six adjacent metal sites
F_Z	An F centre modified by a divalent impurity
F_2 (M)	A pair of adjacent F centres
F_3 (R)	A triangular array of three F centres, probably in the (111) plane
F_4 (N)	Four F centres; at least two configurations exist
F_2^-, F_3^-, F_4^-, etc.	F_2, F_3, F_4 centres with additional electrons
H	This is an interstitial halogen atom centre; the defect is stabilized by a hole which is trapped by a linear array of four halogen ions spread over three lattice sites in the $\langle 110 \rangle$ direction
V_k	A hole trapped by a pair of halogen ions which relax along a $\langle 110 \rangle$ direction
I	Interstitial halogen ion
I_A	Interstitial halogen ion adjacent to an alkali impurity
H_2 or H_M (H', V_4)	Di-H centre
$H_A(V_1)$	A trapped H centre next to an alkali impurity
H_Z (H-N)	A trapped H centre close to a divalent impurity

showed that a rise or fall in surface conductance was not always consistent and that the only permanent effect of low energy electron irradiation in silicon can be explained in terms of irradiation induced surface charge accumulation. His experiments showed that misleading surface-dominated irradiation effects can easily appear in heteroepitaxial p-type silicon samples but not necessarily in similar n-type samples or in bulk samples. This can be explained in terms of the formation of a weak inversion (or depletion) layer beneath the irradiated contacts. This could occur, for example, if a thin SiO_2 layer were buried beneath them.

Electron irradiation effects below the threshold in SiO_2 are probably the most important degradation phenomena in practical applications of the EVE in Si because silica is quite often used as a passivating layer in any silicon p–n junction device. The results of electron irradiation of SiO_2 at energies between 2 and 18 keV suggest that two distinct processes occur. One is associated with knock-on atomic displacements and F-centre formation, and the other is associated with ionization induced relaxation of inherently strained Si–O bonds (Norris and EerNisse, 1974).

The main results of these sub-threshold effects are to increase the reverse bias leakage current in electron bombarded silicon p–n junctions and to increase the surface recombination velocity. The radiation induced space charge build-up in oxide layers on silicon was studied by Mitchell (1967). His model assumed that electron–hole pairs are created within the SiO_2 by the primary electron beam, and that some of the generated electrons drift out of the SiO_2 under the action of an applied electric field. The generated holes become trapped at defects, and the dependence of the positive space charge on electron irradiation dose (D) is approximately of the form $(1 - e^{-\beta D})$ and linearly dependent on the potential across the oxide. In most practical cases examined it was found that the space charge (assumed positive in his example) accumulated within a distance less than 200 Å from the silicon–oxide interface.

Studies of this topic were also undertaken by Snow *et al.* (1967) and Zaininger and Holmes-Siedle (1967). They also found that sub-threshold irradiation of a passivated silicon diode has two effects: (i) the trapping of holes in the SiO_2 layer; and (ii) the creation of fast surface states at the Si–SiO_2 interface. The first of these effects is reversible and it does not occur unless the electric field exists across the silica during bombardment. The second effect is permanent and it occurs whether a field is applied or not. This is the cause of the higher surface recombination velocity and a higher dark current. The effect begins to appear (with a 1 μm thick SiO_2 passivating layer) after bombardment by 10^{10} electrons cm^{-2} at 20 kV (10^4 rad; 1 rad corresponds to an absorbed dose of 10^{-2} J kg^{-1}), and it saturates after further bombardment lasting about 10^3 times longer. This saturation agrees

2. ELECTRON BEAMS AND THEIR INTERACTION WITH MATTER

approximately with the maximum density of F-centres in the alkali halides, which is 10^{19} cm^{-3} (Seitz, 1946). Choisser (1976) found that his diode leakage current at this saturation value (and for this thickness of oxide) was about 15 times higher than its initial value. After this point the diode leakage did not increase on further bombardment.

Further implications of the effects just described, when the EVE is applied in electron devices, will be discussed in Chapter 7.

At high electron beam densities, and very short pulses, it is possible to momentarily melt or nearly melt the crystal within a closely defined volume. If the electron beam pulse is short enough, the crystal is merely annealed. Thus, in contrast to the case of cumulative damage caused by long electron pulses or continuous beams, in compounds this can be used for repairing such damage, especially if the intensity and duration of the electron pulse are appropriately chosen.

At $V_p = 20$ kV and a beam intensity of about 350 A cm^{-2}, Tandon et al. (1979) were able to anneal ion-implanted GaAs with pulses of about 10^{-7} s duration. Similar results for ion-implanted InP are available (Davies et al., 1979). Here the electron pulses were between 0·4 and $1·6 \times 10^3$ A cm^{-2} at $V_p = 10$ kV and pulse lengths of 5×10^{-8} s.

It is emphasized that, when EBC or the EVE is used for measurements characterizing a material, the electron beam intensities are many orders of magnitude smaller than those used for electron beam annealing. Transient EBC drift mobility techniques use beam currents typically of 10^{-7} A cm^{-2}, and the scanning electron microscope in the charge collection mode 10^{-4} A cm^{-2}.

F. Cosmic rays

Although the components (mainly fast ions) of cosmic radiation are all capable of causing ionization in solids and also radiation damage, their intensity at ground level is too small to be significant. However, galactic cosmic rays are important when electronic equipment is flown in a spacecraft, and especially if the craft passes through regions of high intensity of ionized particles such as the van Allen belts or the Atlantic Anomaly. The former are of toroidal shape in the equatorial plane and have a maximum for electrons at an altitude of 15 000 km and for protons at an altitude of 6000 km. The latter covers a broad area over the south Atlantic ocean between America and Africa at an altitude of 370 km. An underestimate of the influence of such radiation on polarization effects to the SiO_2 insulated gate field-effect transistors in Telstar I (the first commercial transatlantic communications satellite) was a cause of irreparable damage to the on-board electronics system; as a result, no insulated gate field-effect transistors

Table 2.11. Components of galactic cosmic rays; notice the fairly high proportion of particles having atomic numbers Z in the vicinity of the Fe, Co, and Mn group (very heavy ions). (Curtis, 1976)

Particle	% of total
Protons	37
He ions	28
Light and medium ions ($3 < Z < 9$)	15
Light heavy ions ($10 < Z < 14$)	10
Very heavy ions ($26 < Z < 28$)	10

are now flown in spacecraft. The components of galactic cosmic rays are given in Table 2.11.

The maximum energy of galactic protons is about 300 MeV during solar quiet times and 500 MeV during periods of solar activity. The corresponding figures for helium ions are about the same although of course there are fewer of them. At an intensity about three orders of magnitude below the peak, helium ion energies up to 300 GeV have been measured (Shapiro and Siberberg, 1974). Outside the earth's magnetosphere, a cosmic ray particle flux of four particles cm^{-2} s^{-1} has been measured inside the spacecraft responsible for these measurements.

The distribution of radiation trapped by the earth's magnetic field is shown in Fig. 2.25.

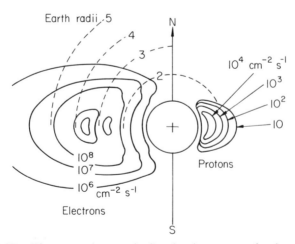

Figure 2.25. Electron and proton isoflux density contours for the trapped radiation belts round the earth, shown on the left for electrons and on the right for protons; they extend round the earth in a toroidal shape. (Curtis, 1976)

G. Nuclear radiation

Some of the earliest observations of the influence of certain kinds of radiation on the conductivity of insulators or semiconductors were made before the electron was discovered. The influence of light was discovered by E. Becquerel (1839; photovoltaic effect) and by W. Smith (1873; photoconductive effect). Nuclear particles and γ-rays from radium and X-rays were found to be the cause of an increase in conductivity in an insulator at around the end of the nineteenth century. It was found that β-rays (as opposed to cathode rays) increase the conductivity of a dielectric before cathode rays were found to have a similar effect.

If we leave out for the moment ionizing radiation which can only be efficiently generated by a cyclotron, a linear accelerator, or an atomic reactor, the only kinds that need to be considered here are light (both visible and ultraviolet), γ- and X-rays, fast electrons, β-particles (i.e. electrons whose source is the nucleus of an atom), and α-particles. All these rays deposit energy along their path according to an exponential absorption law, apart from fast electrons and α-particles, the ionization along their track being proportional to their energy loss. The constant of proportionality is almost independent of the type of radiation; the energy required for pair creation was discussed in Section A and is shown in Fig. 2.24. This is smaller than the electron–hole pair mobilization energy which is a function also of the number of carriers that recombine at their place of origin.

Generation of γ-rays on a laboratory scale for experimental studies of materials is more convenient than that of X-rays, although their influence for the same photon energy is identical. Their relevance in the context of this book is mainly with regard to the extent that X- and γ-rays can be detected by solids, how they may degrade the performance of an EBC device, and how X-rays produced when a solid is bombarded by fast electrons can spread the extent of ionization beyond the depth of electron penetration.

The general properties of X-rays and γ-rays are well documented. The spectrum of generated bremsstrahlung when a fast electron of energy E strikes a solid is very broad, with a short wavelength limit given by $\lambda = hc/E$ (where c is the velocity of light) and an average photon energy at about or slightly more than $\frac{1}{2}E$. The characteristic radiation arising from discrete electron transitions between atomic orbitals accounts for only a minute proportion of the emitted energy. The efficiency with which X-rays are generated is given approximately by $\eta = (1 \cdot 1 \times 10^{-6})ZV$, where Z is the atomic number of the element stopping the electron and V is its accelerating voltage in units of kV.

In the energy range below about 50 keV the X-rays are absorbed to a large extent by the production of photoelectrons, but as their energy is increased

Compton processes gain prominence and eventually become dominant. As a result, high energy X-rays may be scattered more than once before their photon energy has decreased sufficiently for the photoelectric absorption to become significant. These processes help to spread the region of ionization but the total number of electron–hole pairs generated is still proportional to the incident energy of the primary electrons. The most convenient laboratory sources of X-rays or γ-rays are radioisotopes. They are not often used in the measurement of electrical properties of non-metals because of their exponential absorption with depth and the greater difficulty of control compared with other modes of excitation. However, pulsed megavoltage X-rays from a linac have been used for liquids. Convenient radioisotopes are Cs-137 (γ energy 660 keV) or Co-60 (γ energies 1·17 and 1·33 MeV). Gas-filled ionization chambers excited by X-rays provide a model for the theory of steady-state EBC in insulating solids, and bremsstrahlung X-rays are an intermediary in the excitation of solids by fast electrons.

Table 2.12. Range of α-particles of various energies in some elements.

Energy (MeV)	Range (g cm^{-2}) of α-particles in						
	H	C	Al	Cu	Pb	U	
2·0	2·84	11·6	18·5	31·1	61·5	66·2	$\times 10^{-4}$
6·0	1·70	5·40	7·53	11·6	22·8	24·7	$\times 10^{-3}$
10·0	4·18	12·2	16·5	23·9	45·4	49·6	$\times 10^{-3}$
22·0	1·74	4·83	6·07	18·18	14·2	15·3	$\times 10^{-2}$
50·0	7·85	20·7	25·1	32·1	51·1	54·5	$\times 10^{-2}$
100·0	2·80	7·16	8·50	10·6	15·8	16·8	$\times 10^{-1}$
200·0	9·98	24·9	29·0	35·2	50·6	53·5	$\times 10^{-1}$
500·0	5·21	12·7	14·6	17·3	23·9	25·1	$\times 10^{0}$

Radioactive sources such as Am-241 emit α-particles. This isotope emits α-particles with a mean energy of about 4 MeV and a maximum of 5·5 MeV; the half-life is 433 years. The penetration law is not exponential, and the rate of energy deposition reaches a maximum at the end of the range (Bragg peak). The range of α-particles in some elements is given in Table 2.12. Americium-241 also emits low energy γ-rays, but these are relatively unimportant unless the source is fairly thick.

H. Persistent polarization

In addition to creating ion pairs in traversing solids, cathode rays themselves of course carry a charge. If the fast electron has exhausted its range inside a specimen, it can either move towards an electrode, or become trapped, leaving the specimen charged. A space charge established by such residual charges differs from that arising from the separation of pairs in two respects: (1) the negative space charge is located at the end of the range (the "chord" range of the electron; see Fig. 2.10), which could be in or near the middle of a sheet or a film; (2) a positive charge equal to that of the retained electron is attracted towards the surface of the specimen, and this either stays at the electrode or infiltrates into the dielectric. The existence of such charges was spectacularly demonstrated by Gross (1957), by irradiating discs of borosilicate glass with 2 MeV electrons in a Van de Graaff accelerator. The part of the glass penetrated by the electrons acquired a deep yellow colour owing to the production of colour centres. Then, when the samples were touched with a pointed piece of metal, a breakdown occurred which led to a visible pattern of destruction in a clearly marked plane which defined the depth at which the primary electrons had come to rest. This breakdown can be produced months after the irradiation. Similar effects have been observed in Perspex. Sessler and West (1973) observed permanent charges in films of polymers such as Teflon after their bombardment with cathode rays of energy between 15 and 35 keV. These charges decay at room temperature very slowly, with a half-life of over 100 years, and there is no, or very little, initial leakage. They are extreme cases of the absence of induced conductivity. These "electrets" have become technically important in such applications as electrostatic microphones; they are also described by Sessler and West.

3
Steady state EBC of thin insulating films

By the term "thin films" we mean those that are sufficiently thin to allow the primary electrons to penetrate them completely at the highest accelerating voltage used for the experiment. Of crystals, only mica and selenite can be cleaved thin enough to qualify as films; neither shows a strong EBC response. Attempts to lap down crystals usually result in cracking them when they are less than 50 μm thick. Several authors have described attempts to wash down single crystals of alkali halides by dissolving selected regions by a small jet of saturated aqueous solution, but generally such methods still caused cracking.

I. Preparation of films

In recent years many advances have been made in developing new techniques for producing thin films of insulators, semiconductors, and metals. It is beyond the scope of this book to describe these in sufficient detail to be useful, and the reader is referred to the excellent book edited by Vossen and Kern (1978). The few methods about to be described were well established in about 1948 when studies of EBC were started, and they are included here to provide a background to the available techniques used by most experimentalists in the years between then and about 1970. One exception is the glow discharge decomposition of silane (SiH_4) which has become of importance for producing high resistivity amorphous thin films of silicon; this will be described below in Section D.

A. Evaporation

The most common process for producing thin films is by evaporation in a good vacuum. The evaporation "boat" can be made from a 40 mm length of tantalum or molybdenum sheeting about 0·1 mm thick × 10 mm wide; a heavy current transformer is used to pass the heating current through the boat. The residual gas pressure in the evaporation chamber should be less than about 10^{-5} torr during evaporation, and much spitting and showering of the evaporant from the boat is avoided if it is preheated gradually while the vacuum pump is reducing the pressure. The boat should be provided with a "dimple" in its centre to accommodate the material being evaporated, which is usually in a pure finely powdered dry state, although single crystals or a compressed pellet can be used.

Suitable substrates for the films can be glass microscope slides carefully selected for the absence of small surface scratches. Even better substrates are made from ground and polished glass discs cleaned in a mixture of sulphuric acid and hydrogen peroxide, washed in distilled water, and dried in hot trichlorethane vapour. Occasionally the substrate should be heated in order to improve stoicheiometry in those cases where decomposition takes place during evaporation; examples of compounds to which this applies are CdS and ZnS. Masks to limit the extent of the layer can be made from stainless steel, and occasionally mechanical masking arrangements are used so that they can be changed without breaking the vacuum if electrodes are also evaporated.

B. Sputtering

Some layers can be deposited by sputtering in an inert atmosphere. The material is compressed into a thin disc about 0·5 mm thick and held on aluminium or steel by a pressure bond. The metal disc, hung horizontal by a hook on a vertical conducting wire which enters from the top of a glass bell-jar, is kept at a distance of about 10 cm from the substrate. The bell-jar is first evacuated; while still pumping, argon gas is allowed to enter the bell-jar through a needle valve and the equilibrium pressure is adjusted to lie between about 1 and 0·01 torr. This continuous flow method tends to sweep away any impurities that might be desorbed from the electrodes during deposition.

Sputtering takes place when a high voltage is applied between the substrate and cathode. A dc voltage of around 5–20 kV is sufficient, and this should be applied to the metal disc (which is negative) through a resistor of about 1 megohm; the substrate is at earth potential. The discharge glow can be limited to a particular volume by surrounding the space between anode and cathode by a glass cylinder.

C. Glassy bubbles

Glasses can be blown as thin bubbles. For example, molten As_2S_3 can be blown as a bubble on the end of a Pyrex tube, using a controlled flow of dry nitrogen inside a glove-box filled with nitrogen. When cold, such bubbles can be used by cutting small regions which are not too far from being flat. Ordinary glass such as Pyrex is too conductive to show a response to electron bombardment but thin bubbles blown in air were used successfully by Spear and Mort (1963) to make blocking contacts to CdS under pulsed electron excitation.

D. Deposition by glow discharge

An rf or dc glow discharge in an atmosphere of silane (SiH_4) can be used to produce layers of amorphous silicon (a-Si) up to about 3 μm thick. The excitation may be either from an rf coil or between parallel planar electrodes; the latter method provides larger area films of greater uniformity. Suitable apparatus is shown in Fig. 3.1. The gas pressure is reduced by means of a vacuum pump fitted with a throttle valve so that the equilibrium pressure is about 0·1–1 torr. Stimulation is provided by a 500 W oscillator operating at 1 MHz; although the rated power is much higher, the actual discharge is run at a power of 10–20 W. Substrates placed in the glow so produced, which remains in close contact with the specimen surface, are coated at a rate of about 1 Å per second. The substrate is usually heated because the properties of the amorphous silicon depend critically on this temperature.

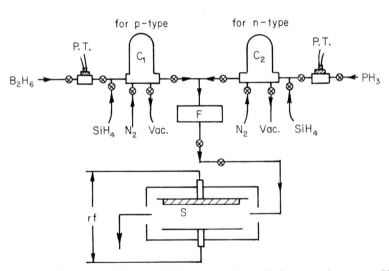

Figure 3.1. Schematic diagram of the preparation unit for n- and p-type a-Si. C_1 and C_2, glass cylinders; P.T., pressure transducers; F, flow meter; S, substrate. (Le Comber and Spear, 1979)

Glow discharge a-Si can be doped with phosphorus or with boron by adding controlled amounts of PH_3 or B_2H_6 to the starting gas. It is thus possible to produce substitutionally doped n-type or p-type a-Si. The equipment for this process is also shown in Fig. 3.1 (Le Comber and Spear, 1979). Glow discharge a-Si is in fact a hydrogen-compensated amorphous semiconductor, the hydrogen effectively saturating (or nearly saturating) the dangling bonds.

E. Anodizing and chemical deposition

Oxide films can be grown on some metals such as aluminium and tantalum. The anodized film on aluminium has received some attention because it shows appreciable induced conductivity; such films have a complicated structure. Tantalum oxide films are also fairly good EBC insulators.

The method employed to form good EBC films of Al_2O_3 was described by Ansbacher (1950) as follows. The pure aluminium foil about 40 μm thick was cleaned by immersing it for about 5 min in a nearly boiling solution of 20 g of CrO_3 in 35 ml of H_3PO_4 (specific gravity 1·72) made up to 1 litre with distilled water. After rinsing in distilled water, the samples were anodized by the chromic acid process in a 3% Cr_2O_3 solution at 40°C. The cathode was a carbon rod and all the wires and other fixtures were of aluminium. The current density was 5 mA cm^{-2} at the start but it rapidly died down. During the first 15 min the voltage was raised from zero to 45 V and then kept at 45 V for a further 40 min; finally it was raised to 50 V or more for 5 min. Samples produced with up to 135 V final anodizing voltage were used. The thickness of the oxide layer could be determined from optical interference fringes at the edges.

After anodizing, the metal-based specimen was washed, dried, and fitted with a thin vacuum-evaporated aluminium electrode over its surface. Alumina films produced by the chromic acid anodizing process have a higher EBC gain than those produced by the tartaric acid process or Al_2O_3 films produced by sputtering.

Films of tantalum oxide may be grown by anodizing pure tantalum sheet in 0·02 M phosphoric acid solution, the ultimate thickness being determined fairly precisely by the anodizing voltage; Aris *et al.* (1976) describe the EBC of such films in the thickness range 72–360 nm.

If thermal decomposition is severe, vacuum evaporation cannot be used. Chemical methods can be employed to prepare thin films in some instances. Pensak (1948) prepared thin films of SiO_2 by heating Nichrome base plates to about 700°C in the presence of ethyl silicate vapour. It is interesting that even today the highest purity synthetic silica is still produced by hydrolysis of ethyl silicate vapour in a similar way.

Thin films of ZnO can be prepared by evaporating zinc in a good vacuum and subsequently oxidizing by baking in air or oxygen for 5 min at 400°C, and thin (polycrystalline) layers of PbO can be produced by evaporating PbO from a platinum crucible through oxygen at a pressure of 10^{-2} torr on to a substrate held at 110°C.

F. Other methods

Under this heading we include techniques that were little known, or even unknown, when studies of EBC were started seriously. Even though some of these techniques evolved while such studies were in progress, the older methods were often still employed, so a direct comparison of different properties of similar films produced by the same techniques could be made in the same laboratory. The available technologies for producing thin films now include chemical etching, plasma deposition of inorganic films, glow discharge polymerization, ion beam deposition, magnetron sputtering, sputter gun deposition, glow discharge sputter deposition, and electron beam evaporation. For details of these, the reader is referred to the book edited by Vossen and Kern (1978).

II. Specimens for measurements

Specimens of an insulator for EBC measurements are usually prepared in the form of a metal–insulator–metal sandwich. The contact on the side of the thin film remote from the electron source is usually a conducting film of metal deposited on a glass substrate, although in some instances it can be the substrate itself (e.g. Ta or Al), or self-supporting (or weakly supported) films can be prepared for electron beam access from either side.

A typical arrangement is shown in Fig. 3.2. In this example, a gold or aluminium back electrode is vacuum evaporated on to a cleaned glass microscope slide using a brass or stainless steel mask to limit the size of the conductive area. Good electrical contact can be made to this at a later stage if a strip of Aquadag is painted on one edge of the glass beforehand and, after drying the Aquadag in a warm place, the metal layer is evaporated on to this layer also. Such contacts are more reliable than those made with Aquadag after metallizing.

The specimen (insulator or semiconductor) is deposited over an area slightly larger than that of the bottom electrode, again using an appropriate mask. If the edges of this film are slightly diffuse it is often possible to measure the thickness by the Tolansky interferometric method.

The top electrode is now evaporated on to the insulator, but this time the contact point is brought out on the opposite edge of the glass slide so that

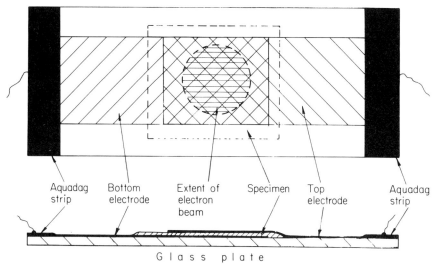

Figure 3.2. Diagram showing a mounted specimen prepared for measurements of steady-state EBC; the contact wires at each side are of 0·1 mm diameter copper and are joined to the Aquadag with silver paste.

short-circuits between the top and bottom metal electrodes are avoided. Of course, if the top contact is an electron beam the surface of the insulator is left free.

If the insulator is to be weakly supported it can be vacuum evaporated on to a thin alumina membrane made by the tartaric acid anodizing process (Harris, 1955).

Films completely open on both sides can be made by vacuum evaporation on to copper foil; the copper can then be removed, if the insulator is not attacked, by carefully dropping a spot of concentrated nitric acid on to the foil (Spear, 1955). An alternative method if the insulator is not damaged by heat is to evaporate it on to a thin "skin" of nitrocellulose supported on a fine-pitch electroformed metal grid (e.g. 20 threads per mm). Such substrates can be made by dropping a spot of nitrocellulose in ethyl acetate (Collodion) on to the surface of water in a vertical-sided trough. When the solvent evaporates, the floating nitrocellulose membrane can be picked up on the metal grid, which is mounted on a frame immersed in the water, by slowly raising it at an angle of 45° through the surface. After drying it and evaporating the insulator, the nitrocellulose can be destroyed by baking the specimen in air or in a vacuum at 400°C for 30 minutes. Other workers have reported a technique whereby the insulator is evaporated on to a smooth rock-salt surface which is subsequently removed by dissolving it in water.

A. Solid electrodes

If the primary electrons penetrate the front (top) electrode of a thin metal–insulator–metal sandwich film, this contact should be electrically continuous and have very little stopping power at the bombarding voltage (V_p) used. For this reason aluminium is often chosen; the thickness should be about 1000 Å. This metal can be readily evaporated in a vacuum better than 10^{-5}–10^{-6} torr from a helix about 8 mm in diameter made from 1 mm tungsten wire or, better still, from a multiple strand of three tungsten wires 0·5 mm thick twisted together. Pure and clean aluminium wire 1 mm in diameter is cut into pieces about 10 mm long and bent into horseshoe shapes, and these are crimped around the tungsten wire at 5–10 mm intervals with pliers. The aluminium evaporates from the tungsten in a high vacuum when it is electrically heated to about 1000°C by passing a current through it. The substrate is typically at a distance of 15–25 cm from the source.

Alternatively an evaporated gold contact is used; this should be 350–400 Å thick. A suitable source is a 5 mm diameter helix, 5 cm long, of stranded wire made by twisting together three lengths of 0·5 mm tungsten and one of gold wire. The tungsten wire is cleaned before use by an electrolytic method using the wire as anode and a solution made up from 25 g of NaOH and 100 g of Na_3PO_4 in 2 litres of water as electrolyte at about 85°C.

If aluminium or gold are unsuitable, an Aquadag electrode can often be painted on. Wetting of the specimen is aided by adding a wetting agent such as Teepol to the aqueous dispersion or by selecting one of the dispersions that are available in an organic medium such as ethanol.

B. Electron beam contacts

Electron beam contacts can be explained by reference to a particular example. Suppose a target made of some insulating material is mounted at one end of a tube facing an electron gun; surrounding the target is a collector electrode by which secondary electrons emitted by the target may return to earth. Such a tube is shown diagrammatically in Fig. 3.3(a). Consider what happens when the cathode of the electron gun is at earth potential and the collector potential (V_{coll}) is made progressively more positive. For values of V_{coll} less than the bombarding voltage (V_{p1}) at which the external secondary emission yield σ of the target first attains a value of unity, the number of secondaries reaching the collector electrode will always be less than the number of primary electrons incident on the insulator. This will occur whatever the initial positive value of the insulator potential and the initial value of σ, since the collector cannot accept secondaries emitted with small

3. STEADY STATE EBC OF THIN INSULATING FILMS 87

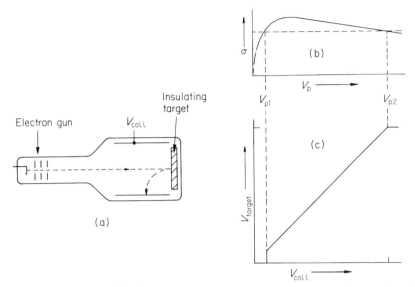

Figure 3.3. The stabilization potentials of an insulator when its free surface is bombarded by electrons. (a) Tube used to explain the text; (b) secondary emission yield curve; (c) surface potential of target as a function of collector potential. (Gibbons, 1966)

energies from a target at a potential more positive than V_{coll}. Thus the potential of the insulator will float in a negative direction until it attains cathode potential. This is known as cathode potential stabilization.

Qualitatively the shape of the secondary emission yield curve is shown in Fig. 3.3(b); this shape is representative of all materials. The potentials V_{p1} and V_{p2} are respectively the lower and upper target potentials for which σ has the value unity. Tables giving values of σ, V_{p1}, and V_{p2} for a number of materials are given in the Appendix.

As V_{coll} is increased beyond V_{p1} the target may remain at cathode potential but this is not the only stable state. Any increase of the target potential, even momentary, beyond V_{p1} will cause the emission of more secondaries than the number of primaries arriving, and the target potential will float positively. This potential will continue to rise until the target attains a potential slightly more positive than V_{coll}. Stabilization occurs at the potential where the number of primary electrons incident on the target is exactly balanced by the number of secondaries that can overcome the slight retarding field of the collector. This may be a few volts positive with respect to V_{coll}. This process is known as anode potential stabilization.

The target potential may now follow V_{coll} as it is increased until the higher potential (V_{p2}) at which σ attains a value of unity is reached. The target potential can no longer follow to higher values because, beyond V_{p2}, the number of secondaries will always be less than the number of primaries and the target can only charge negatively. Thus, when V_{coll} is greater than V_{p2} the target potential will stabilize at V_{p2}. For this reason V_{p2} is often referred to as the sticking potential of the insulator.

These three stabilization conditions are shown diagrammatically in Fig. 3.3(c). In general, the condition of cathode potential stabilization can be achieved only if the stabilizing beam lands everywhere on the target orthogonally.

The free surface can be scanned by an electron beam in the form of a television raster. In this case a charge is induced in the back conducting contact if the beam replaces those charges that flowed through the insulator in the interval between scans. If the surface is stabilized at V_{coll} both positive and negative potential excursions can be discharged, but if at cathode potential ($V = 0$) only positive charges can be neutralized. The former method was used by Pensak (1949, 1950) to measure the EBC of thin insulating films. Diagrams of his apparatus are shown in Figs 3.4 and 3.5. His, or similar methods, have been used to measure the steady EBC of insulators in self-supported or weakly supported thin films.

There is good experimental evidence to support the belief that if an electron beam contact is used, and even if the contacting beam arrives at the specimen surface with such a low velocity that it can do little more than deposit a surface charge, it can act as an injecting contact for electrons. Thin films with either type of contact can be compared when it is known that the source material and deposition conditions were identical. Decker and Schneeberger (1957) measured films of As_2S_3 and found that values of EBC

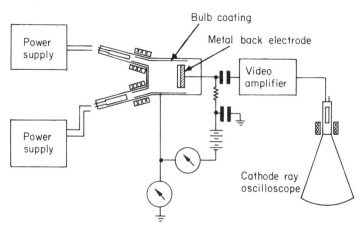

Figure 3.4. Experimental arrangement used to demonstrate the EBC of thin films of chemically deposited silica. (Pensak, 1949)

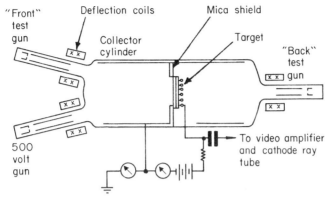

Figure 3.5. Experimental tube used to measure the steady-state EBC of lightly supported thin films of amorphous selenium when excited by electrons from either side. (Pensak, 1950)

gain when an electron beam was used were higher than those with a film having contacts of thin metallic layers. Bowman (1972) found similar behaviour for ZnS. This results in secondary EBC gain and it is thus accompanied by higher g values and increased very low-level conductance "tail", especially for long times after the end of excitation.

III. Transfer of excitation

A. X-rays

Firmin and Oatley (1955) first drew attention to the fact that when an insulator sandwiched between metallic electrodes is excited by electron bombardment (as in EBC) the influence of the excitation may greatly exceed the electron penetration depth. Since then, others have studied the influence of processes that could extend the depth of excitation, and the main results have been summarized by Bleil (1962) and Sturner and Bleil (1963).

The first experiment was performed on electron bombarded diamond crystals with thin silver electrodes of different thicknesses. The EBC was not measured directly, but the change in conductivity was inferred from measurements in which the specimen was placed in one arm of a Blumlein ac bridge (see Clark and Vanderlyn, 1949). This bridge uses closely coupled coils as ratio arms, and very high ratios of the measured capacitance to the known capacitance can be used. The change in the effective specimen series resistance Δr, or capacity ΔC, was used as a measure of the increased conductivity. It was found that the change in conductivity became greater as the thickness of the top (bombarded) electrode was increased, provided that the beam energy exceeded a threshold. Firmin and Oatley further found that the value of Δr was 4 times greater for the same i_p and V_p if a piece of tungsten foil 1 μm thick was mounted in front of, but not touching, the diamond. This thickness of tungsten was enough to stop the 20 keV incident electrons so that none reached the specimen. It was inferred from this that X-rays generated in the front electrode seriously affected the excitation, because a small number of electron–hole pairs liberated throughout the bulk of the specimen would have had as big an effect as a much larger number generated directly near the surface. The greater range of X-rays compared with electrons has already been mentioned; this is more important if the specimen is too thick for the primary electrons to penetrate it completely.

The effect of X-rays was also studied in a detailed set of experiments on CdS by Bleil and his coworkers. Basically their experiment was to use two single-crystal specimens of CdS mounted in close proximity to each other. The two crystals were not quite touching, and they were optically screened from each other by a thin aluminium foil. The change in conductivity in both crystals was measured when only one of them was bombarded by electrons. It was shown that characteristic X-rays generated when electrons struck the Cd and the S atoms of the first crystal are capable of exciting conductivity in the other (non-bombarded) crystal. However, characteristic X-rays constitute only a small proportion of the total energy emitted. The total generation efficiency of X-rays by 50 keV electrons in CdS was calculated as 0·2% but the results showed that only a small part (of the order of 10^{-5}) of the energy available in the X-rays was needed to explain the observed changes in conductivity.

The experiment of Firmin and Oatley was repeated with a copper foil a few tenths of a mm thick in a series of experiments with CdS using beam voltages up to 90 kV. As expected, the role of X-rays in producing a change in conductivity was confirmed.

It was subsequently found that the role of some specific X-rays, such as the characteristic rays emitted from the constituent atoms, is more pronounced in CdSe than in CdS. In particular, the importance of the double conversion electron–X-ray–electron was found to be significant and that the importance of individual characteristic X-ray wavelengths due to Cd_K and Se_K can be clearly demonstrated (Sturner and Bleil, 1963).

B. Excitons

If the primary electrons generate excitons or electrons and holes which diffuse through the lattice by ambipolar processes, energy can be deposited at distances from the point of entry many times the penetration depth of the beam. The energy carried by the primary beam may thus be spread by mechanisms of this kind. In organic or molecular crystals there is also migration of molecular excitons—the equivalents of excited states of isolated molecules. In any of these cases, energy is transported without the net movement of charge, and energy is liberated in the form of a phonon or a number of phonons when the electron and hole recombine or, in the case of the molecular exciton, when it reverts to its ground state. The localized region of high phonon density can now lose its energy either by liberating electron–hole pairs or by diffusing through the lattice.

Exciton diffusion in CdS crystals when they were illuminated by light at some distance remote from the electrodes was studied extensively; the same considerations would apply of course in the case of EBC. Large values of the diffusion length were shown to be due to excitons, by a combination of measurements involving the spectral response of photoconductivity, the temperature dependence of the diffusion length, and the PEM voltage. A thermal activation energy of about $0 \cdot 1$ eV was found for exciton migration in CdS, and diffusion lengths ranging from a fraction of a μm to several hundred μm were observed (Diemer and Hoogenstraaten, 1957; Diemer *et al.*, 1958). Using these results and others, Bleil assigned an exciton diffusion length of $0 \cdot 4$ mm to his CdS crystals excited by electrons, and an ambipolar diffusion length of 1 μm or less. This compares with a mean attenuation length of about 1 cm for X-rays of 10 keV energy in a typical crystal.

One of the most extensively studied organic crystals is anthracene. In this material, singlet excitons were found to have a diffusion length of about 1000 Å but there is a complication in the mechanism of energy transfer. The singlet exciton may undergo what is termed inter-system crossing, by changing from the excitonic structure corresponding to the singlet excited states of the molecule to that corresponding to the triplet states. The triplet exciton has a much longer life and hence a longer diffusion time. In anthracene its lifetime is typically 6 ms and the diffusion length can be as long as 2 μm. The excitons decay to their ground state and, with the assistance of a small amount of additional energy from collision with a surface state or an impurity, free carriers may be produced.

Thus, in the particular case of some molecular crystals where the lifetime of triplet excitons is long, this mechanism of energy transfer is often accompanied by a process of delayed generation.

C. Fluorescence

A solid such as CdS, which can be fluorescent as well as photoconductive, is capable of exhibiting the influence of excitation by an electron beam at some distance from the point of measurement owing to its own light emission and absorption. This was studied by Diemer et al. (1958) who concluded that, for separations greater than about 4 mm, the visible light fluorescence of a copper-activated CdS sample could be responsible for 80% of the change in conductivity. Taking into account the accuracy of the measurements, they concluded that exciton diffusion did not play a measurable part at this distance in the crystal they used.

However, an interesting and unexplained effect was discovered. It was found that infrared light ($\lambda \approx 850$ nm) was effective in quenching photoconduction. This might have been expected, since they were observing secondary photocurrents, but they found that, if infrared light was allowed to fall on part of the crystal containing the electrodes while a nearby region was illuminated by a narrow patch of green light, the quenching effect was about 10 times greater than when that region of the electrodes was illuminated directly. The authors suggested that the occupancy of the centres with which the excitons interact giving rise to photoconduction might be changed by the infrared light.

A different situation applies to organic rigid glasses and to plastics (Förster, 1959). In these a process called resonance transfer can take place; this has the same basic requirements as radiative transfer just discussed for CdS in so far as the absorption spectrum of the accepting molecule overlaps the fluorescence spectrum of the emitting molecule. Normally, in the isolated molecule, Stokes's law states that the absorption peak is at a shorter wavelength than that of the fluorescence. However, in condensed phases, neighbour interactions will broaden the spectral line widths, and overlap is more likely. In the case of resonance transfer, no emission or absorption of photons takes place. Dipole–dipole interaction between neighbouring molecules is mainly responsible for direct coupling of the mutual radiation field of two excited molecules. In practice, the resonance condition is usually soon broken because the accepting molecule loses thermal energy to the lattice as soon as it receives enough energy to raise it to an excited state. Further diffusion therefore stops and the excited molecule now decays to its ground state. This process may transfer energy over a range of only 5–100 Å.

IV. Conductivity induced by electrons

The conductivity of films measured through their thickness is many orders of magnitude greater than that in their plane. Since 1948 conductivity changes under electron bombardment have been studied in a variety of insulators that might be highly resistive in the absence of bombardment.

A. Influence of bombarding voltage

1. The threshold

When a space-charge-free film of a typical insulating solid appropriately biased by an applied voltage across its thickness is exposed to a short burst of electrons, a charge due to internally generated carriers is induced on the electrodes. We then define the gain as the ratio of this charge to the bombarding charge. On repeating the excitation the gain decreases and finally vanishes unless the energy qV_p of the impinging electrons exceeds a threshold qV_0. As a result, the response to *steady* bombardment commences only at electron energies above qV_0. The gain then rises to a maximum g_m at $V_m \approx 2V_0$ beyond which it declines because the electron energy lost in the film decreases. The gain above the threshold increases with bias but is normally not more than about 100. Figure 3.6 shows a typical family of curves giving the EBC gain (g) of a thin film of amorphous As_2S_3 as a function of bombarding voltage (V_p), with applied bias (v) as a parameter. It is seen that, even at 105 V bias, about 1000 eV are dissipated at V_m for each

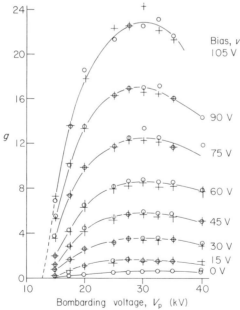

Figure 3.6. The EBC gain g of a thin film of amorphous As_2S_3 as a function of electron bombarding voltage V_p, with applied bias voltage v as a parameter (○ negative gain; + positive gain). The shape of these curves is typical of those for all amorphous insulating thin films in which there is no carrier injection from the electrodes. (Ansbacher and Ehrenberg, 1951)

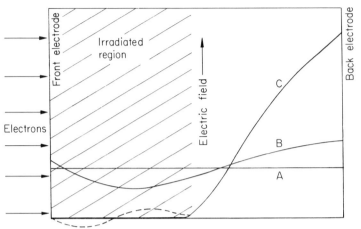

Figure 3.7. Schematic representation of the electric field in a dielectric bombarded by electrons that do not penetrate the slab; the left electrode is positive. A, before bombardment; B, after a brief pulse of electrons; C, equilibrium under steady electron bombardment. The broken line refers to the case where the positive carriers are less mobile than the negative carriers. (Ehrenberg and Hidden, 1962)

carrier moved across the film. The threshold, and V_m with it, moves to higher voltages as the thickness of the film increases.

The behaviour at low electron energies has been explained on the basis of the presence of a very large number of traps with escape times long compared with the intervals between the bursts of excitation. Figure 3.7 shows an insulating lamina fitted with thin front and back electrodes. Of the carriers generated near the front electrode, those with the polarity of this electrode are pulled out of the plasma and may be trapped in the non-irradiated regions shown on the right-hand side of the diagram. The process continues until the charge in this region is so large that it takes up the whole potential drop across the specimen. This progressive change is shown by the sequence of potential curves: A, initial distribution uniform; B, slight space charge build-up; C, final stable potential. The plasma region then becomes field-free and carriers are no longer encouraged to leave it, and recombination in the generation region becomes severe. Since the threshold applies equally to positive and to negative carriers it must be concluded that both are mobile and liable to be trapped. In the non-irradiated region no recombination can take place; after delays depending on the depths of the traps, the carriers will eventually again become mobile.

The existence of a threshold is typical for amorphous films but not displayed by all films. A number of experiments have been carried out in order to examine the nature of this threshold. Ansbacher and Ehrenberg (1951) measured the charge trapped in an As_2S_3 film bombarded with electrons having energy near the threshold by interrupting the bombardment when the induced current had dropped to zero. After reducing the bias to zero they connected the electrode through a ballistic galvanometer; the stored charge is discharged on renewed bombardment. On a specimen 2 μm thick of As_2S_3, $1\cdot4 \times 10^{-7}$ C cm^{-2} was discharged; this would have caused a field of about 2×10^5 V cm^{-1} (taking the dielectric constant of the material as 7). Even if this charge is distributed over half the film, it requires 10^{16} traps cm^{-3}; this estimate is a lower limit but it is within the range commonly found for amorphous solids.

An estimate of the density of filled traps can also be obtained directly from the bias at which the threshold is observed. Thus, Ehrenberg and Hidden (1962) gave a lower limit of 10^{18} cm^{-3} for evaporated ZnS films, assuming that the filled traps occupy half the specimen. A similar consideration led these authors to an estimate of the energy required to create a mobile pair. The dose, at an energy near qV_0, was measured after the response had been reduced to half that of a virgin film. Since the gain is proportional to the bias for ZnS films, the charge must then cancel half the applied bias. This was found to correspond to the creation of one mobile pair for each 2000 eV dissipated by the beam.

On his silica films Pensak had observed a rapid decline of the conduction effect with decreasing bombarding voltage, and had suggested that it occurs when the penetration ceases to be complete. Ansbacher and Ehrenberg also interpreted the threshold as the beginning of complete penetration of the specimen by fast electrons. Spear (1955), examining transmission and EBC on the same specimens, found however that electrons commence to be transmitted only when their energy reaches qV_m, and that the range of the bombarding electrons at the threshold is about half the specimen thickness. This was confirmed by Bowlt and Ehrenberg (1969), directly and by comparison with Spencer's electron range calculations.

All experiments have confirmed that the threshold effect is related to the build-up of space charge although details of the process are still obscure. It is not explained why sufficient space charge to compensate the applied bias occupies half the specimen, so that for example 1 μm is sufficient in a 2 μm specimen but insufficient in a 4 μm specimen.

The response to a pulse can be taken as a measure of the existing space charge; for example, a steady response indicates that the space charge has reached its limiting value for the particular bombarding conditions. This state was first established in a series of experiments carried out by Bowlt and Ehrenberg (1969), illustrated in Fig. 3.8. The top surface of a self-supporting film of As_2S_3 was bombarded with electrons; the controlling space charge was then expected to reside in the lower portion of the film. Then the specimen was bombarded from the bottom at zero bias in order to

96 ELECTRON BOMBARDMENT INDUCED CONDUCTIVITY

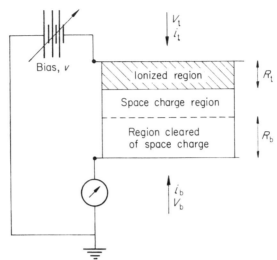

Figure 3.8. The two-beam experiment. Schematic diagram of the ionized region and the space charge region produced by top bombardment, with an applied bias; the region cleared of space charge by bottom irradiation is also indicated, and R_t and R_b are the practical ranges of bombarding electrons from the top and from the bottom respectively. (Bowlt and Ehrenberg, 1969)

remove the space charge in a region adjoining the bottom surface; this created a strong ionization in which the trapped carriers were liberated and conducted away through the bottom electrode. A strip of space charge remained intact in the middle region, unless the cathode rays from the bottom reached the plane at which the rays from the top had stopped. In the third stage of the experiment, the response to a single shot from the top was measured at the original bias and electron energy. It indicated how much of the space charge established in the first stage was left; it would have been equal to the response of a virgin specimen if the space charge had been completely removed.

Results for an arsenic trisulphide film with a threshold of 23 kV are shown in Fig. 3.9. The sensitivity of the method is limited to the cases where the space charge decay in the interval between first and second stages mentioned above is less than the irradiation induced clearing. The natural decay accounts for the larger than steady-state gains at low values of the bottom bombarding voltage and especially in the absence of irradiation from the bottom.

In the second curve the pulses of electrons from the top have an energy of 23 keV which is the threshold energy for this film. It is seen that a small effect is observed even at $V_b = 10$ kV, but that it requires electrons of about nearly threshold energy impinging on the opposite side to restore the full response.

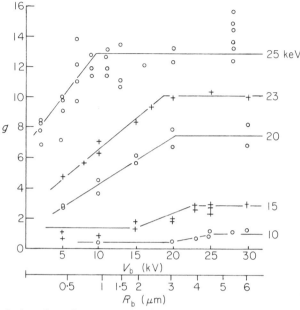

Figure 3.9. Induced conductance gain of the first pulse of fast electrons from the top gun at various energies. The pulses of electrons from the top gun (V_t) were interrupted by irradiating the specimen with electrons from the bottom gun (V_b). The lower scale gives the penetration depth (R_b) of the electrons. a-As_2S_3 film thickness 6·6 μm; $i_t = 5 \times 10^{-9}$ A (pulsed); bias $v = 200$ V (top negative). (Bowlt and Ehrenberg, 1969)

Thus, as V_b (and hence the penetration of the neutralizing beam) increases, more and more of the space charge is swept away until at $V_b = V_{b2} = 20$ kV the whole space charge layer is neutralized and g ceases to increase. The limits of penetration R_t and R_{b2} of the 23 kV and 20 kV electrons are 3·9 and 3·1 μm respectively; the sum is 7 μm, slightly more than the thickness of the film. The results show that the space charge extends from about the middle of the film, i.e. the end of the range of the bombarding electrons, to very near the bottom electrode.

At higher V_t (top curve) the space charge layer commences nearer to the bottom of the film, so the neutralizing electrons have to penetrate less deeply in order to release all the space charge. But for $V_t = 15$ kV and 10 kV (third and fourth curves) the irradiation induced clearing of the space charge is small or absent when the penetration of the bottom electrons is small. Accurate measurements are therefore difficult in this region but the results suggest that at low values of V_t very little charge resides near the bottom electrode. The difference ($R_{b2} - R_{b1}$) of the penetration depths at which the upward trend of the curves commences and ends (indicating clearing of the space charge) showed that the thickness of the space charge layer ($R_t + R_{b2}$) should equal the thickness of the film. This was indeed found to be the case, within a reasonable margin of error.

In another series of experiments a specimen was exposed to repetitive bursts of bombardment on one side or on both sides. With As_2S_3 2·6 μm thick, at 100 volts bias, gains of 80 and 25 were observed at V_p = 18 and 13 kV respectively (ranges 2·6 and 1·5 μm) when one side only was bombarded. The same gains were recorded for 12·7 and 9·2 kV respectively (ranges 1·3 and 0·8 μm) when both sides were bombarded. This shows that the gain depends on the sum of the ranges, and has its full value when the formation of a space charge is prevented. Top and bottom pulses need not be exactly simultaneous. Second only to As_2S_3 among insulating films, ZnS has received the most attention. Tests have been made on glass-supported sandwich films (Ehrenberg and Hidden, 1962; Benoit *et al.*, 1969) and on lightly supported films (Didenko *et al.*, 1959; Guillard and Charles, 1966; Bowman, 1972) which were stabilized on one side by a low velocity scanned electron beam. Didenko *et al.* exposed this non-metallized side to both the "writing" gun and the stabilizing gun, operating at a voltage causing high secondary emission, whereas Guillard and Charles, and Bowman, let the "writing" gun operate from the supported metallized side of the film, and stabilized the other side by a beam of nearly zero velocity electrons.

The experimentalists who examined glass-based films found clearly marked thresholds and gain values between 10 and several hundred. Benoit *et al.* found that X-ray diffraction photographs indicated very disordered lattices which could be improved by annealing at elevated temperatures; initially the films might have been amorphous. Didenko *et al.*, on the other hand, kept their mica-Al bases at 120–150°C *during deposition* and reported gains of up to 360 but no clear threshold. Diagrams given by Guillard and Charles indicate similar gains and also the absence of a threshold. The targets examined by Bowman fall into two groups, giving gains of up to about 500 and of more than 1000 respectively; only the former displayed a well defined threshold. Figures 3.10 and 3.11 summarize Bowman's results. One explanation of this behaviour is that ZnS, quite different from As_2S_3, is very easy to crystallize but tends to have disordered lattices with an interplay between the wurtzite and the blende form. It is therefore reasonable to associate the differences in EBC behaviour with different degrees of crystalline perfection, which is completely lost in the amorphous form. Ehrenberg and Hidden had found that thin crystal platelets gave gains without threshold that were about 1000 times higher than those given by amorphous films deposited from the same material.

The idea that the existence of a threshold is due to the super-abundance of traps in the amorphous state is confirmed by experiments carried out by Bril

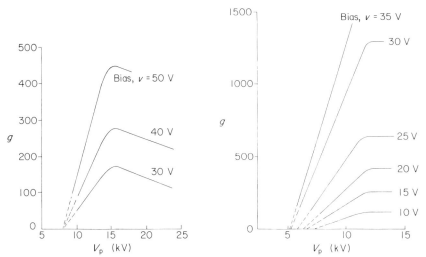

Figures 3.10 and 3.11. The steady EBC gain of thin lightly supported films of ZnS with an electron beam contact on one side. (Bowman, 1972) LEFT "low gain" film; RIGHT "high gain" film.

and Gelling (1962) who examined layers of cadmium selenide 5–25 μm thick sandwiched between glass (made conductive by a coating of tin oxide) and a thin film of aluminium. The film was bombarded through the aluminium with electrons of 10, 15, and 20 keV energy. The authors estimate that 20 keV electrons penetrate only 2 μm. In spite of this, a steady gain of about 2000 was observed in a film 15 μm thick, at a negative bias of 20 volts. The interesting point is that the temperature of the substrate during evaporation has a decisive influence on the structure of the evaporated layer; the higher the temperature the greater the opportunity for the condensing material to arrange itself into a neatly crystalline solid. The authors also report that the gains of their films, and their dark resistance, were appreciably enhanced when the films after deposition were exposed to copper vapour so as to give them a Cu content of 0·1 mol%, and heated at about 400°C in the presence of sulphur vapour to compensate for any excess Cd. The holes were found to be immobile, so the steady current must have been maintained by injection; this is confirmed by the rapid increase of gain with bias.

Selenium evaporated in a vacuum is deposited in the red amorphous modification. Pensak (1950) found that the gain of these films is exceptionally large and that considerable currents are induced with positive bias at less than 2 keV electron energy in films 1 μm thick. This means that there is no threshold. These results were confirmed by Spear (1956) who also found that, for low bombarding energies and currents, the induced current satu-

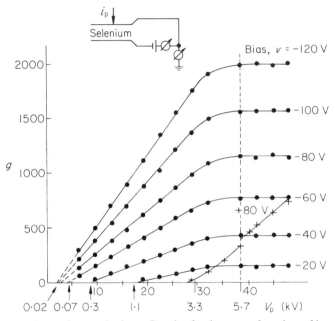

Figure 3.12. Amorphous selenium. Graph of gain g as a function of bombarding voltage V_p, for different values of applied bias v; $i_p = 1.4 \times 10^{-8}$ A, specimen thickness 5·7 μm. Numbers below the V_p axis are calculated values of the average penetration depth of the bombarding electrons in μm. (Spear, 1956)

rates with bias when an average of about 20 eV energy is used for the creation of a pair (see Fig. 3.12).

Broerse (1966) produced polycrystalline films of PbO about 3 μm thick that were highly photosensitive and cathode ray sensitive by evaporating the material in the presence of oxygen; the density of these films was about half that of the bulk material. There was no threshold; the gain increased linearly with bias initially and then appeared to saturate above 50 volts bias at about $g = 100$ with 10 keV electrons and $g = 700$ with 20 keV electrons.

2. The maximum

If there is no entry of carriers from the electrodes (i.e. "primary" EBC), as the energy of the primary beam is increased, the gain increases rapidly and comes to a maximum (g_m) at about twice the bombarding energy of the threshold. At qV_m the ionization of the specimen is roughly uniform, and the gain varies only slowly with electron energy. Spencer's calculations show that, at qV_m, about half of the energy of the electrons entering the platelet is dissipated in it (i.e. the energy of the impinging electrons less the energy of those back-reflected). It is common for qV_m to be chosen as the energy for EBC studies of films.

At $V_p = V_m$ the induced current should be the same for either polarity of bias, neglecting the charge carried by the cathode rays. This is not always exactly the case, perhaps because the ionization is not strictly uniform, and the two surfaces are not quite equal. The existence of a maximum and its variations with film thickness were first reported by Pensak (1948) who observed gains of up to 70 for very thin films of silica deposited on Nichrome plates by thermal decomposition of ethyl silicate vapour, at the highest bias the films would tolerate. For aluminium oxide, values of g up to 15 have been reported (Ansbacher and Ehrenberg, 1951), for As_2S_3 values up to 60, and for Sb_2S_3 higher values. Sb_2S_3 was also examined in more detail by Oksman and Tikhomirov (1959) who used films deposited on thin metal-lized organic foils and stabilized on the other side by secondary emission; the much higher gain (up to 600) found by these authors is probably related to the difference in measuring techniques.

All such figures are of course of restricted significance. They do not take into account the increase of gain at elevated temperatures, the injection of carriers, the effect of thickness, etc. It would therefore serve no useful purpose to compile and present a list of observed gains.

3. Space charges in the steady state

Ansbacher and Ehrenberg (1951) suggested that, for uniform ionization, Thomson's and Mie's idea of the ionization chamber should apply, thus allowing for carriers firmly trapped near the electrodes. In the central region of the specimen the applied field is then reduced by the charges near the electrodes. Following Newton's (1949) analysis of McKay's (1948) work on diamond, which showed that the electric field in the centre of the specimen was less than about 200 V cm^{-1}, Ansbacher and Ehrenberg related the gain at $V_p = V_m$ to the mobility (μ) of the carriers, idealizing the field as having one value in the two boundary regions and another in the central region. A mobility of 6×10^{-4} cm^2 V^{-1} s^{-1} was suggested. Alcock (1962) and Ghosh (1967) attempted to estimate the charges stored near the electrodes for As_2S_3 and Spectrosil films (SiO_2) as follows. After exposure to electrons at $V_p = V_m$, at given applied bias, the bombardment was interrupted and, after a short interval, re-started with zero bias. The field due to the charges near the electrodes then provides a bias which caused an induced current to flow in the opposite direction. The initial current depended on the interval, but its value at zero interval was found by extrapolation. The bias responsible for this current was found to increase with the charging current but never to exceed 10–20% of the bias at which the specimen was charged—a percentage low enough to justify the procedure. Ghosh pointed out that a reverse current is observed after the bias has been reduced to zero on As_2S_3 film even without renewed bombardment. This reverse current must be allowed for when estimating the effective bias. The curves given show that in the absence of irradiation the charge decays initially to $1/e$ in about 30 s, and to $1/e^2$ in about 100 s. Gibbons (1974a) studied the stored charge by recording the discharge current on successive bursts, after the removal of the bias, in thin films of As_2S_3 and ZnS. It was found that substantial discharging currents could be detected and that, with successive pulses of electron excitation under a finite forward or reverse bias, the amount of trapped charge reached a saturation value. The size of the discharge pulse under zero bias is shown in Fig. 3.13 as a function of the number of equal duration excitation pulses.

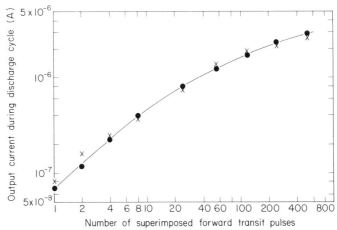

Figure 3.13. Discharge pulse amplitude as a function of the number of superimposed equal forward induced current pulses (● positive bias, X negative bias; discharge bias zero). Zinc sulphide evaporated film 0·9 μm thick, $V_p = 14$ kV. (Gibbons, 1974a)

The amplitude of the discharging pulse could be fitted to a Hecht curve from which values of the schubweg (w) could be deduced. Values of w/bias field were found to be of the order of 10^{-13} m^2 V^{-1} for evaporated ZnS. The deduced values of w ranged from 0·08 to 3·1 μm for specimens in the thickness range 0·4–2·0 μm.

It is worth noting that in these charge storage experiments the EBC effect is responsible for both charging and discharging the specimen, according to the value of the bias. With the open film technique (Pensak's method), in particular with ZnS films, a different kind of charge storage has often been employed and has become of technical importance. The film acts as the dielectric in condenser elements of which the electrodes are the free surface and the metal backing. Areas of the surface are charged by scanning it with an electron beam which stabilizes it by secondary emission at the potential of a neighbouring grid. Electrons from a different source and of higher energy make the dielectric leaky so that the condensers are discharged; the charge required to bring the surface back to its original value is then a measure either of the gain at given current density or of the density and duration of the cathode ray current at given gain.

4. Secondary EBC

Secondary EBC gain can be explained by reasoning analogous to that used in the case of secondary photoconductivity, but it should be emphasized that the word "gain" has slightly different meanings. In EBC, gain is the ratio of circulating current to bombarding current; in photoconductivity, gain is the ratio of the number of circulating charge carriers to the number of incident photons. The secondary effect in photoconduction arises through one sign of charge carrier being trapped near an injecting electrode which thereby increases the electric field in its vicinity and enhances injection. These trapping centres are known as activation centres; they have a high trapping cross-section for one sign of carrier and thereafter a low recombination cross-section for carriers of the opposite sign. If t' is the carrier lifetime in the presence of filled activation centres, and T is the transit time, the secondary photoconductive gain is given by $G = t'/T$. In EBC the explanation of the secondary effect is similar; the secondary EBC gain is given by $g' = gG$, where g is the EBC gain for the primary effect.

Activation centres can be introduced into CdS, CdSe, or ZnS by adding impurities such as copper or silver. Secondary EBC gain values in the region of several thousand at room temperature can be observed in stoicheiometric and non-stoicheiometric formulations of Se–S, Se–O, Se–O–S, Se–Cd–S, As–Se–S, Cd–As–S, and Zn–Cd–S. Secondary gains up to 40 000 have been seen for As_2S_3 at high temperatures; they can be up to 2000 for amorphous selenium at room temperature and as high as 10^{10} for single crystals of CdS activated with copper and chlorine. Secondary gain is always associated with time lag effects, and usually the higher its value the longer the response time. The cited targets having g' values in the region of 1000 had a response time of about 1 s, and the very high gain of 10^{10} was measured on a CdS target with a time constant of many minutes.

B. Gain

The gain at qV_m, at a bias just below breakdown, is a measure of the EBC sensitivity of the material. Normally materials with large gains are also photoconductive in some region of the spectrum. If the gain is of the order of unity or even less, of course it matters whether its definition does or does not include film leakage (i_l) and the primary current (i_p). In these cases gain must be defined as either

$$g = i_s/i_p \quad \text{or} \quad g = (i_c - i_b - i_l)/i_p$$

where i_s is the secondary (induced) current, i_c is the total conduction current, and i_b is the current due to primary electrons reaching the bottom electrode. In practice i_b can be measured by determining the current flowing in the circuit containing the specimen when the bias is reduced to zero.

1. Gain as a function of thickness

Observations on specimens having different thicknesses have supplied valuable evidence as to how the applied bias increases the EBC gain in thin films. However, there is uncertainty about such results because even values for gain measured on equally thick samples often vary by perhaps 30% or more if the technique for preparing them is not rigorously controlled. Pensak (1949), comparing films of silica between 0·25 and 1·5 μm thick, found that the gain at V_m is approximately proportional to the energy dissipated. In the light of Hecht's formula, this suggests that the schubweg of the carriers in these films is much longer than 1·5 μm. However, this was not Pensak's conclusion; he was puzzled as to how his results could be reconciled with a gain much smaller than the ratio of qV_m and the band gap. Ghosh (1967) and Ehrenberg and Ghosh (1969) also found that the gain for As_2S_3 between 2 and 6 μm thick was proportional to V_m. At a field of 10^5 V cm^{-1} the corresponding value of the mobilization energy E was 650 eV. An independent series of samples led to the same result, and a third series gave $E = 215$ eV with a field of 2×10^5 V cm^{-1}. The calculated schubweg varied considerably, from 6 to 50 μm, but was never small compared with the sample thickness.

These authors also examined films up to 22 μm thick excited by β-rays from a Sr-90 (+ Y-90) source. Under this excitation the gain reached its final value only after 2 hours, when it was 2–3 times greater than after 30 s. Although the thickness fitted the Hecht pattern reasonably well, the schubweg was only 1–2 μm after 30 s with $E = 54$ eV at a field of 10^5 V cm^{-1}; after 2 hours the schubweg was even shorter and the mobilization energy had fallen to 1·8 eV at 3×10^5 V cm^{-1}.

At first sight it might appear that the mobilization energy is much greater than the radiant ionization energy on account of Shockley–Read recombination. But the electron bombardment results pretty well exclude the possibility that this alone is responsible for such a large discrepancy; it could be due to initial or columnar recombination—or both. The authors decided that columnar recombination was mainly responsible because of the higher value of E for cathode rays than for β-rays, the density of ionization along a track being much greater for cathode rays than for β-rays. Some later results of Ing et al. (1971), who found the photoconductive quantum efficiency of As_2S_3 films to be of the order of unity for blue light at fields in the neighbourhood of 10^6 V cm^{-1}, also suggest that initial recombination plays only a minor role. The initial recombination for photons can barely be smaller than that for cathode rays.

The different values for β-rays were attributed to their tracks being straight and parallel to the field, so the carriers moved in the wake of the tracks. This interpretation does not explain the apparent reduction of the schubweg with increasing exposure. A high rate of injection of carriers caused by a slow build-up of the field near the electrodes would make the gain just as independent of the thickness of the specimen as would a short schubweg.

Recombination of carriers explains in general why the gain does not come up to expectations. However, on As_2S_3 films at high values of bias (and especially at elevated temperatures) the gain rises above the value that could not reasonably be expected without any tendency to saturate. This effect has been related to the injection of carriers from the electrodes, and no other explanation seems to be possible.

Mott and Gurney (1940) first drew attention to the possibility that the work function of the electrode material with respect to a dielectric can be so small that a significant cloud of electrons exists in equilibrium in the dielectric near the electrode. An electric field will draw this cloud to the opposite electrode. Mott termed this effect space charge limited current. The effect here referred to is of a similar kind.

Trapping of carriers near the electrode can increase the average field acting near it by a factor of the order of 10, and more due to statistical fluctuations. In the experiments, fields up to 10^6 V cm^{-1} are applied. Field emission rises according to the Fowler–Nordheim formula from 10^{-11} to 10^{-9} A cm^{-2} when the field increases from 10^6 to 3×10^6 V cm^{-1}. Carrier injection is therefore a predictable effect. Because gain rises with bias owing to reduced recombination and because of rising or falling injection of carriers, any region of saturation must be expected to be masked.

2. Gain as a function of applied field

The current induced in films always increases with bias. Of course, even the best films break down at fields of about 10^6 V cm^{-1}, but experimental evidence suggests that the absence of saturation is not due to weakness of the field.

At moderate fields the gains are so low that no tendency to saturate could be expected; they have been shown to be only a small percentage of what is energetically possible. It is seen from diagrams such as Fig. 3.6 that, as a function of V_p, the gains recorded for different values of bias have the same shape and can be made to coincide by multiplying them by a factor $f(v)$. Ansbacher and Ehrenberg (1951) found that gain increases linearly with bias at low temperatures and becomes quadratic in bias as the temperature increases. They provided a curve showing the gain at fields between 5×10^4 and 6×10^5 V cm^{-1} for various specimens of As_2S_3 film; it increases in this range from 1 to 40. Ghosh found that, at room temperature over a wide range of bombarding currents and specimen thicknesses, the gain for As_2S_3 films is proportional to $v^{1.4}$; for Spectrosil and fused quartz the gain is proportional to $v^{1.25}$. For sandwiched ZnS films, Ehrenberg and Hidden found that the gain was proportional to the bias. Benoit's curves indicate a superlinear increase. For open ZnS films Didenko et al. and Bowman found that the induced current can be represented as a power of the bias, with an index not exceeding 2. Guillard and Charles claimed an exponential increase of gain with bias for their open films, but this was based on a rather small range of values.

We have to ask why the bias influences the generation or recombination of carriers so that only a minority of them contribute to the circulating current. In order to recombine, a carrier must find another carrier of opposite sign; it can normally do this in an insulator only within the plasma. The longer the carriers remain in the plasma the greater is their chance to recombine—whether the plasma is restricted to a thin layer near the bombarded electrode or extends over the whole film. We would therefore expect the fraction of carriers escaping recombination and contributing to the gain to be greater the higher the drift velocity of the more mobile carriers. Hence the gain should be proportional to the bias if it is small, and a measure of mobility and trap density.

In the scheme referred to as Shockley–Read recombination, two stages are involved, the creation of a pair by the incident radiation, one partner of which is either initially localized or becomes localized after some diffusion. If now a mobile charge of opposite sign (one of a pair created elsewhere) comes near to this localized charge, it can neutralize it so that, of two mobile pairs, one has disappeared. In the first process a carrier is trapped and hence can become free again. Only the second event terminates the career of a carrier.

In order to gauge the effect of recombination on the circulating current, Hecht (1932), studying field dependence in photoconductivity, introduced the concept of schubweg (w). This is the mean distance a carrier travels in the direction of the field before it recombines or becomes trapped. If the

3. STEADY STATE EBC OF THIN INSULATING FILMS 107

specimen is L units thick, the carrier contributes the charge $(w/L)q$ to the charge measured in the external circuit. The actual path of the carrier is devious and much longer than w, because the field has only a slight effect on the carrier's movements. As a result, w is proportional to the field. If $w \ll L$, and the ionization is uniform, the gain will be w/L times the number n of carriers set free per primary electron, and hence proportional to the bias and independent of the thickness of the specimen. If $w \gg L$, each carrier pair will contribute the charge q and $g = n$; the gain is proportional to the energy lost by the incident electrons, i.e. to V_m if the electron is re-adjusted to V_m for varying thickness. Hecht's formula for intermediate cases will be referred to in Chapter 5.

A carrier released by radiation can also disappear by initial recombination, i.e. by returning to its parent atom. This can happen because it loses the kinetic energy with which it is ejected by collision and can be left with only thermal energy while still within the range of Coulomb attraction. If it is far enough, of course, the bias field will prevent the return. The distance from the parent atom at which the applied field can be equal and opposite to the Coulomb attraction is, neglecting screening, given by $q^2/r^2 = q\mathscr{E}$ or $r = \sqrt{(q/\mathscr{E})}$. For $\mathscr{E} = 10^5$ V cm^{-1}, $r \approx 1 \cdot 3 \times 10^{-6}$ cm ($q = 1 \cdot 6 \times 10^{-19}$ C). This distance is of the same order as that at which a carrier is normally expected to have become thermalized. Onsager (1938) estimated the effect of a bias field on initial recombination; fields of 10^4, 10^5, and 10^6 V cm^{-1} will reduce the probability by $1/2$, $1/10$, and $1/100$ respectively.

Columnar recombination stands between initial recombination and recombination in Shockley–Read centres.

It was Langevin (see Moulin, 1908) who first drew attention to the special nature of the ionization by α-particles, viz. the high linear density of ion pairs along the track (columnar ionization) which must greatly increase the rate of recombination and the field necessary for saturation. A detailed experimental study by Moulin fully confirmed Langevin's suggestion. Moulin confirmed in particular that an electric field normal to the columns produces a higher current than a field parallel to the track. The recombination in the columnar plasma is so effective that it swamps any initial recombination of freshly separated pairs. Jaffé (1913, 1914) studied the directional effect in greater detail and also investigated the ionization of liquids. For example, in hexane at about 1000 V cm^{-1}, the currents caused by α- and β-particles respectively were found to be $0 \cdot 1\%$ and 10% of the saturation currents produced by the same source in air. For α-particles the ionization current increased in proportion to the component of the field normal to the track. Jaffé also made calculations based on the idea of a critical radius, and obtained theoretical values for the reduction of current due to columnar recombination. Columnar recombination is more effective the higher the (bimolecular) coefficient of recombination, the greater the density of ions along the track and the smaller their mobility. From this it must be inferred that the efficiency of cathode rays in inducing conductivity increases with their energy and depends critically on the trap-controlled mobility of the carriers.

Jaffé's figures for the ion density in a column in hexane are $4 \cdot 3 \times 10^7$ cm^{-1} for α-particles and $1 \cdot 46 \times 10^5$ cm^{-1} for β-particles. Figures for the energy loss in arsenic trisulphide suggest, with 10 eV per pair created, densities of 8×10^6 cm^{-1} for cathode rays and 8×10^5 cm^{-1} for β-rays—values quite comparable to those found in hexane.

Whilst the mobilities of electrons and holes in many crystals are far greater than those of ions in liquids, the mobility of carriers in amorphous films is trap-controlled or associated with an intermolecular hopping process and probably smaller than that of ions in hexane. The generation of carriers in arsenic trisulphide films by cathode rays and β-rays should therefore have all the features of columnar ionization. For β-rays, as opposed to cathode rays, columnar recombination is reduced owing to the greater separation of the ions and enhanced by the directional effect. The number of carriers released from columnar recombination increases in proportion to the field.

3. Effect of current density

Except at very low temperatures, or for very low current density, the gain decreases as the current density increases. The effect is not great. For example, the gain for a specimen of As_2S_3 4·63 μm thick, at a field of $2 \cdot 2 \times 10^5$ V cm^{-1}, reaches its upper limit of 60 at a current density of 5×10^{-10} A cm^{-2}. At a current density of 5×10^{-7} A cm^{-2} the gain is about 15. Ehrenberg and Ghosh (1969) attribute the reduction of gain to the incidence of Shockley–Read recombination.

4. High temperature effects

The EBC gain in a film increases with increasing temperature. Recombination by any one of the three processes referred to is reduced at elevated temperature, and the drift mobility rises, so an increase in gain must be expected. For As_2S_3 a bend occurs in the gain vs. temperature curves above room temperature as shown in Fig. 3.14; a second feature comes into play at about 100°C.

Below 100°C the gain varies reversibly with temperature, i.e. if the temperature returns to a previous value the gain comes back to its original value. Also, on reversal of bias the current changes sign without change in magnitude after a short transition period during which the space charge settles down to a new equilibrium arrangement.

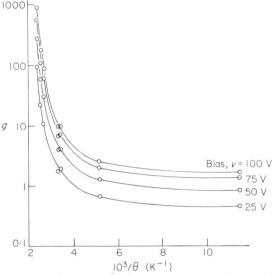

Figure 3.14. The EBC gain g as a function of temperature for an a-As_2S_3 film, with specimen bias v as a parameter. The upturn in this family of curves at about 100°C ($10^3/\theta = 2 \cdot 7$) can be clearly seen; this is accompanied by a rapid increase of g with rising temperature caused by entry of carriers from the electrodes. (Ansbacher and Ehrenberg, 1951)

Above 100°C the gain becomes dependent on the history of the sample. At room temperature, for a field of 9×10^5 V cm^{-1}, the maximum gain measured on a new sample is about 50. After the sample has been heated beyond 100°C and allowed to cool, a gain of several hundred is measured for the same field. Under bombardment this eventually falls by $1 \cdot 3$–$1 \cdot 5$ times, but this is restored to its original value after bombardment at liquid air temperature. Also, above 100°C the increase is accompanied by a kind of switching effect. The current increases initially under bombardment until it stabilizes after seconds or minutes at a "high" value several times its initial value. Once the "high" gain is established for a particular bias and temperature, the EBC current instantaneously follows any variations of the beam. Now, if the polarity of the bias is reversed while the bombardment continues, the gain (in the reverse direction) is initially a small fraction of the previous "high" gain. It then rises slowly to about the magnitude that it had before the reversal of bias. The "high" gain is reached more quickly if the bombarding current is repeatedly interrupted while the bias is kept on. Such "conditioning" of a specimen does not take place without bombardment. Oscilloscope traces showing some of these effects are given in Fig. 3.15.

The difference between low gain and high gain increases with increasing bias and increasing temperature, and it decreases with increasing bombard-

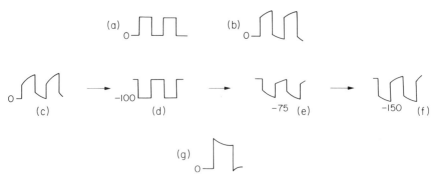

Figure 3.15. Pulsed EBC. (a) represents the bombarding current as a function of time; (b) conduction pulse in a specimen of As_2S_3 at 26°C, with large bias; (c)–(f) conduction pulse in a specimen at 120°C, $I_p = 1.4 \times 10^{-7}$ A cm^{-2}. The conduction pulse shape changes on bias reversal: (c) high gain with positive bias; (d) bias merely reversed; (e) negative bias on for 1 s; (f) negative bias on for some time. (g) Al_2O_3 film at 300°C, primary current pulse length 5 ms, $I_p = 2.7 \times 10^{-7}$ A cm^{-2}, $V_p = 5$ kV.

ment current and with increasing thickness of the sample. For example, at $0.27\,\mu$A cm^{-2} there is no difference between low and high gain below 100°C, but for the very low current density of 1·4 nA cm^{-2} a small effect is already noticeable at 20°C. Also, above 100°C and at low primary energy, the negative high gain can be double the positive gain. This means that the holes travel with less hindrance than the electrons or that the recombination of a mobile electron with an immobile hole is easier than the reverse process. The highest gain on As_2S_3 films is 40 000 at 163°C with $V_p = 24$ kV; this would allow only 0·5 eV for the creation of a pair of carriers.

Appreciable increase of gain with temperature has also been reported for ZnS films. Benoit *et al.* give 25, 60, and 150 as relative values of gains at -100, 50, and 150°C. Ehrenberg and Hidden found that the gain increases 2½-fold between room temperature and 330°C. No secondary effects are mentioned. The threshold, when found, is independent of temperature.

5. *Rise and decay of induced current*

If cathode rays are switched on as a step function, the induced current rises instantly to perhaps half its final value which may be reached after a few milliseconds or hours. It takes longer the lower the bombarding current. For example, with weak β-rays, rise times of up to 4 hours have been observed for As_2S_3 films, and a fraction of an hour for silica. An example is shown in Fig. 3.16. The break between the "prompt" and the "slow" rise is often quite conspicuous. If a build-up of space charge takes place during the slow rise, the increase can turn into a decrease. A reverse process takes place when the bombardment current is switched off.

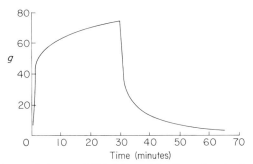

Figure 3.16. Rise and decay of EBC in a thin film of a-As_2S_3 (10·8 μm) when bombarded at a low current density; bias $v = \pm 108$ V, $I_p = 5 \times 10^{-14}$ A cm^{-2}. (Ghosh, 1967)

For the beginning of the slow rise, Alcock (1962) found that the gain for β-rays acting on an As_2S_3 film was proportional to log t for 20 ms $< t <$ 30 s. Hirsch found $g \propto c \tanh(L/t)$ for cathode rays. He suggested (1966) that initially the liberated electrons are trapped by V centres which then become V^- centres capable of neutralizing holes. This process makes up the prompt rise. As the number of V^- centres increases, the density of holes reaches its limiting value during the slow rise. Hirsch carried out his calculation assuming that 10 eV is the average energy required to create a pair, and arrived at the result that the schubweg is small compared with the thickness of the specimen, using likely values for the other parameters—a result that justifies the omission from his calculation of a term denoting the loss of carriers by drift into the electrodes. This omission is no longer justified if, as shown above, the mobilization energy is several hundred volts and the length of the schubweg becomes comparable to the thickness of the specimen. A calculation allowing for this is not at present available.

Figure 3.17(a)–(d) shows how the rise and decay of the EBC under conditions of pulsed beam and steady bias depend on the bias voltage and on the bombarding electron current density in a film of amorphous selenium. Except at low bias and at low current densities, the two-component behaviour is easily seen, and also the change from a rise with time to a fall if the value of V_p is less than the penetration voltage (Spear, 1956). Similar behaviour can be found in thin films of amorphous As_2S_3, and Fig. 3.17(e) is a tracing from a typical oscillograph; in this case the two components cannot be distinguished (Hirsch, 1966). Figure 3.18 shows how the two components can be analysed separately using a log-log plot of the magnitude of the prompt components of gain (g_0) and the slow component (g_s) as a function of bombarding current density (I_p) when the beam is totally penetrating. The slopes of the two double-logarithmic plots are ½ and 1, so indicating that a bimolecular recombination process is responsible for the kinetic behaviour

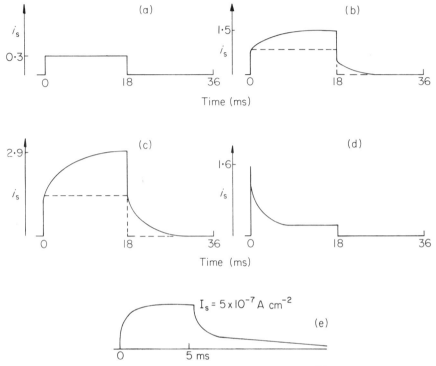

Figure 3.17. Growth and decay of induced current in a-Se [(d) $V_p = 10$ kV, $i_p = 4 \times 10^{-9}$ A; (a)–(c) $V_p = 45$ kV, $i_p = 8 \times 10^{-10}$ A]: (a) bias $v = 30$ V, (b) $v = 70$ V, (c) $v = 100$ V, (d) $v = 45$ V. (Spear, 1956) (e) Trace of typical oscillogram for a 4·3 μm film of a-As_2S_3 at room temperature when $V_p = 50$ kV, $I_p = 10^{-8}$ A cm^{-2}, and the bias field was $2·5 \times 10^5$ V cm^{-1}. (Hirsch, 1966)

of the slow component, whereas a monomolecular process is applicable to the prompt component (Hirsch, 1966). These two can therefore be attributed to the primary and secondary EBC processes respectively.

6. The nature of EBC materials

There is no doubt that the conductivity induced by electron bombardment varies from material to material. Mica shows no induced conductivity whereas some plastics films retain the bombarding charge of electrons almost indefinitely at the location where they originally came to rest. Normally, materials that do not show induced conductivity are not very sensitive to impurities; this is in conformity with the often quoted statement that amorphous semiconductors cannot be doped (Mott and Davis, 1971). However, Spear and Le Comber (1975) have shown that, in one amorphous solid

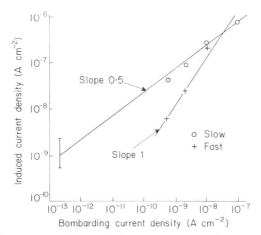

Figure 3.18. Separation of fast and slow components of rise and decay of EBC. Double-logarithmic plot of induced current density as a function of bombarding current density in As_2S_3 evaporated film 4·3 μm thick bombarded by 50 keV electrons, except for the lowest value where β-particles were used. Notice the bimolecular nature of the slow component compared with the monomolecular nature of the fast component. (Hirsch, 1966)

at least, it is possible to add impurities substitutionally to form either n-type or p-type amorphous silicon. It should be mentioned that this does not make it less necessary to use pure and well defined materials, since certain impurities may make films leaky and liable to breakdown. Normally the experimentalist will use the purest material commercially available, or try to improve on it; for example, As_2S_3 is easily purified by molecular sublimation (Bowlt, 1967), and ZnS can be purchased as "phosphor quality".

A notable dependence of EBC gain on impurities was observed for fused silica by Ehrenberg *et al.* (1966); the specimens had the shape of microscope cover slips and were exposed to β-rays from a Sr-90 radioactive source. The electrons lost only a fraction of their initial energy in the specimens and thus caused uniform ionization; no induced conduction was observed in silica-rich glass, but this may have been masked by its very high leakage. Natural fused silica gave a gain of unity at 500 V bias, Vitreosil (natural silica purified by electrolysis but retaining 50 parts per million Al) gave a gain of about 5, and Spectrosil WF (produced by vapour phase hydrolysis of silicon tetrachloride, with 500 parts per million Cl as main impurity) gave a gain of about 15; an alternative process gave Spectrosil with 4000 parts per million OH, but much less Cl, which showed a gain in the neighbourhood of 100. It should be remembered that Pensak previously had observed gain of the same order on thin silica films produced by decomposition of ethyl silicate vapour.

Films of amorphous selenium behave in many respects differently from "normal" films but share with them the property of giving two rates of rise. At low bias the "slow" rise is absent, as shown in Fig. 3.17(a). As the bias is increased, the "prompt" rise grows to a limiting value whilst the "slow" rise develops and grows. Spear suggested that the prompt rise represents the current due to pairs created by the primary electrons, and that the slow rise is due to injected carriers. He reports that the limiting value of the fast rise leads to the same value of E^* (energy of pair creation) as the saturation value of the positive gains at low electron energy, a coincidence confirming this value for E^* and the interpretation of the fast rise. If we accept this interpretation we are tempted to ask if the same idea could apply to materials like As_2S_3. Figure 3.17(e) shows that in As_2S_3 the fast rise is also absent at high bias.

The general equations for the rise and decay of induced conductance in the case of a bimolecular recombination process can be derived from the general kinetic equations. If the generation rate is β pairs cm^{-3} s^{-1}, and the recombination law is of the form αnp, we can put the recombination rate as αn^2 if one sign of carrier is not immediately trapped. Thus, during generation,

$$dn/dt = \beta - \alpha n^2 \qquad (3.1)$$

This has the solution, given the boundary conditions $n = 0$ when $t = 0$,

$$n = (\beta/\alpha)^{1/2}\tanh(\alpha\beta)^{1/2}t \qquad (3.2)$$

In the steady state (i.e. $t \to \infty$), $n = (\beta/\alpha)^{1/2}$. The decay law and its solution are given by the equations

$$dn/dt = -\alpha n^2 \qquad (3.3)$$

$$n = 1/(1/n_0 + \alpha t) \qquad (3.4)$$

where n_0 is the initial concentration of carriers. This relationship approximates to $n \approx 1/\alpha t$ as $t \to \infty$.

Thus it is seen that, where the conduction mechanism is dominated by bimolecular recombination, the rise and decay process is approximately hyperbolic. The tangent to the current rise curve at $t = 0$ meets the tangent for $t = \infty$ at $t_r = (\beta\alpha)^{1/2}$. Similarly, the tangent to the current decay curve meets the line $n = 0$ at a time t_r after the excitation has been turned off.

Variations in equations (3.2) and (3.4) occur if trapping, trap distributions, and detrapping are taken into account.

C. Lateral induced conductivity

If the direction of incidence of the bombarding electrons is perpendicular to the direction of induced currents, we can describe this EBC effect as lateral

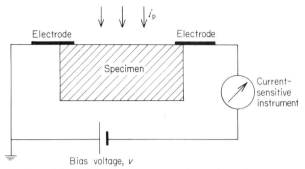

Figure 3.19. Disposition of electron beam, surface of specimen, and electrodes in the case of lateral EBC.

induced conductivity; the arrangement of the beam and the specimen is shown in Fig. 3.19.

Selenium, and the photoconductivity of its grey (metallic) modification, were discovered in the 19th century. Simple cells are made by painting or pressing molten selenium just above its melting point on to a glass plate provided with electrodes; such cells remained more or less a curiosity because their response to light is sluggish and unreliable. It occurred to Ralph de Laer Krönig (1924) that electron bombardment should have the same effect as illumination. He placed a cell of 3500 ohm dark resistance, forming an arm of a Wheatstone bridge, in the position of the anode of a thermionic triode and examined the change in resistance as a function of anode voltage and current; the resistance was reduced to half the dark value at 90 V anode potential and 400 μA anode current.

Although the grey metallic form had been known for a long time to be photoconductive, it was Weimer (1950) and Bixby and Ullrich (1951) independently who discovered that amorphous selenium also possesses this property. The amorphous form is produced by vacuum evaporation on to a cold substrate, or a substrate certainly no warmer than 50°C. The EBC of this form was described by Pensak simultaneously with Weimer's description of the photoconductivity.

1. Surface layers

Unless the bombarding electrons are very energetic, the excitation of a thick specimen is limited to a shallow region near the surface. The path of the electrons and holes liberated internally by the primary beam is limited to a region determined by their free lifetime.

Lateral EBC in crystals of CdS, CdSe, CdTe, and ZnSe, and in thin layers of CdSe and CdTe, was studied by Kot and Simashkevich (1962). The thin layers of CdSe and CdTe were "sensitized" by heating the layers in an oxygen atmosphere; this reduced the dark current and increased the EBC gain by forming "traps" (activation centres). The entire gap between the electrodes was bombarded with electrons. It was found that the most responsive crystals were those with smooth surfaces, and the least sensitive those with transverse grooves. Typical values for the EBC gain are given in Table 3.1, from which it is seen that the sensitizing procedure on the last

Table 3.1. Lateral EBC gain for a variety of specimens when the entire gap between the electrodes is bombarded; $V_p = 3$ kV, $i_p = 5 \times 10^{-8}$ A.

Specimen	Form	Gain
CdS	crystal	10
CdSe	crystal	10
CdTe	crystal	0·5
ZnSe	crystal	0·04
CdTe	layer	0·5
CdSe	layer	2×10^3

specimen mentioned clearly had a pronounced effect. A similar effect in thin evaporated films of CdS was reported by Gibbons (1974b); a (secondary) EBC gain of several thousand could be obtained by doping with Cu followed by an air-bake. Kot and Simashkevich found that the induced conductance varied linearly with i_p at low values of current, and that with increasing V_p the gain rose approximately hyperbolically. This was interpreted as due to an increase in the carrier lifetime as the generated carriers were then further away from the surface.

Soon after the discovery by McKay (1949) that the region near the contact of a germanium point-contact diode was sensitive to bombardment by α-particles, similar effects were found when the experiment was repeated but instead using electron bombardment (Moore and Hermann, 1951). This is a very important forerunner of a valuable number of effects associated with electron bombardment of semiconductor p–n junctions which will be described in detail under the heading "electron voltaic effect" in Chapter 4, and it provides the basis for obtaining images from a scanning electron microscope when used in the EBIC mode (Holt et al., 1974).

It is interesting that Ansbacher (1950) was unable to detect lateral induced conductivity in electron bombarded layers of ZnS, glass, mica, natural ruby, electrolytic Al_2O_3, and coloured or uncoloured alkali halide crystals.

3. STEADY STATE EBC OF THIN INSULATING FILMS

Further measurements of the lateral EBC of single crystals of CdS with Aquadag contacts were made by Archangelskaya and Bonch-Bruevich (1951). The entire gap between the electrodes was bombarded, and the induced currents were measured with a steady dc bias. The dark conductance of the CdS was 5×10^{11} ohm-cm and $V_p = 2$ kV for all the experiments; the bombarding current density was between 1 and 15 μA cm^{-2}. All measurements were made after an initial conditioning period of electron bombardment lasting 15 minutes; after this process there was no hysteresis in the conductivity decay rates when the beam was interrupted, and reproducible results could thereafter be obtained for periods of several days provided that i_p was not increased beyond that needed for conditioning.

It was found that approximately $\Delta\sigma_0 \propto i_p^{1/2}$ where $\Delta\sigma_0$ is the steady value of increased specimen conductance in the presence of bombardment. Also, when the beam was interrupted, the induced conductance decayed with a time constant $\tau_m \propto i_p^{1/2}$. It did not decay according to a bimolecular model, but an accurate expression

$$\Delta\sigma = \Delta\sigma_0/(1 + at)^\alpha \tag{3.5}$$

was found to apply, where $a = 500$ s^{-1} and $\alpha = 0.65$. The increase in decay time constant with time, as shown by equation (3.5), is often found in a large number of photoconductors and of phosphors.

The presence of an insensitive surface layer on single crystals of CdS when subjected to electron bombardment was demonstrated by Ryvkin et al. (1954). It was found that in single crystals from different sources, and in thick polycrystalline layers, the beam had to penetrate a thickness of between 100 and 1000 Å before the induced conductance rose rapidly with increasing V_p (see Fig. 3.20). The authors suggested that the fall in sensitivity near the surface was connected with an increase in recombination rate due to defects associated with an adsorbed surface layer.

Experiments on thin layers of ZnO, produced by oxidizing evaporated zinc, have provided valuable evidence for the kinetics of the process by which this compound loses oxygen under the influence of exciting radiation, as well as incidental data on the EBC; Heiland (1952) distinguished them as the "slow" process and the "fast" process respectively. Figure 3.21 shows how the induced current ("fast" process) responds to a step-function of bombardment. The thickness of the layer before baking in air at 400°C for 5 min to form the oxide was between 0.06 and 0.25 μm for different specimens, and the bombarding electrons had a set energy in the range of 1–6 keV. The results show EBC gain of about 100 at i_p about 6×10^{-7} A; it rose to a steady value within about 0.5 ms. The primary beam irradiated the entire space between the contacts as well as a large proportion of the contacts themselves (zinc contacts). As mentioned, the "slow" process

118 ELECTRON BOMBARDMENT INDUCED CONDUCTIVITY

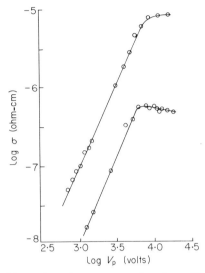

Figure 3.20. Induced lateral conductance as a function of bombarding voltage for two different single crystals of CdS. (Ryvkin *et al.*, 1954)

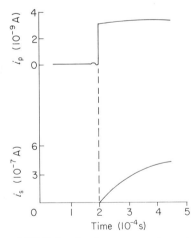

Figure 3.21. The lateral EBC for the fast process in a zinc oxide layer, showing (above) the exciting electron beam current waveform and (below) the corresponding rise in induced current with time. (Heiland, 1952)

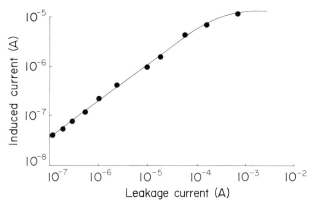

Figure 3.22. Dependence of the lateral induced current on the leakage current in a thin film of ZnO when the increased leakage arises through loss of oxygen; bias 10 V, $I_p = 3\cdot 2 \times 10^{-7}$ A cm^{-2}, layer thickness 0·06–0·25 μm, $V_p = 5$ kV. (Heiland, 1952)

represents a loss of oxygen from the lattice under the influence of excitation; the resulting oxygen vacancies act as donors and the leakage current then rises. Figure 3.22 shows how the induced current depends on the leakage for a constant electron excitation intensity of $1\cdot 6 \times 10^{-3}$ W cm^{-2}; it was found that the gain depends on the leakage current as $g \propto i_l^{0\cdot 77}$ and on the bombarding current as $g \propto i_p^{-0\cdot 60}$ approximately. These results agree substantially with the photoconductivity results by Mollwo (1948).

Sverdlova and Rokakh (1964) also exposed films of ZnO (0·4–0·8 μm thick) and of CdS (0·4–4 μm thick) to electron bombardment currents of 10^{-5}–10^{-8} A, at energies up to 5 keV. The ZnO films again were prepared by evaporation of Zn and subsequent oxidation; the gap between the electrodes was about 7 mm. Changes of resistance by factors of 10^4–10^5 were observed, some of which were due to chemical changes such as loss of oxygen.

Only one measurement of the lateral EBC of a thin-layer photoemissive cathode has been published. An antimony–caesium photocathode was studied under electron excitation with $V_p = 600$ V (Safratova-Eskertova, 1955); it was found that $i_s \propto i_p^{1/2}$, and in the range 10^{-10}–10^{-11} A gains of several thousand were obtained.

The earlier experiments by de Laer Krönig on the EBC of a selenium photocell were repeated by Rittner (1948) using electrodes made of tungsten wire spaced by 5 mm. Under a dc bias of 225 V he measured steady values of g equal to 63 and 123 respectively when the bombarding voltage was 1·0 and 2·0 kV; the bombarded area was only 1 mm in diameter but the published data make no reference to the influence of the position of the beam.

2. Position of the beam and the influence of contacts

A very important feature of the lateral EBC effect in insulators is the way in which the gain varies with the position of the beam between the electrodes. Experiments have provided valuable evidence for the existence of space charges due to trapped carriers or ionized impurities in the immediate vicinity of the contacts. The lateral induced conductivity of single crystals of CdS between 20 and 30 μm thick was measured by Benda (1951) using gold or silver electrodes between 1 and 3 mm apart for the various specimens, with bombarding voltages in the range of 800 V to 3 kV. The induced conduction current i_s varied with the position of the beam on the specimen; it increased markedly when the beam moved from the centre to the edge of the gap between the electrodes, and maxima were observed when the beam actually touched them. The maximum was higher when the electron beam hit the positive side, and this indicates that diffusion of excitons and production of light or X-rays is not a likely explanation for the higher gain near the contacts. Additional measurements in this series showed that $i_s = k i_p^a V_p^b$ where the constants a and b have values of 0·61–0·88 and 1·1–3·7 respectively and k is an arbitrary constant.

The experimental results of Kot and Simashkevich are summarized in Fig. 3.23 in the case where the contacts of gold or silver to CdS are rectifying. If ohmic (injecting) contacts are used the influence of the position of the beam is the opposite of that just described; for example, if indium or gallium

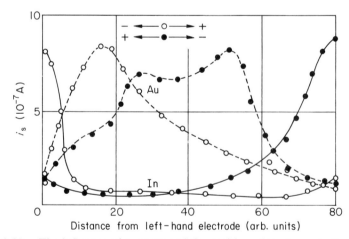

Figure 3.23. The influence of contacts and the position of the beam in the case of lateral EBC, for specimens of CdS or CdSe with In or Au electrodes; note the enhanced EBC when the beam strikes the In contact when it is biased negatively whereas there is a marked drop in the EBC if the beam approaches a negative Au contact. (Kot and Simashkevich, 1962)

contacts are used, i_s rises to a maximum when the beam strikes the negative side. If mixed contacts are employed, such as gold or silver with indium or gallium, the dark current is unsymmetrical and it is found that the highest value of i_s occurs when the region in the vicinity of the negative contact is excited, although when the "diode" is reverse biased the maximum value of i_s occurs when the position of the beam is slightly nearer the centre of the gap than in the case when it is forward biased. The latter experimental results are shown on the same figure for comparison.

Ehrenberg and Shrivastava (1973) showed that surface and volume induced currents could be detected simultaneously in 0·5 mm thick single crystals of ZnS. The surface currents were carried by "channels", and the volume currents were due to carriers with a bulk drift length of about 200 μm. The surface currents arose mainly as a result of built-in fields, and their influence could be suppressed by using a guard ring electrode.

D. The influence of contacts in normal EBC

As might have been expected, the influence of the contacts (and the relative position of the beam) is more important in lateral EBC than in the case where the exciting electrons penetrate the sample completely and we are interested only in the conductivity through its thickness. Obviously, if the electrode metal is thick (in units of mass/area) it will absorb more energy from the beam, and in this way the variation of g with V_p as a function of electrode thickness can be easily understood. This will be discussed in Chapter 7 where the influence of the thickness of a gold top (bombarded) electrode will be shown (Fig. 7.15).

If the specimen is capable of exhibiting significant EBC without the need for complete beam penetration, the contacts can have an important effect. An example showing this is single-crystal CdS; aluminium or indium form injecting contacts whereas graphite or gold form rectifying contacts because current transport is mainly by electrons. If crystals about 30 μm thick are fitted with two aluminium, gold, or indium contacts, on opposite faces, the gain at bombarding voltages of about 30 kV is about the same for any contact metal. However, if the specimen is not symmetrical, such as may be the case when one contact is of indium and the other of gold or graphite, the "dark" current characteristic is asymmetric and the EBC gain is higher if the bombarded contact is negative with respect to the other (Gibbons, 1974b). Trodden and Jenkins (1965) also found that symmetrical specimens of CdS with Au–Au or In–In contacts yield higher g values if the bombarded contact is negative with respect to the unbombarded one; this result is the opposite of that found by Benda for lateral EBC.

These results confirm that, even in the dark, conduction currents in single crystals of CdS are determined by the electron injecting properties of the metallic contact. The ability of the negative contact to supply electrons is promoted by electron excitation, and this explains why a normally non-injecting gold electrode can be rendered effective as an injector of electrons when it is bombarded. A similar promotion occurs if the bombarded contact is made from graphite (Aquadag). Electron bombardment of a negatively biased graphite contact can make it so effective that the EBC gain is then higher than when the specimen is turned so that the indium contact faces the beam; this is probably because near the negative contact the space charge barrier (which also limits the dark current) is nearer the contact when the graphite is biased negatively.

From these experiments it is clear that contacts (if they are thin) affect normal EBC only in the case of materials like CdS where appreciable EBC in the steady state can occur with a non-penetrating beam. In these circumstances a higher EBC is obtained when the bombarded contact is biased in such a direction that this contact injects the more mobile carrier. A higher ratio of induced current to dark current is obtained if the electrodes are different, the specimen is back-biased, and the non-injecting contact is bombarded. This seems also to apply to evaporated thin films of CdS, because in this material intercrystalline barriers are not a serious impediment to steady current flow beyond the penetrated region.

4
The electron voltaic effect (EVE)

I. Semiconductor junctions

In a semiconductor, near a metal electrode, or where differently doped regions come together, i.e. at a junction, the potential energy of carriers changes; this difference in potential prevents holes from crossing into the n-region and electrons into the p-region. The potential necessary to establish thermodynamic equilibrium is increased if the junction is back-biased. The potential energy diagrams for biased Schottky-barrier and p–n junctions are shown in Figs. 4.1 and 4.2 respectively. A depletion region is a result of all the donors or acceptors being fully ionized, and the free carriers liberated from them being drawn out of the region near the junction. An accumulation region represents the opposite situation—carriers are attracted to the junction and the donors or acceptors revert to their unionized state. Such junctions are rectifying, and are sources of an emf when irradiated by light which produces a photoelectric effect in the material. Units designed for this purpose (Fig. 4.3) are called photoelements or photovoltaic cells, but all p–n junctions respond to ionizing radiation.

The behaviour of photovoltaic cells under light excitation has been briefly mentioned in Chapter 3. The equivalent phenomenon when electron excitation is used (known as the electron voltaic effect or EVE) differs from the photovoltaic effect mainly owing to the non-exponential absorption of energy from the beam and complete freedom of choice of the energy of the ionizing particle (photon or electron energy). For reasons such as these, the beam energy can often be chosen to be a high multiple of the mobilization energy. Selenium photoelements were shown by Becker and Kruppke (1937) to be sensitive to bombardment by fast electrons. McKay (1951) reported the decrease in resistance of the barrier associated with a point contact to germanium when bombarded by α-particles. In such devices the values of g can be very high. The EVE is of importance in device applications, especially when the junction is back-biased to reduce its capacitance and to increase the width of the most sensitive region. Such applications will be discussed in Chapter 7.

124   ELECTRON BOMBARDMENT INDUCED CONDUCTIVITY

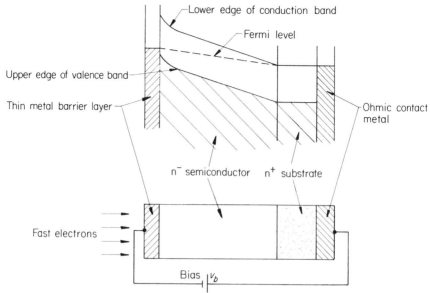

Figure 4.1. Energy band diagram for a depleted Schottky barrier diode. (Siekanowicz *et al.*, 1974)

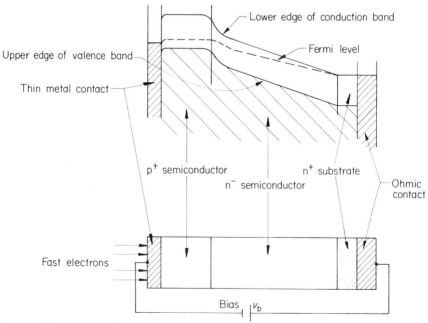

Figure 4.2. Energy band diagram for a depleted p–n junction under reverse bias. (Siekanowicz *et al.*, 1974)

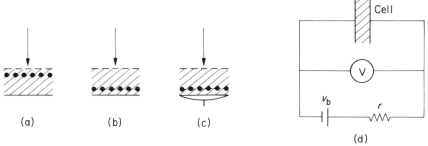

Figure 4.3. Photovoltaic cells (the semiconductor is shaded); the broken lines represent electrodes that are transparent to electrons, and the position of the junction is indicated by a row of dots. (a) selenium cell or silicon solar cell; (b) copper oxide cell; (c) Rappaport's silicon cell; (d) circuit containing a voltaic cell; bias v_b will be 0 if the cell is used as a source of emf.

A. Point contacts

Moore and Hermann (1951) measured point contacts on n-type germanium when bombarded by electrons having energies between 10 and 20 keV. The sensitive area surrounding the contact was found to have a diameter of 5×10^{-2} cm. The increase of current flowing through the specimen during bombardment was very much greater than the bombarding current both under forward and under reverse bias, although more so under reverse bias. The saturation value of g for forward and reverse bias respectively was 10^3 and 1.8×10^4 at $V_p = 10$ kV. The deduced values for production of an electron–hole pair in the two cases correspond to 5·2 eV and 0·55 eV. Since the latter is less than the band gap of germanium, and also much less than the value now accepted of $E^* = 2.9$ eV, this is strong evidence for secondary induced currents or avalanche multiplication in the germanium used for these measurements.

McKay observed pulses due to individual α-particles equivalent to about 10^6 electronic charges per particle with reverse bias across an n-type germanium point contact rectifier. Pulses were also observed under zero bias due to the influence of the barrier field alone. The sensitive region had a diameter between 10^{-3} and 10^{-2} cm surrounding the contact.

B. Schottky barriers

Ehrenberg *et al.* (1951) observed that copper oxide rectifiers showed a response to electron bombardment only after the thickness of the oxide layer was reduced by grinding and the energy of the cathode rays increased to 70 keV. Selenium cells however responded even at low electron energies, and Ehrenberg and Lang (1954) showed them to respond to a small reverse

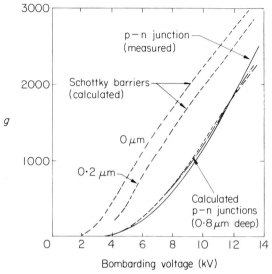

Figure 4.4. Current gain g as a function of electron bombarding voltage for silicon Schottky diodes and p– junctions. (Siekanowicz *et al.*, 1974).

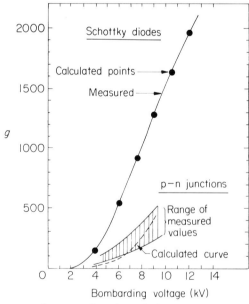

Figure 4.5. Current gain g as a function of electron bombarding voltage for two gallium arsenide cells. (Siekanowicz *et al.*, 1974)

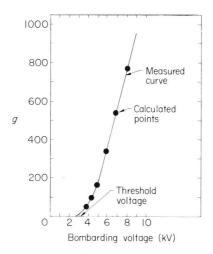

Figure 4.6. Current gain g as a function of electron bombarding voltage for $GaAs_{0.7}P_{0.3}$ Schottky barrier diode. (Siekanowicz *et al.*, 1974)

bias with greatly increased gain. On the one hand, the variation of gain with energy confirmed that the sensitive junction in the selenium cells is at the front, whereas that of the copper oxide cells is where the oxide has grown on the copper face. On the other hand, the variation suggested the use of cathode rays as probes for barriers or the use of semiconductor junction devices for cathode ray detection.

At one time selenium photoelements were thought to be Schottky barrier devices, but recent evidence suggests that the barrier acts as a p–n junction (which will be discussed in the following section). Silicon, GaAs, and $GaAs_{0.7}P_{0.3}$ barrier layer cells, made by using contacts of evaporated chromium or Nichrome, were measured by Siekanowicz *et al.* (1974) with a certain amount of reverse bias applied so that the short-circuit current was proportional to i_p at all values of bombarding voltage used. The current gain for such cells as a function of electron beam voltage is given in Figs 4.4–4.6 respectively. The way in which g depends on the back-bias for a GaAs Schottky barrier diode is shown in Fig. 4.7. The metallic contact constituting the barrier metal had a mass per unit area of about 25–50 $\mu g\,cm^{-2}$ and this yielded diodes with a critical voltage of $V_p \approx 3$ kV. Beyond this voltage (which corresponded to the mean loss in electron beam energy in penetrating the metal), the gain rose with increasing V_p linearly and with a slope that represented radiation ionization energies E^* of 3·6 eV for Si, 4·6 eV for GaAs, and 5·2 eV for $GaAs_{0.7}P_{0.3}$. It is seen that a biased Schottky barrier junction provides a convenient method for determining E^* since the mobil-

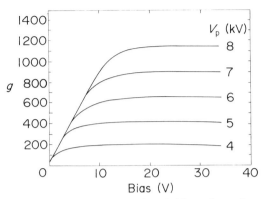

Figure 4.7. Current gain g as a function of back-bias voltage for a GaAs Schottky barrier diode employing a 400 Å chromium barrier, with bombarding voltage V_p as a parameter. (Siekanowicz et al., 1974)

ization energy then has the same value as E^* if the electron–hole pairs are generated in a high field region such as the space charge volume near a junction.

C. p–n junctions

The junctions of this type that have been studied extensively are those in grey Se, in Si, and in GaAs. Early published work on silicon p–n junction diodes under bombardment by β-rays from a Sr-90 source was done by Rappaport et al. (1956). Their main aim was to produce a source of emf, and for this reason no external bias was applied across the diode. A bias in the first instance does not affect the value of the induced conduction current because this is already saturated as a result of the junction field alone. However, second-order effects due to a bias may be important in certain circumstances. For example, a bias causes the width of the space charge region near the junction to grow, and thus the thickness of the sensitive region will increase. This may be important if the value of V_p is high. Similarly, if i_p is high the value of the induced current will be smaller at a given value of V_p if the applied bias is zero. The latter effect will be discussed in Section III.A.

The current gain of p–n junction cells is typically equal to V_p/E^* after an allowance has been made for an insensitive surface layer; the latter might be the highly doped diffused region. A typical g versus V_p curve for silicon p–n junctions is shown in Fig. 4.4 where a direct comparison can be made with a silicon Schottky barrier diode. It can be seen that the insensitive layer of 0·2 μm on the Schottky barrier diode absorbs less energy from the primary beam than the p⁺ diffused layer $0.8\,\mu m$ deep on the p–n junction diode. Typical EVE gains are respectively 2500 and 2000 at $V_p = 12\,kV$ (Siekanowicz *et al.*, 1974).

Analogous studies were carried out on selenium photovoltaic cells by Billington and Ehrenberg (1960, 1961).

In experimental studies, one difficulty with selenium elements is their tendency to fatigue under electron bombardment, and little progress was made before it was found that this phenomenon could be suppressed by exposing the cells to suitably timed pulses. When the cells were exposed to 6 ms square pulses interrupted by 54 ms off periods, the response clearly followed the primary current pulse and did not change with time. For all cells, families of curves were obtained such as are shown in Fig. 4.9. The linear rise, at the rate corresponding to $E^* = 16.5\,eV$, is due to electrons liberated in the p-material of the bulk; the junction is very close to the surface.

II. Practical junctions

A. Theory of the EVE

The theory can be simplified if we consider a p–n junction near a surface of a platelet of, say, n-type Ge or Si, and carrier pairs generated near the junction [Fig. 4.3(a)]. The minority carriers, e.g. the holes, will diffuse away from the place of their generation according to Fick's law. At the junction they will be drawn by the junction field into the p-region so that near the junction in the n-region the density of holes $p = 0$. But not all carriers generated will reach the junction since alternatively they may recombine or diffuse to the other electrode. Rappaport *et al.* therefore introduced the term "collection efficiency" (Q) for the ratio of carriers crossing the junction to those generated. In view of their technological problem—creating a source of emf—they did not consider the use of an external bias which could establish a field in the n-region and thus modify Q. We shall first consider the case without external bias.

Let I be the current in the circuit due to carriers generated near the junction and drawn into it, and let V be the voltage across the junction [Fig. 4.3 (d)]. I will be independent of V because the rate of generating carriers is determined by the radiation, and under all practical conditions (i.e. except at a forward bias which would cause a very large current to flow) the p-region remains at a lower potential than the n-region, so *all* holes in the n-region near the junction will move into the p-region. If $V = 0$, no other current passes the circuit, i.e. I is the short-circuit current.

In the absence of radiation-generated pairs, rectifying junctions satisfy the current–voltage relation

$$i_+ = i_s(e^{V/b} - 1) \qquad (4.1)$$

where in practice for silicon b is normally between $2k\theta/q$ and $3k\theta/q$, and i_s is perhaps 10^{-7} A cm^{-2}. i_+ is here taken as positive in the forward direction, i.e. in the direction of the majority carrier flow in the junction. Hence, if the generation current is I, the total current is

$$i = I - i_s(e^{V/b} - 1) \qquad (4.2)$$

On the other hand, in the absence of a bias (external emf, $v_b = 0$),

$$i = V/r \qquad (4.3)$$

where r is the resistance of the circuit.

For $r = 0$, $V = 0$, then I/i_p is the gain if i_p is the current carried by the impinging electrons. Given I, then currents, voltages, and power output can be calculated. Rappaport et al. found that the overall efficiency of their device, i.e. the ratio of the electric power available from the cells to the power supplied by the radioactive nuclei, was 0·1% for Ge and 0·5% for Si cells. The authors concluded that under optimum conditions the efficiency could be raised to 2·5%, but even if this low initial efficiency is accepted there remains the difficulty that the β-rays damage the crystal lattice by throwing silicon atoms into interstitial positions owing to their high energy, so the efficiency is halved in about 25 hours.

While obviously an external bias has no place in the circuit of a voltaic cell used as a source of power, its insertion can be advantageous if this cell is used as a particle detector or to monitor incident electrons. Equations (4.1) and (4.2) remain valid, but (4.3) must be replaced by

$$V = ir + v_b \qquad (4.4)$$

Rappaport et al. also excited their Ge cells with electrons from a Van de Graaff generator, and in a cathode ray tube with electrons of 50, 80, and 100 keV. In the latter case the energy dissipated in the platelets was accurately known, so the energy E^* used to produce a circulating electron could directly be determined, as 5·7 eV. This is rather more than the value of 2·94 eV obtained by McKay and McAfee (1953) with α-particles. The results with the Van de Graaff generator suggested a value of $E^* = 3$ eV. (It is not clear if these values are corrected for reflection losses.) The authors attribute the high value for E^* obtained with cathode rays to a low value of the collection efficiency, since with their arrangement [Fig. 4.3(c)] the cathode rays which do not penetrate deeply into the material produce holes only a long distance away from the junction.

In general, the calculation of the collection efficiency Q presents a formidable problem, even in the one-dimensional case. Let $g(x)$ be the number of minority carriers generated per cm inside the semiconductor. Then if D and τ are the diffusion constant and lifetime respectively, we have to solve the differential equation

$$-g(x) = p/\tau - D(\partial^2 p/\partial x^2) \qquad (4.5)$$

with the boundary conditions (if the junction is at $x = l$)

$$p = 0 \quad \text{at} \quad x = l$$

and

$$\partial p/\partial x = \text{const. at the near boundary } x = 0$$

There is no genuine analytic expression available for $g(x)$. Rappaport et al. have given a solution of equation (4.5) assuming that the β-rays from their source are absorbed exponentially, having entered the specimen from the face opposite the junction. With $(\partial p/\partial x)_{x=0} = 20 \text{ cm}^{-1}$ they find for Q values between 0·7 and 0·9.

The EVE in Si p–n junctions was also examined by Billington (1960), Billington and Ehrenberg (1961), Rosenzweig (1962), and Bowlt and Ehrenberg (1962, unpublished). Here the junction was close to the face irradiated. Figure 4.8 includes the short-circuit current for a typical cell (D_4) with a diffusion length greater than the penetration depth of the electrons. The current commences to rise only when the range of the bombarding electrons exceeds 2·6 μm, which was the depth of the junction below the surface. It may be concluded that the thin p-layer is damaged with the result that its minority carriers have a very short lifetime. The slope of the straight portions corresponds to $E^* = 3\cdot1$ eV, which is rather less than McKay's value of 3·6 eV. Figure 4.8 shows four curves measured on cells obtained from the Bell Telephone Co., of which D_1, D_2, and D_3 had been damaged by exposure to β-rays from a radioactive source; these cells had a thin n-layer on a p-matrix. The slope of the straight portion near the origin is $E^* = 4\cdot1$ eV. The reason for E^* being greater than that of the Ferranti cell as determined from the slopes of (a) and (b) is not clear; some of the difference is probably due to a grid which had been deposited on its surface in order to reduce the resistance of the conducting film. The damaged cell D_3 saturated at a gain of 6800 and an electron energy of 30 keV, corresponding to a range of electrons of 7 μm.

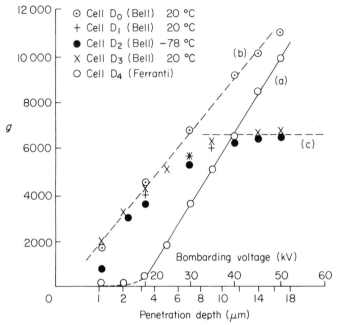

Figure 4.8. Gain as a function of electron bombarding voltage for five silicon p–n junction diodes. The penetration depth of the electrons was calculated from $d = 0.025 V_p^{1.65}$ where V_p is in kV and d is in μm. The open circles refer to measurements on a Ferranti cell; the remainder were on damaged and undamaged cells from Bell laboratories. (Billington, 1960)

B. Diffusion length of conduction electrons in silicon solar cells

A simple solution of the diffusion equation suggests that the density of carriers decreases by a factor 1/e for each distance $L = \sqrt{(D\tau)}$ travelled, where L is the diffusion length and D and τ are the diffusion constant and lifetime respectively. This feature has been approximated by interpreting L as a distance to which all carriers diffuse. Consequently, as the depth of penetration of the cathode rays increases with their energy, the number of carriers coming into circulation increases until the penetration equals L, and then stays constant. Accordingly, 7 μm should be the diffusion length of the minority carriers in the damaged cell [see Fig. 4.8(c)].

The same damaged cell was examined by Rosenzweig (1962) using a reasoning and technique which, slightly simplified, can be described as follows, adapting equation (4.5) for electrons, with boundary conditions

$$n = 0 \text{ at the junction}$$

and
$$\partial n/\partial x = 0 \quad \text{at} \quad x \to \infty$$
This has the solution
$$n = g_0\tau(1 - e^{-x/L}) \tag{4.6}$$
if g is taken as constant equal to g_0, over a range of the order of L. Then
$$I = -qD(\partial n/\partial x)_{x=0} = g_0 Lq \tag{4.7}$$

For high energy electrons, such as are supplied by a Van de Graaff accelerator, the excitation will vary along the path, but so slowly that it can be taken as constant over distances well exceeding L. Starting now with a beam of known energy qV_p and then placing, in small steps, increasing thicknesses of silicon as absorbers in front of the cell, Rosenzweig measured I over the whole range of the electrons. He then plotted I as a function of the thickness of the absorber, or, what is the same, of the location of the junction within a specimen. Then
$$\int I\,dx = Lq\int g_0\,dx = LqV_p/E^* \tag{4.8}$$
because $\int g_0\,dx$ is the total number of pairs that can be liberated by an electron from the accelerator. Now, we take g_o as *slowly* varying over the range of the electron. Hence, using the known value for E^*, a graphical integration over I yields the diffusion length L for the particular specimen. Therefore, with equation (4.7), g_0 for the material of the specimen and L as a function of the thickness of the absorber can be determined. Thus, in all subsequent tests on specimens of this material, and for the same voltage of the generator, one reading sufficed for the determination of L. In this way Rosenzweig obtained a diffusion range of 6·9 μm for the cell D_3 (private communication) in remarkably good agreement with the value given above.

III. Particular semiconductor junctions

A. Selenium

Some of the more general properties of the EVE in selenium p–n junction photovoltaic cells have been mentioned in Section I.C; the gain characteristics are summarized in Fig. 4.9.

The diffusion length of the generated carriers depends here on the current density. In order to extrapolate the diffusion length to zero primary current, Billington and Ehrenberg (1961) suggested a relation between the lifetime (τ) of the carriers and that under zero excitation (τ_0), of the form
$$1/\tau = 1/\tau_0 + 1/\tau_1$$

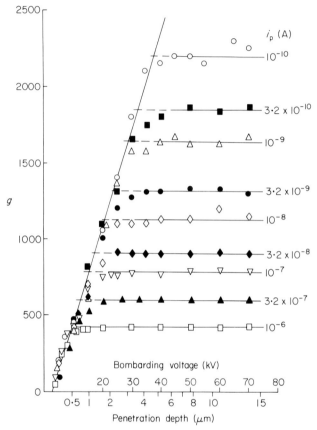

Figure 4.9. Gain g as a function of electron bombarding voltage for selenium photoelements, with beam current i_p as a parameter; g ceases to rise with increasing V_p when the penetrating power of the impinging electrons is within a carrier diffusion distance of the junction. The diffusion length depends on i_p. (Billington and Ehrenberg, 1961)

where τ_1 is the lifetime if irradiation were the only cause of decay. This is equivalent to

$$1/L^2 = 1/L_0^2 + f(i_p)$$

The measured values fitted a relation

$$f(i_p) = 2\cdot 8 \times 10^{11} \, (i_p)^{1/2} \text{ cm}^{-2}$$

and $L_0 = 8\ \mu$m. At a current density of 10^{-6} A cm^{-2}, L was less than 1 μm.

B. Gallium arsenide

For use as the source of an emf, in contact with a β-emitter, GaAs cells are preferable to Si cells since, owing to their heavier constituents, they are less sensitive to radiation damage (see also p. 292).

Gallium arsenide voltaic cells exposed to cathode rays were studied by Pfister (1957) who found that his cells, which had a 10 μm thick p-layer on n-type material, satisfied equation (4.2) and that the short-circuit current was proportional to the bombardment current. If the gain is plotted against the bombarding voltage it becomes evident (Fig. 4.10) that the gain

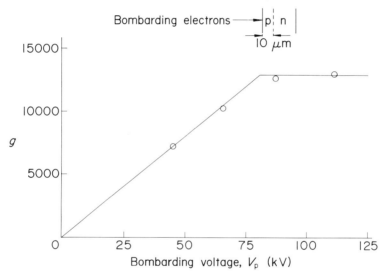

Figure 4.10. The electron voltaic effect in a GaAs p–n junction as a function of the bombarding voltage. (Pfister, 1957)

saturates at $V_p = 80$ kV, equivalent to 17 μm penetration. There is no insensitive layer, and the p- and the n-type material contribute carriers corresponding to $E^* = 6 \cdot 3$ eV. The diffusion length of the electrons in the p-layer cannot be less than its thickness; that of holes in the n-layer is 7 μm if the thickness of the junction is neglected.

C. Silicon

Advances in silicon planar technology in the period since about 1967 have made the EVE in biased diffused silicon p–n junction diodes very important. Only a brief account is given here because most of the practical device applications will be discussed in Chapter 7. In any event, the detailed

136   ELECTRON BOMBARDMENT INDUCED CONDUCTIVITY

behaviour of such diodes depends strongly on the precise diode design and manufacturing methods.

The primary beam of electrons may be incident either on the front (diffused) face of the diode or on the back face. In the latter case the diode is etched to a thickness over the bombarded area of about 10–15 μm so that the liberated holes (or electrons) do not have far to diffuse to reach the edge of the high-field region associated with the space charge near the junction. In some cases it is possible to make this high field extend right through to the back face. In those diodes where the incident beam penetrates the surface through which the dopant was diffused (the front face), a shallow junction is obviously desirable for high values of g at low V_p since the field-free highly doped side is invariably damaged by the diffusion process. Because of the low field and the lattice damage, this does not contribute many carriers to the sensitive region near the junction.

Curves showing g as a function of V_p for a typical back-face irradiated diode are shown in Fig. 4.11, and for a typical front-face irradiated diode in Fig. 4.4. Gain versus bias curves for silicon junctions when the direction of the impinging electrons is parallel to the junction are shown in Fig. 4.12; it is seen that the gain can rise with increasing bias—which is abnormal (Takeya and Nakamura, 1958). A discussion of "abnormal" behaviour was published by Guldberg and Schroder (1971).

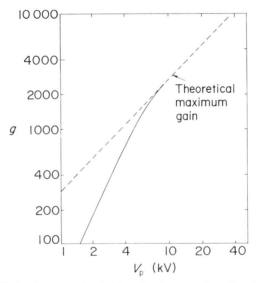

Figure 4.11. EVE gain g on a back-biased silicon p–n junction diode when bombarded by fast electrons on the back face, i.e. the face opposite to that through which the p-type impurity was diffused. (Engstrom and Rodgers, 1971)

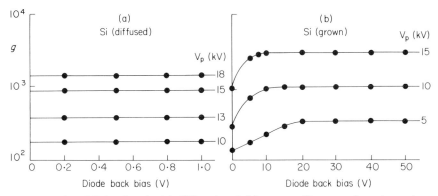

Figure 4.12. EBC gain g in (a) diffused and (b) grown silicon p–n junctions when bombarded by electrons travelling substantially parallel to the plane of the junction. (Takeya and Nakamura, 1958)

D. EBC avalanche effect

1. Silicon

When a bias voltage is applied across a semiconductor junction, two additional effects are possible: (a) secondary currents may flow owing to a modification of the space charge associated with it, or (b) avalanche multiplication of generated carriers crossing it may occur in the vicinity of the junction. The first of these effects in single crystals of CdS with indium contacts was described in Chapter 3. Experimental results obtained for the second were described by Gibbons *et al.* (1975). A silicon avalanche photodiode was mounted on the stage of a scanning electron microscope (SEM), and the normal microscope image was used to line up the irradiated patch accurately just inside its guard ring. The primary electrons were thus scanned over a patch about 350 μm square. The beam current was measured by directing it into a small Faraday cup provided on the stage beside the diode; this was again accurately positioned with the aid of the normal SEM image. The beam current and the diode current were measured on a battery-driven electrometer, and the bombarding voltage was measured directly using the meter supplied with the microscope. Results for a front-face bombarded silicon avalanche photodiode having a junction at 2·5 μm below the irradiated face are given in Table 4.1.

It is constructive to calculate the response time expected. Measurements of electron velocity in the drift region of an avalanche diode by Duh and Moll (1967) indicate complete velocity saturation. The saturation drift velocity of hot holes in silicon is 10^7 cm s^{-1}, so the response time of

Table 4.1. Experimental results for the EBC of a silicon avalanche photodiode. (Gibbons et al., 1975)

Beam current, i_p (A)	Diode bias, v (V)	Increase of diode current, i_s (A)	EBC gain, g	Avalanche gain, g'	Overall gain, G	Normalized G for $g' = 100$
\multicolumn{7}{c}{Bombarding voltage, $V_p = 10$ kV; penetration, $r = 1 \cdot 1$ μm}
0	50	7×10^{-8} (leakage)	—	—	—	—
790×10^{-12}	50	$1 \cdot 63 \times 10^{-7}$	206	1	206	
790×10^{-12}	193·6	$5 \cdot 36 \times 10^{-6}$	206	32·9	$6 \cdot 7 \times 10^3$	$2 \cdot 06 \times 10^4$
\multicolumn{7}{c}{Bombarding voltage, $V_p = 15$ kV; penetration, $r = 2 \cdot 2$ μm}
0	50	7×10^{-8} (leakage)	—	—	—	—
250×10^{-12}	50	$5 \cdot 3 \times 10^{-7}$	2120	1	2120	
250×10^{-12}	194·7	$4 \cdot 3 \times 10^{-5}$	2120	81	$1 \cdot 7 \times 10^5$	$2 \cdot 12 \times 10^5$

the EBC avalanche effect, which is equal to the depletion width divided by the saturation velocity, is $(1 \cdot 1 \times 10^{-3})/10^7 = 1 \cdot 1 \times 10^{-10}$ s for this diode. It is thus seen that the EBC avalanche effect is a high-gain phenomenon and also very fast. However, measurements on this diode by Varol (1978) show that the response time is strongly dependent on minority carrier storage in the heavily doped n-type region near the bombarded surface. Varol showed that this could account for a slowing of the response to about 10^{-9} s with this particular construction. The effect can thus not be divorced from the details of junction manufacture; the order-of-magnitude calculation just given can thus be used mainly as a guide showing the potentialities of attainable speed.

Detailed studies of the avalanche multiplication in an electron bombarded silicon p–n junction were reported by Donolato (1977). The diodes were of similar construction to those just described but the onset of avalanche breakdown occurred at 30 V, mainly owing to the choice of starting material which was 0·5 ohm-cm p-type silicon; again a scanning electron microscope was used to make the measurements. Figure 4.13 shows how the avalanche multiplication factor (M) depends on the diode bias (v) for different values of electron bombarding voltage. It is plotted as log (v/v_B) versus log $(1 - 1/M)$, since this provides a family of straight lines through the origin with slope n by virtue of Miller's law,

$$1 - 1/M = (v/v_B)^n$$

The bias (v) has been normalized here to the junction breakdown voltage (v_B) which marks the onset of avalanche multiplication.

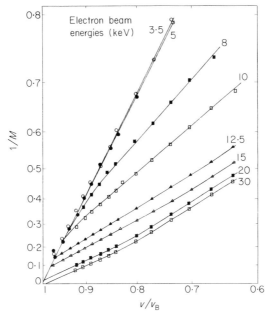

Figure 4.13. EBC avalanche effect in a silicon diode. Experimental multiplication factor (M) as a function of applied reverse bias (v_B), for different values of bombarding voltage. (Donolato, 1977)

By analysing a step-junction approximation to the actual diode used, and by assuming that both the generated holes within the strongly n-type (surface) region and the generated electrons in the p-type bulk contributed to the observed total conduction current, Donolato was able to account for the main features of the experimental curves, and especially for the dramatic change in slope of the curves between $V_p = 5$ and $12 \cdot 5$ kV. It is clear that there is a possibility of measuring useful multiplication data by use of electron excitation in a similar way to that which has been undertaken using light (e.g. Urgell and Legucrrc, 1974), and since fast electrons have a fairly well defined range this method of excitation has an advantage over absorption of electromagnetic radiation which leads to an exponential rate of carrier generation with depth.

2. Gallium phosphide

A number of GaP avalanche diodes were measured by Oelgart (1979) using excitation by a scanned electron beam. The starting material was p-type single-crystal GaP grown by liquid-phase epitaxy or vapour-phase epitaxy, and an n^+–p junction was prepared by diffusing a narrow stripe of zinc to a depth of about 10 μm across one surface; this formed a diode structure which was scanned by an electron beam in a direction parallel to the crystal surface and perpendicular to the junction. The behaviour of this diode was dominated by surface effects, especially at low values of V_p. For example, in diodes that had been stored in air for an extended period, the influence of surface recombination could be detected for bombarding voltages below 20 kV, whereas for carefully prepared structures these effects were not observable.

As expected, the spatial extent of the contribution to the diode current by carriers generated on either side of the junction could be measured by monitoring the amplitude of the current flowing when the electron beam was scanned across it. The contribution extended about 2 μm on either side and, from the change in shape of the signal as V_p was increased, it could be shown that the rate of ionization by energetic holes in bulk GaP is greater than that of electrons. It was also shown that, in the neighbourhood of breakdown, an avalanche multiplication gain of about 100 could be detected.

A difficulty with this type of diode, compared with those measured by Gibbons and by Donolato, was that, apart from powerful surface effects, the magnitude of the useful number of generated electron–hole pairs depended on the reverse bias applied across the diode. It was found empirically that

$$i_0 \propto 1 + \exp(-a/V_p^b)$$

where i_0 is the diode current in the absence of multiplication, and the constants $a = 2 \cdot 55$ and $b = 0 \cdot 73$.

5
Transient EBC time-of-flight measurements

I. Time-of-flight technique

The specimen, in the form of a parallel sided lamina and with thin conducting electrodes on either face as indicated in Fig. 5.1, has a bias field provided across its thickness by a voltage between the electrodes F and B. A short pulse of ionizing radiation (such as fast electrons, α-particles, or a light flash) penetrates the front electrode (F) and generates a cloud of electron–hole pairs near the adjacent surface. These are drawn out of the generation

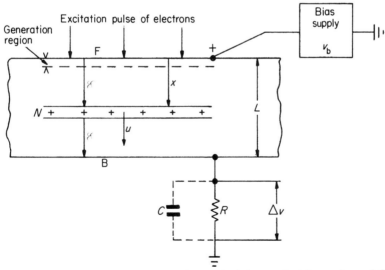

Figure 5.1. Diagram illustrating the principle underlying time-of-flight measurements on a resistive specimen. The excitation may be a short pulse of high velocity electrons which penetrate the front conducting electrode F. A narrow cloud of N carriers then drifts with velocity u towards the back electrode B under the influence of the electric field produced when the bias voltage v_b is applied. (Spear, 1969)

region by the field and, depending on its polarity, either electrons or holes are drawn across the specimen towards the back electrode (B). The passage of the charge carriers from one face to the other can be detected in the circuit containing the specimen. Either the induced charge or the current can be displayed on an oscilloscope, and if a break in the pulse shape can be determined marking the arrival of carriers at the far electrode, a transit time T can be measured. This chapter is devoted to the determination of materials properties from observations of T and the shape of the transient EBC pulse, mainly in the case where the ionizing radiation is a short pulse of fast electrons.

A. Background of similar work prior to 1957

Before the development of the cathode ray oscilloscope as a measuring instrument and advances in electronic circuitry, little could be done to resolve details of a carrier transit across an insulator when excited by a short pulse of ionizing radiation. The early results of Gudden and Pohl (1923) and of Hecht (1932) made use of relatively slowly responding galvanometers, and these workers thus concentrated on determining the total charge transported. The first indications that progress towards faster measurements were being made came with the work of Kosmata and Huber (described by Stetter, 1941) who were close to resolving details in the millisecond time

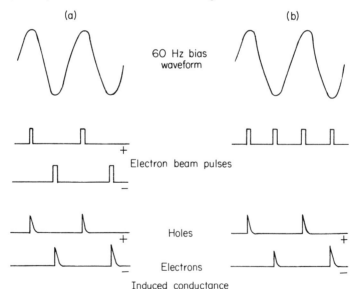

Figure 5.2. The relative position of the electron pulse and the bias waveform used by Moore (1949): (a) 60 Hz system; (b) 120 Hz system.

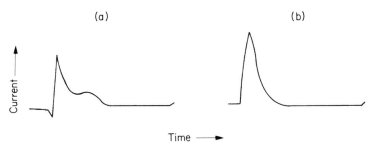

Figure 5.3. Historic oscillograms showing (a) the transit of electrons and (b) the movement of holes generated in a single crystal of AgCl by short pulse electron excitation. The slight rise in current towards the end of pulse (a) can be attributed to a static space charge due to trapped electrons, whereas the decay to zero in (b) is due to a short hole free lifetime. (Moore, 1949)

range by using electronic pulse amplifiers and a fast mechanical oscilloscope with a bifilar suspension. Further progress was made in the laboratories of Cornell University under Smith and Sproull around 1947 (private communication), and a thesis by Moore (1949) showed photographs of an oscilloscope trace of the transit of electrons and the movement of holes in a single crystal of AgCl excited by a short electron pulse. At about the same time, Haynes (1948, see Shockley, 1950, page 210) compared the Hall and drift mobilities in AgCl using photoexcitation. In Moore's experiments a space charge was prevented from developing in the AgCl by his using an alternating field and pulsing the electron beam at the peak of both the positive and negative half-cycles (see Figs 5.2 and 5.3). Warfield (1950) dealt with a similar aspect of the transient EBC time-of-flight technique, but little more of significance followed until independent developments led to the first full paper (Spear, 1957). Since then, this technique has been extensively developed and used by experimental materials scientists to provide information about the fundamental physical properties of many insulators, semiconductors, amorphous materials, and even some non-ionic liquids.

B. Measurements on a crystal in its virgin state

Different specimens of the same type of crystal have identical molecules arranged in the same order, so they might be expected to be indistinguishable from each other in all respects. However, crystals of the same kind can vary greatly in some properties which have been described as structure sensitive. The differences are due to small amounts of impurities, to Frenkel and Schottky defects (atoms in interstitial positions or missing respectively), to a mosaic structure of real crystals, or to a variety of dislocations. Some disorder is built into the crystal when it is growing; others vary with time and disclose some of the history of the crystal. This distinction is not definite but it has an important bearing on EBC effects.

Any irregularity of the crystal lattice provides a location where the potential energy of a carrier is different from the energy that it has in undisturbed regions; if it is lower, the location will be a trap. Clearly, the more perfect a crystal, the fewer the number of traps. Largely under the influence of transistor technology, the art of crystal growing has advanced tremendously during the past few decades, and a large number of crystals are now available having many orders of magnitude fewer irregularities than occur in natural crystals, but none is perfect. Regardless of its degree of perfection, a crystal can be in a state of thermal equilibrium in which a certain fraction of the traps are filled. Such a crystal is, with respect to EBC, in a virgin state. After exposure to cathode rays, the state is altered; the ratio of filled to empty traps in some parts of the crystal will be changed, with the result that its response to a second dose of cathode rays will differ from that to the first.

Left to itself, it may take milliseconds, or hours, or a century before the virgin state is restored; often the return to the virgin state can be artificially accelerated by annealing or by infrared light. Obviously, the more perfect a crystal, the more slowly will significant deviations from the virgin state arise. Anyway, with the advance of electronic switching, it has become possible to measure crystals using transient EBC techniques with less intense electron beams of shorter duration on sampling or storage cathode ray oscilloscopes. As a result, the study of virgin crystals has become possible; the more perfect the material, the longer it remains virgin, and improved techniques permit more observations to be made in shorter times. For example, it is now possible by the transient EBC technique to take measurements of the transit of as few as 10^6 electronic charges across an insulating or some semiconducting specimens typically 200 μm thick and of cross-sectional area 0·1 cm^2.

C. Types of transient EBC measurements

1. Space-charge-free transits (SCF)

An SCF transit would be appropriate to the condition $\sigma \ll Cv$, where σ is the space charge due to the packet of generated carriers and any static charges that might exist between the electrodes, and C is the capacitance between them. This corresponds to the field within the specimen being uniform and equal to v/L between the electrodes, where L and v are the interelectrode separation and the applied bias potential respectively.

The charge $Q(t)$ induced in the electrodes by the generated carriers may or may not be a linear function of time after a short excitation pulse, but it ceases to rise when the cloud of all the drifting carriers arrives at the remote electrode. Alternatively, the induced current $i(t)$ can be observed, and this falls towards zero at the same time. In either event, if a discontinuity in the $Q(t)$ or $i(t)$ curve can be seen, a transit time T can be defined. This is clearly a function of the drift velocity u of the moving carrier and thus of the applied field $\mathscr{E} = v/L$. Thus

$$u = L/t \qquad (5.1)$$

and if a drift mobility μ can be assigned,

$$\mu = L^2/Tv \qquad (5.2)$$

One limitation arises when this basic method is used. If the mean free carrier lifetime is too short, it is no longer possible to measure a discontinuity in the oscilloscope trace at $t = T$. Under these conditions, the transient EBC has fallen to zero before this can be seen, as in Fig. 5.3(b). However, special techniques can be used in some circumstances; these will be described in Sections I.E, II.E, and II.F.

The shapes of transient EBC pulses after short-pulse electron excitation are shown in Fig. 5.4. In practice, because the equivalent circuit element of the specimen is a constant current source shunted by its own capacitance (i.e. the capacitance between electrodes F and B in Fig. 5.1), the charge display (a) is

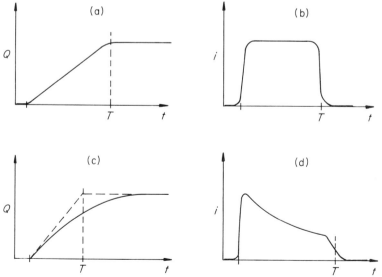

Figure 5.4. (a) Transient EBC pulse shape expected when there is little or no trapping and the charge is displayed on the oscilloscope; (b) the appearance when the current is displayed; (c) and (d) are the equivalent charge and current pulses expected if the mean free carrier lifetime is short.

obtained by employing a large value of input resistance (R) of the preamplifier or oscilloscope, and (b) with a low input resistance. When the mean free carrier lifetime is short, one of the most commonly observed pulse shapes is that shown in (c), and the transit time is obtained by measuring the intersection of tangents to the curve drawn at $t = 0$ and $t = \infty$, as shown in the figure. Exactly the same information is thus obtained as would otherwise have been seen by observing the $i(t)$ curve as shown in (d). In general, a value of transit time can be assigned so long as the duration of the transient pulse remains field-dependent.

The original analytical model of Hecht (1932) yields expressions [(5·4) and (5.5) below] for the externally measured total charge. This takes into account the proportion of carriers in the initial cloud that become trapped in the bulk before arriving at the far electrode, and the proportion that traverse it completely and escape. Without sacrificing generality, we assume in the following analysis that the drifting carriers are electrons.

Figure 5.5 shows a rectangular crystal on the two opposite faces of which some means of electrical contact is made so as to apply an electric field across it. When the front face of the crystal is bombarded by high-energy primary electrons, a number of electron–hole pairs are released in a region of the crystal determined by the penetrating depth of the primary beam.

Suppose the excitation pulse is short, and N electrons are excited into the conduction band at time $t = 0$. If a number n_1 of these terminate their paths within the crystal, and if w is the mean free path of a carrier in an

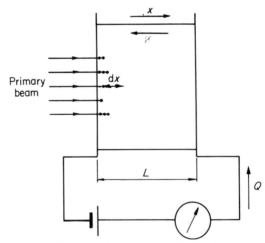

Figure 5.5. Diagram to illustrate the basic experimental arrangement for a crystal bombarded by a short pulse of electrons, where the total charge transported from face to face is measured.

infinite crystal (the schubweg), $n_1 = N(1 - e^{-L/w})$. Similarly, the number that leave via the far electrode is $n_2 = Ne^{-L/w}$. The average distance travelled by the n_1 electrons trapped within the crystal is given by

$$w' = (N/n_1)\int_0^L (x/w)e^{-x/w}\,dx = (N/n_1)[w(1 - e^{-L/w}) - e^{-L/w}]$$

Thus, the average distance travelled by the initial pulse of N electrons is

$$\bar{x} = (1/N)(n_1 w' - n_2 L) = w(1 - e^{-L/w}) \tag{5.3}$$

By Ramo's theorem (Appendix, Section IV), the total charge induced in the external circuit is therefore

$$Q = (1/L)Nq\bar{x} = (1/L)(Nqw)(1 - e^{-L/w}) \tag{5.4}$$

The opposite limit corresponds to uniform excitation throughout the thickness. In this case,

$$Q = (1/L)(Nqw)[1 - (w/L)(1 - e^{-L/w})] \tag{5.5}$$

Equations (5.4) and (5.5) are commonly known as Hecht laws for surface and bulk excitation respectively.

The expressions given by equations (5.4) and (5.5) are plotted in Fig. 5.6 (a) and (b) respectively. The amplitude of both has been normalized to equal measured total charges for $t = \infty$. This implies that a total of twice as many carriers must be generated for the same externally measured charge if these are uniformly distributed compared with the situation when they are all generated near the front face.

McKay (1948b) modified the equation (5.4) Hecht law for the special case of electron pulse excitation as follows. His experimental arrangement made

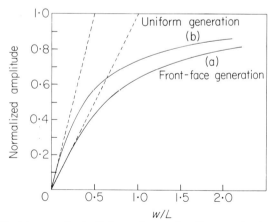

Figure 5.6. Hecht curves for both front-face and uniform excitation; L = specimen thickness and w = mean free path of mobile carrier.

use of an electron beam having a penetrating power that was only a small fraction of the insulator thickness. If each primary electron excites N electrons into the conduction band, the rate at which electron–hole pairs are generated is Ni_p where i_p is the primary current. If a fraction b of the internal secondaries are capable of contributing to the conduction current (the remainder being trapped or recombining at their place of origin), the internal yield may be defined as the ratio of the charge traversing the crystal, as measured externally, to the primary charge. Thus

$$g = Nbi_p\bar{x}/i_pL = (bNw/L)(1 - e^{-L/w}) \tag{5.6}$$

and this is normalized to

$$G = W(1 - e^{-1/W})$$

where $G = g/bN = g/g_\infty$, and $W = w/L$. The yield at infinite field strength is denoted by g_∞ and this corresponds to the case where all electrons capable of contributing to the conduction process arrive at the back (i.e. non-irradiated) face of the crystal.

If the primary (non-penetrating) beam is pulsed for an interval of time τ, and N electrons are excited into the conduction band, the number n remaining in this band after a time t measured from initiation of the primary pulse is given by the charge balance equation

$$dn/dt = N/\tau - n/t' \tag{5.7}$$

where N/τ is the rate at which electrons are excited into the conduction band, and n/t' is the rate at which electrons are lost to traps. For the present, we assume that once an electron is trapped it plays no further part in the conduction process for the duration of observation. The effect of carrier release will be considered later.

Using the initial conditions that $n = 0$ when $t = 0$ and $i_s = \Delta Q/\Delta t$, we obtain the solution in the case of no trap-release:

$$i_s = \begin{cases} (Nt'qu/L\tau)(1 - e^{t/t'}) & \text{for } 0 < t < \tau \\ (Nt'qu/L\tau)(e^{\tau/t'} - 1)e^{-t/t'} & \text{for } T > t > \tau \\ 0 & \text{for } t > T \end{cases} \tag{5.8}$$

These equations can often be simplified if the excitation pulse is much shorter than the transit time. In this case,

$$i_s = \begin{cases} (Nqu/L)e^{-t/t'} & \text{for } t < T \\ 0 & \text{for } t > T \end{cases} \tag{5.9}$$

Similar measurements of the motion of generated holes through the body of the crystal may be made by reversing the polarity of the applied bias voltage.

If traps release their carriers with a mean "emptying time" of t'' which is short compared with T, the charge balance equation is

$$dn/dt = N/\tau - n/t' + (Nt/\tau - n)/t'' \quad \text{for} \quad 0 < t < \tau \quad (5.10)$$

Thus the externally measured conduction current is given by

$$i_s = \begin{cases} (Nqut_0^2/L\tau t')(1-e^{-t/t_0}) + Nqut_0/L\tau t'' & \text{for } 0 < t < \tau \\ (Nqut_0^2/L\tau t')(e^{\tau/t_0} - 1)e^{-t/t_0} + Nqut_0/Lt'' & \text{for } t > \tau \end{cases} \quad (5.11)$$

where $1/t_0 = 1/t' + 1/t''$.

If the excitation pulse is short compared with t_0, equation (5.11) becomes

$$i_s = (Nqut_0/L)[(1/t')e^{-t/t_0} + 1/t''] \text{ for } t > \tau \quad (5.12)$$

When the first carriers arrive at the back electrode the current falls sharply by an amount given by

$$i_s = (Nqut_0/L)e^{-T/t'} \quad (5.13)$$

where the transit time $T = L/u$.

The shape of the conduction pulse for times greater than the transit time, but in which trapping and detrapping occur within a time comparable to it, is complicated; the results of computer calculations by Baron have been quoted by Mayer (1968). The shapes of the current transients for representative values of the parameters are shown in Fig. 5.7. On the diagram the charge induced or integrated current pulses are also shown for the same experimental conditions.

It should be emphasized that these pulse shapes are those to be found in an insulating specimen, i.e. one in which the fundamental relaxation time $\varepsilon\varepsilon_0/\sigma$ of the material is greater than the carrier transit time. It does not apply to typical low resistivity semiconductors where strict charge neutrality must always be maintained and only the drift of minority carriers can be studied (Haynes and Shockley, 1951). Also, the case of trapping and detrapping shown in Fig. 5.7 refers to fairly deep traps, i.e. those where the times spent in traps and the mean free time between trapping events are comparable to T. However, if many trapping and detrapping events take place during the carrier transit, the effective mobility is now trap-controlled. A detailed discussion of trap-controlled mobility will be given in Section III.B.

Transient EBC time-of-flight measurements are used extensively for measuring the drift velocity of generated electrons or holes in insulating solids. If the thickness L is known, the drift mobility can be derived from the slope of a simple plot of $1/T$ against bias v. The reason for doing this will be explained later but, apart from trapped space charges, a short carrier free

150   ELECTRON BOMBARDMENT INDUCED CONDUCTIVITY

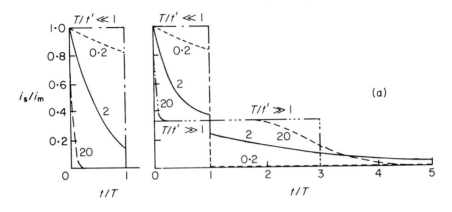

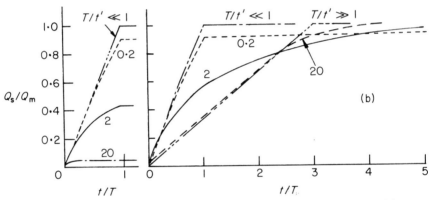

Figure 5.7. Transient EBC pulse shapes in the case of carrier trapping and detrapping: (a) current pulses; (b) charge pulses. The nomenclature is as follows: T = transit time in the absence of trapping; t' = mean free carrier lifetime; t'' = mean release time of carriers from traps. The left-hand curves in each row correspond to the case where there is trapping but no detrapping, and the right-hand curves to the case where $t''/t' = 2$. (Baron, quoted by Mayer, 1968)

lifetime can distort a plot in such a way that it may not pass through the origin. An SCF transit can thus be used for measuring the mobility (μ) or the trap-controlled drift mobility (μ_t), and since the latter is temperature dependent the depth of the trap which dominates carrier transport can also be derived.

The shape of the SCF pulse also allows the mean free lifetime t' and the mean carrier detrapping time t'' to be determined in some cases. Also, the trap density N_t can be derived if specimens of the material of differing purity are available, so the true microscopic lattice mobility μ_0 and its temperature dependence can be deduced.

In molecular crystals or amorphous semiconductors, SCF transits can be used to measure many of the materials parameters characterizing hopping transport or other mechanisms applicable to low-mobility solids.

2. Space-charge-perturbed transits (SCP)

Usually the SCF condition can be fulfilled by using an excitation pulse having a sufficiently small amplitude. The build-up of a static space charge in an insulator can also be prevented by employing one or a number of additional excitation pulses under zero applied bias. This section is devoted to a discussion of the transient EBC pulse shape when the space charge field due to the drifting charge cloud is not small compared with the applied field.

If the size of the cloud of drifting charge is large enough, the self-field of the drifting carriers is sufficient to perturb the applied field. This problem was dealt with by Keating and Papadakis (1964). Further theoretical work was published by Papadakis (1967), and this model was fitted to experimental measurements of electron transport in orthorhombic sulphur by Gibbons and Papadakis (1968). In qualitative terms, if the charge in the cloud of carriers is of magnitude Q per unit area, the electric field at the leading edge exceeds that at the trailing edge by an amount given by Poisson's law, $\Delta \mathscr{E} = Q/\varepsilon\varepsilon_0$. This results in the carriers near the front of the cloud travelling faster than those at the back, and the centre of gravity of the pulse moves increasingly rapidly as the carriers approach the far electrode. The externally measured current pulse is modified by this effect, and the current now rises to a peak at a time T_1, which is less than T, before falling to zero at times equal to about twice this. The shape of the space-charge-perturbed current transit (SCP) is shown in Fig. 5.8 by the full lines.

The SCP transit time, which may be defined as the time for the measured current to reach its peak, is computed as a function of the parameter β which is the ratio of the applied field to the self-field of the drifting carriers. The value of T_1 as a fraction of T is given in Fig. 5.9. This graph also shows how the peak value of the current at $t = T_1$ depends on β. It is seen that, in the asymptotic case of Q approaching zero ($\beta = \infty$), $i(T) = i(0)$ which is the case described earlier for small induced currents.

3. Space-charge-limited transits (SCL)

This model was used by Many and Rakavy (1962) in drift experiments designed to provide a pseudo-infinite reservoir of carriers near the excited electrode. Carriers excited by a very intense flash gas-discharge tube produced a situation in which carriers were injected into the crystal bulk under

152 ELECTRON BOMBARDMENT INDUCED CONDUCTIVITY

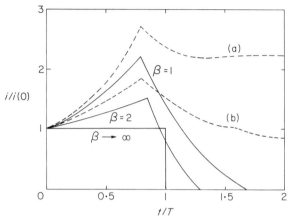

Figure 5.8. Shape of the transient SCP current pulse after surface generation, for different values of the parameter β which is the ratio of the applied field to the field perturbation due to the travelling charge packet. (Gibbons and Papadakis, 1968) The dashed curves are the shapes predicted (a) by Many for SCL transits, and (b) by Many and Rakavy for SCL currents with trapping.

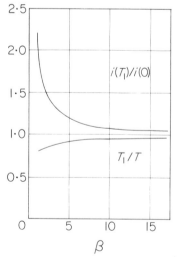

Figure 5.9. Theoretical variation of the normalized time to the cusp, T_1/T, and of the current, $i(T_1)/i(0)$, with the parameter β for SCP transients. (Gibbons and Papadakis, 1968)

Table 5.1. Comparison of space-charge-limited (SCL) and space-charge-perturbed (SCP) current in the absence of deep trapping.

	SCL current	SCP current
Initial current	$i(0) = 0·5 i_0$	$0·5 \leq i(0)/i_0 \leq 1$
Ratio of maximum current to initial current	$i(T_1)/i(0) = 2·718$	$2·208 \leq i(T_1)/i(0) \leq 1$
Transit time of first carriers to arrive at opposite electrode, i.e. time to reach maximum value of current	$T_1/T = 0·787$	$0·792 \leq T_1/T \leq 1$
Current during period $0 \leq t \leq T_1$	Current rises as $i/i(0) = (1 - t/2T)^{-2}$	Current rises as $i/i(0) = \exp(t/\beta T)$
Current for $t > T_1$	Current decays to its steady-state SCL value at $t \approx 1·5T$	Current decays to zero at $t = T_2$ where $T \leq T_2 < 1·688 T$
Total charge collected	Very large	Equal to Q_0, the total generated charge drawn into the bulk

T = carrier transit time in the absence of space charge perturbations
T_1 = time taken for the current pulse to reach its peak value
Q_0 = total generated charge drawn into the bulk
$i(0)$ = initial value of the current
$i(T_1)$ = maximum value of the current
i_0 = Q_0/T
β = $\varepsilon \mathscr{E}_0 / Q_0$ where $\mathscr{E}_0$ is the applied field and ε is the permittivity of the insulator.

space-charge-limited conditions. In this case, the space charge conditions are defined precisely as soon as the induced current is established. Table 5.1 and Fig. 5.8 show the differences between the expected transient induced current pulses in the cases of SCP and SCL. The figure also shows the expected pulse shape in the case of SCL current transients with trapping. It can be seen that the SCL transient current pulse shape in the presence of trapping is very similar to that of the SCP pulse shape. To distinguish between the two cases it might not be sufficient merely to examine the shape of the pulse; it is often necessary to examine critically the experimental conditions used. An SCL current transient demands a fairly long intense excitation pulse; an SCP current transient needs a relatively short pulse of intense excitation. Figure 5.10 shows how the induced transient current

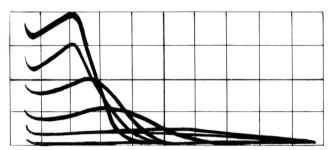

Figure 5.10. Oscillograms of transient SCP currents in a single crystal of orthorhombic sulphur excited by high intensity flash illumination, for different applied fields; the applied bias increased from the lowest to the uppermost trace. Vertical scale, 5×10^{-8} A/div.; horizontal scale, 2 ms/div. (Gibbons and Papadakis, 1968)

pulse for electrons in a single crystal of orthorhombic sulphur changes shape when the bias voltage is altered.

The physical situation analysed by Many for transient SCL currents in insulators has not been applied in the case of electron beam excitation, but three cases of space charge perturbation using this mode have been investigated. Experimentally this has advantages over the space-charge-free case ($\beta = \infty$), especially when the usual fall in current at $t = T$ is difficult to determine either because of a poor signal/noise ratio or because of severe carrier trapping. The cost paid for this advantage in detection ($t = T_1$) is that, unless β can be determined, the relationship between T_1 and T is not exact. To within about 5%, a value $T_1 = 0.85T$ can often be taken so long as the observed induced current pulse shape provides qualitative evidence for space charge perturbation.

When the carrier reservoir generated by an excitation pulse near the electrode is intense but short compared with T_1, the trailing edge of the transient current pulse can be used to determine the lifetime of generated carriers at the surface. According to an analysis by Simhony and Gorelik (1965), the trailing edge of the current pulse falls to zero linearly after $t = T_1$, so defining a reservoir emptying time (t_e). Thus, the current falls to zero at a time $t = T_1 + t_e$. The carrier lifetime at the surface (t'_s) can then be derived from

$$t'_s \exp(t_e/t'_s) = N_0/j \qquad (5.14)$$

where $t_e = t_f - \frac{1}{2}T_1$, and $t_f + T_1$ is the value of t for which the current falls to zero.

A similar analysis for transient electron-bombardment induced conduction currents in a depleted semiconductor p–n junction diode was made by Taroni and Zanarini (1969). Experimental confirmation of their analysis

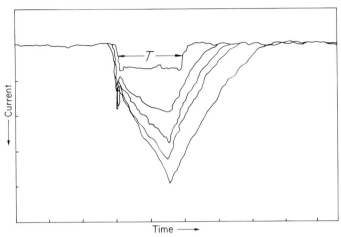

Figure 5.11. Transient EBC current waveform in a totally depleted silicon diode. The shape is explained in terms of the non-uniform field within the diode and the self-field of the cloud of drifting electrons. The lowest amplitude pulse corresponds to a low generated current density, and the transit time is very nearly equal to T. With increasing initial density of generated electrons, the current rises to a cusp at about $0 \cdot 8T$. (Canali *et al.*, 1971c)

was obtained by Canali *et al.* (1971c) for electron motion under transient space charge conditions in a depleted silicon diode; their results are shown in Fig. 5.11. The current is given by the expression

$$J(t) = (q/L)\int_0^L (\mu_e n + \mu_p p)\, \mathscr{E}\, dx \tag{5.15}$$

where p = density of holes generated by excitation pulse, n = density of electrons generated by excitation pulse, q = electronic charge, $\mathscr{E}$ = internal electric field within specimen, w = width of region excited by pulse as measured from the surface. As with all the previous analyses, the transit time as measured to the cusp is at about $0 \cdot 8T$. It is pointed out that the striking analogies between the behaviour of an insulator and that of a depleted semiconductor must not obscure the fact that the former theory is not applicable to semiconductors since the unperturbed electric field is not constant inside the depletion range. Figure 5.12 shows how T_1, the computed time for the pulse to reach the peak of the transient induced current in a reverse biased semiconductor diode, depends on the normalized range R_1 of the exciting electrons, for different values of the parameter v_3 which is the ratio of the external bias voltage to the depletion voltage.

The computation of the transient space-charge-limited induced current pulse shape for a partly depleted and a totally depleted semiconductor junction diode is more difficult than in the case of a homogeneous insulator,

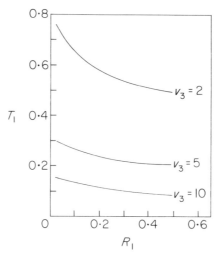

Figure 5.12. Dependence of the position of the peak of the transient EBC pulse in a partly depleted semiconductor junction diode structure. The curves are generalized by measuring the time T_1 in units of the dielectric relaxation time ε/σ, and the bias voltage v_3 in units of the diode depletion voltage v_d; the range R_1 of the penetrating electrons is in units of the depletion width. The incident electrons penetrate the thin heavily doped side of the junction or a Schottky barrier. (Taroni and Zanarini, 1969)

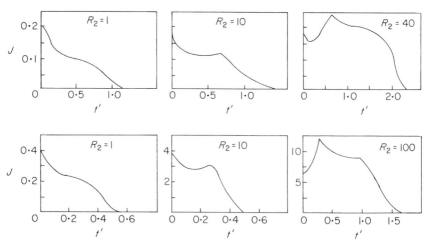

Figure 5.13. Normalized current density as a function of time in an electron-excited p–n junction for two different applied voltages and various initial current densities $J = I/I_0$ ($v/v_d = 2$ in upper row and 5 in lower row); R_1, the range of the exciting electrons in units of the depletion depth, is 0·1. (Taroni and Zanarini, 1969)

and the variety of possible shapes is much greater. The computed shapes are shown in Fig. 5.13 for representative values of the parameters R_2 (the ratio between average density of carriers generated by the ionizing pulse of electrons and the ionized donor density inside the depletion region) and R_1 (the ratio between the range of the ionizing electrons and the depletion width).

4. *Limited-space-charge transits (LSC)*

A slightly different situation was analysed by Schwartz and Hornig (1965) who noted that, since the carrier generation efficiency due to surface photo-excitation in anthracene falls to zero (or very nearly zero) when the electric field at the surface is nil, the situation of an almost infinite reservoir of carriers in the generation region under the SCL conditions suggested by Many is probably not representative of practical situations where the quantum efficiency is field-dependent. This formulation of the SCP problem thus corresponds to the situation where the generation efficiency is so low that a sufficiently large reservoir is not built up for the field in the generation region to fall quite to zero, or to the situation just mentioned where the excitation pulse is short compared with T.

The analysis by Many and Rakavy for completely SCL currents predicts a rise in the transient induced current pulse of the form

$$j(t) = (\mu v^2/2L^3)(1 - t/2T)^{-2} \quad (5.16)$$

and the limited-space-charge analysis by Schwartz and Hornig gives

$$j(t) = (\mu v^2/2L^3)e^{t/T} \quad (5.17)$$

These two analyses lead to qualitatively similar transient current pulse shapes, and either pulse reaches a cusp maximum at $t = 0.79T$.

5. *Microwave time-of-flight technique*

A very fast experimental approach was described by Evans and Robson (1974). The technique, which was particularly aimed at measuring drift velocities in very thin epitaxially grown layers of a semiconductor with short dielectric relaxation times, relies on microwave technology. An electron beam is modulated at microwave frequencies by passing it between parallel transmission lines carrying a UHF current. The sinusoidally deflected beam then passes across an aperture in front of the specimen. The transit time of generated carriers through the specimen causes a phase delay between the externally measured ac conduction current and the sinusoidal excitation. It can be shown that this phase delay is simply $\phi = fL/u$, where f is the microwave modulation frequency of the electron beam, L is the sample thickness, and u is the carrier velocity. Naturally, as the measurements are now in the realm of microwaves, the specimen mounting is a typical microwave varactor header, the bias is applied via a stripline bias tee, and ϕ is measured with a waveguide bridge and spectrum analyser.

D. Comparison with light-flash or α-particle excitation

The benefits of the transient EBC time-of-flight technique over similar methods using α-particle or light-flash excitation were not clear when such methods were introduced. Very often experimentalists were discouraged by the apparent complexity of the equipment. However, it has now become clear that there are many advantages in using electron pulses despite the extra effort that might be involved in producing them. This technique is now the most widely used method for measurement of drift mobilities in a variety of materials as well as other parameters which can be derived from measurements of the transient EBC pulse shape.

When measurements are made on semiconductors, the dielectric relaxation time $\tau_0 = \varepsilon\varepsilon_0\rho$ is often short, the mobility of the generated carriers is high, and the problem of sample heating due to v is considerable. It is thus important to apply the field in the form of a fairly short voltage pulse accurately synchronized with the excitation pulse. The field pulse is applied just long enough before the excitation for any switching transients to have died away; it is then switched off immediately after measurements on the signal. Also, in those circumstances where the transit time is short, sampling techniques must be used. It is important that an accurate trigger be provided, and there should be reproducible excitation pulse amplitudes, positions, and delays. Clearly, some of the time-of-flight measurements described here might be performed using a light flash from a spark or by α-particles, and in several measurements on solids use has been made of such methods to complement those using electron excitation. The main difficulties confronting the user of such methods are that there is no accurate pre-trigger available when α-particle excitation is used, and there is considerable jitter in pulse height and in position when a triggered nanosecond spark or light flash is used. Spear (1969) made the important point that, when measurements are being made on wide gap materials such as noble gas solids (band gap about 10 eV), it is difficult to produce sufficient ultraviolet intensity to generate an adequate number of carriers. Also, light is absorbed according to an exponential law, and the greater density of generated carriers is thus in a region near the surface where surface states may give rise to a high rate of geminate recombination.

The main advantages of electron beam excitation over optical generation are: (i) there is ample intensity, even with nanosecond pulses, for the generation of a sufficient number of carriers; the design and construction of a suitable high-voltage electron gun with magnetic focusing is relatively easy; (ii) the depth of the generation region below the top surface can be varied within wide limits by means of the acceleration potential V_p applied to the gun. A disadvantage of electron beam excitation for materials characterization is that it lacks the selectivity in energy that can be achieved optically by a given photon wavelength. The latter can lead to important additional information, particularly in organic materials, on the transport associated with particular excited states.

It may be remarked here that, although α-particle excitation has been used successfully in similar circumstances for time-of-flight measurements, and the actual determined drift mobilities are often in good agreement with other measurements using optical or electron excitation, the efficiency of carrier generation is usually lower owing to columnar recombination. The physical basis of this is as follows. Along the track of the α-particle there is a very high density of electron–hole pairs. Until these are separated under the influence of the applied electric field, recombination reduces the number drastically. Therefore there is a region corresponding to a high recombination rate, and the average energy absorbed from the α-particle to generate a free electron–hole pair is high. However, the most important result of this effect in high speed pulse measurements in solids is an increase of the transient EBC charge pulse rise time. This is demonstrated markedly in experiments by Canali *et al.* (1971b) on single crystals of CdTe, where a comparison was made between the rise time when α-particle excitation was used and when electron excitation was used (see Fig. 5.14). At an applied

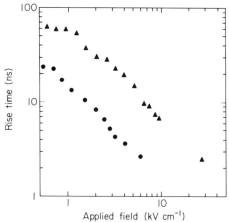

Figure 5.14. Plasma–time effect in CdTe. Transient EBC charge pulse rise time for excitation by α-particles (▲) and by a short electron pulse (●); the larger rise time with α-particle excitation is evident. (Canali *et al.*, 1971b)

field of 500 V cm^{-1} the rise time of the induced charge pulse is 65 ns with α-particles compared with 25 ns for electrons. The main cause is the longer carrier extraction time from the interior of the volume containing generated free pairs, due to a sheath of oppositely charged particles that surrounds them and effectively shields them from the applied field. Thus α-particles are generally not satisfactory for use in evaluation of drift velocity and of carrier trapping times when they are less than the plasma time. In circumstances where none of these effects is likely to influence the results, such as measurements of carrier detrapping times in semiconductors, α-particle excitation may be used because the equipment needed to make these measurements might be somewhat simpler.

E. Insulators having a short carrier free lifetime

If the free lifetime of the carriers in the drifting cloud is much less than T, transient EBC pulse shapes are rather like the one shown in Fig. 5.3(b). In this case the mobile holes have fallen into traps and not been thermally released for the duration of the measurement. This shape of pulse is typical for insulators having a high trap density, and the normal time-of-flight technique cannot be used for measuring mobilities.

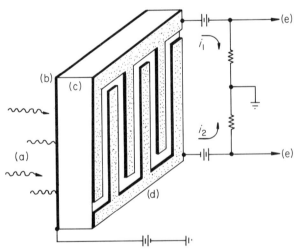

Figure 5.15. Schematic diagram of the experimental arrangement using an interdigital back electrode: (a) incident photons; (b) semitransparent SnO_2 conducting electrode; (c) specimen; (d) interdigital electrodes; (e) connectors to amplifiers and oscilloscope capable of displaying sum and difference of comb currents i_1 and i_2. (Young et al., 1977)

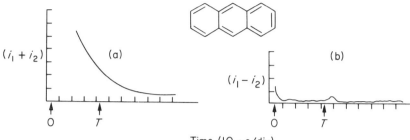

Figure 5.16. Transient electron current in a single crystal of anthracene containing deep traps: (a) total current equivalent to conventional experiment; (b) current difference. The transit time for the generated electrons is T. Crystal 1·04 mm thick. Vertical scale, 500 nA/div. (Young *et al.*, 1977)

A variant of the time-of-flight technique to cope with this situation was described by Young *et al.* (1977, 1979). Rather than observing the current pulse shape, they used a number of interdigital electrodes on the face of the specimen remote from the excited face to detect the arrival of generated carriers. The electrodes were in the form of two interdigitated square combs, the fingers of one alternating with those of the other (Fig. 5.15). They consisted typically of 600 fingers 40 μm wide with 30 μm spaces between them, produced by photolithographic techniques for depositing 700 Å thick chromium on glass. Excess charge carriers arriving at the interdigital electrodes were measured by applying a small voltage (e.g. 48 V) between them and joining the leads to an oscilloscope capable of displaying the sum and difference of the comb currents, i_1 and i_2. The bias field was provided by applying a 700 V potential between (b) and (d).

In the sample bulk the field is uniform provided that its thickness is large compared with the repeat distance of the fingers, and the sum $(i_1 + i_2)$ is the same as the current in the conventional experiment. However, when the charge cloud reaches about $(2\pi)^{-1}$ times the interdigital repeat distance from the electrodes, $(i_1 - i_2)$ is then proportional to the rate of arrival of the charges. The traces $(i_1 + i_2)$ and $(i_1 - i_2)$ for a single crystal of anthracene containing deep traps are shown in Fig. 5.16. The value of T could be clearly measured from the second trace, but a log-log plot of $(i_1 + i_2)$ did not resolve the transit (see below in Section II.E).

F. Semiconductors having a short dielectric relaxation time

An interesting variation to the techniques just described applies when the fundamental dielectric relaxation time, $\varepsilon\varepsilon_0\rho$, is short compared with the carrier transit time. In practice, the relaxation time for many semiconductors is 10^{-12}–10^{-10} s and, if an electron pulse generates a drifting cloud of charged carriers near one of the contacts, charge neutrality within the specimen requires injection of an equal number of carriers of the opposite sign from the opposite electrode.

Such a process occurs within a time equal to about the relaxation time. Su et al. (1970) used this principle for determining the electron drift velocity in p-type silicon, and Neukermans and Kino (1970) used it for electrons in InSb at 77 K. It can be shown, as for example by van Roosbroeck (1953), that the observed drift mobility is the ambipolar mobility of the carriers, which is given by

$$\mu_a = (n - p)/(n/\mu_p + p/\mu_n) \tag{5.18}$$

where n and p are the thermal equilibrium densities of electrons and holes respectively within the bulk of the unexcited semiconductor. If μ_p is much smaller than μ_n, and $\partial u_e/\partial \mathscr{E} \approx 0$, then $u_a = u_n$. Neukermans and Kino give the ambipolar velocity as

$$u_a = -[u_n p(\partial u_p/\partial \mathscr{E}) - u_p n(\partial u_n/\partial \mathscr{E})]/[-p(\partial u_p/\partial \mathscr{E}) + n(\partial u_n/\partial \mathscr{E})]$$

Under the conditions outlined for charge neutrality, although the specimen may be a p-type semiconductor, the measured transit time is that for electrons, and it is seen from this equation that, in a region of velocity saturation, $\partial u/\partial \mathscr{E}$ is small and the ambipolar velocity is essentially equal to the drift velocity of the other carrier. This was used by Neukermans and Kino in a study of velocity saturation effects which are described with the experimental results for InSb.

These measurements are essentially the same as the drift velocity measurements of minority carriers in semiconductors by Haynes and Shockley (1949). However, when it is possible to obtain samples of semiconductor (such as silicon) that are pure enough for p–n junctions to be made with depletion regions sufficiently wide for the time-of-flight to be less than the dielectric relaxation time of the specimen, charge neutrality will not be restored for the duration of the measurements. As an example, if the resistivity of ultra-pure silicon is 10^5 ohm-cm, the dielectric relaxation time is 100 ns, and the transit time of electrons is usually less than this.

Most frequently, measurements by the transient EBC time-of-flight technique on semiconductors are performed on reverse-biased p–n or p–i–n junctions. When such structures are biased, the potential applied is balanced by an internal space charge field due to ionized impurities. With increasing reverse applied field, more and more thermally generated carriers are drawn away from the junction and the space charge due to the residual impurities widens.

The depletion width may be calculated from Poisson's equation. Thus, if we assume that all the donors and acceptors are ionized (which is usually the case), the thickness w_d of a one-sided junction is given by

$$w_d^2 = 2\varepsilon\varepsilon_0 v/q|N_A - N_D| \qquad (5.19)$$

where ε is the dielectric constant of the semiconductor, ε_0 is the permittivity of free space, and N_A and N_D are the densities of acceptors and donors respectively. For a uniformly doped semiconductor, by evaluating ε_0/q, this equation becomes

$$w_d = 1\cdot 05(v\varepsilon/N)^{1/2} \times 10^7 \mu\text{m} \qquad (5.20)$$

where v is measured in volts and the excess impurity concentration $N = |N_A - N_D|$ in atoms cm^{-3}.

When the gap between diffused junction contacts to a semiconductor (or barrier layer contacts) is filled by the space charge due to ionized impurities, the structure is said to be totally depleted; the reverse bias potential at which this occurs is known as the depletion voltage. If the bias is increased beyond the depletion voltage, the applied field will superimpose linearly on the space charge field up to the limit of dielectric breakdown. Since the space charge field gradient is constant, this means that the internal field within the semiconductor between the contacts will not be constant but will assume the shape shown in Fig. 5.17 where the internal field is depicted for a reverse

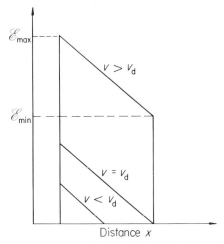

Figure 5.17. Electric field distribution in a back-biased p–n junction or surface barrier on high resistivity material. For applied voltages v, less than the voltages v_d required to deplete the sample completely, the field strength decreases linearly with distance. For applied voltages greater than v_d, the field distribution consists of the superposition of a constant field and the linearly decreasing component. (Martini *et al.*, 1972)

biased semiconductor diode structure as the applied bias is increased beyond the depletion voltage v_d. In the situation where the carrier velocity is proportional to the field, it is easy to show that the carrier transit is still $T = L^2/\mu v$ despite the field non-uniformity. In the more general case, however, when the concept of mobility cannot be applied, this does not hold, and the drift velocity is given by

$$T^*/T = (\mathscr{E}_0/\mathscr{E}_1)\ln[(1 + \tfrac{1}{2}\mathscr{E}_1/\mathscr{E}_0)/(1 - \tfrac{1}{2}\mathscr{E}_1/\mathscr{E}_0)]$$

where T^* is the non-uniform field transit time. Here $\mathscr{E}_1$ is the inherent field variation due to the ionized donors or acceptors within the depletion region, and $\mathscr{E}_0$ is the maximum electric field $\mathscr{E}_{max}$ in Fig. 5.17. In practice it is usually possible to render $T^* = T$ approximately by increasing the applied bias sufficiently.

G. Signal averaging

If the bandwidth of the output amplifier is high and the generated transient pulse is small, the limit of detection is soon reached because the signal/noise ratio of the sampled gain is too small. Details of an experimental arrangement to overcome this problem were first described by Zulliger et al. (1969). Signal averaging is extensively used when measurements are made with very fast transient EBC time-of-flight equipment. This was achieved by a technique similar to the one described for nuclear particle detectors by Aitken et al. (1965). The output from the sampling oscilloscope corresponding to the transient EBC within the specimen was fed into an integrator and time base unit which was connected to the memory of a 1024-channel analyser (Nuclear Data, ND-ITB 180 and ND-FM 181). The memory time base was used to provide the external horizontal input to the sampling oscilloscope. The averaging system worked by comparing the input during each sampling interval with the corresponding signal in each memory channel. If the difference was positive, a count was added to that channel; if it was negative, a count was subtracted. The time interval between samples was 16·7 ms and, with a scale on the averager of $100\,k$, the time-averaged pulse shape could be obtained in 5–10 min. Non-random interference signals were averaged and stored separately, so enabling a virtually interference-free output signal to be obtained by a simple subtraction. The output was displayed on an X-Y pen recorder, an oscilloscope, or computer punched-tape. This technique was extensively used by the group at Modena in Italy throughout the 1970s for materials characterization (Reggiani, 1979).

II. Transient EBC methods for measurements other than drift velocity

A. Transient EBC pulse shape near $t = 0$

When details of the transient pulse shape can be resolved, several materials parameters characterizing the properties can be measured. In many cases, the fundamental principle first demonstrated by Moore (which is suitable for determining drift velocities) is modified to enable other properties to be evaluated. With the added capability of simple averaging computers, it is now possible even to eliminate spurious signal pick-up which often occurs at the onset of switching the electron pulse or the specimen field. Limitations of signal-to-noise ratio can also be overcome as just described. A relatively "clean" transient EBC pulse is obtained by subtracting the coherent interference signal from the observed transient. We are then left with only the fundamental limitation arising from the variance of both. More often than not this complexity is not needed, and many valuable studies of materials have been made on relatively simple equipment. This section is devoted to the kind of measurements that can be made without using a signal averaging computer. Of course, if one is available it could still be used to advantage, but the determinations about to be described do not depend on it.

If the excitation pulse length τ is short compared with the mean free carrier lifetime t', from equation (5.12) the induced charge equation becomes

$$Q = (Nqut_0/L)[(t_0/t')(1 - e^{-t/t_0}) + t/t''] \qquad (5.21)$$

The slope at the origin is given by i_s at $t = 0$, and thus the equation of the tangent is

$$Q_{t0} = Nqut/L \qquad (5.22)$$

Thus the initial rise in charge is the same as in the no-trapping no-release case. The equation of the tangent for t large (but less than T) is

$$Q_{t1} = (Nqut_0/L)(t_0/t' + t/t'') \qquad (5.23)$$

The intersection of the tangents Q_{t0} and Q_{t1} occurs at the point $t = t_0$, where $t_0 = t't''/(t' + t'')$ has the same meaning as given before. However, if the excitation pulse is not short compared with the times t', t'', or t_0, we find the value of Q at $t = \tau$ is $Q = Nqu\tau/2L$. Thus the tangent to the Q_{t1} curve meets the $Q = 0$ axis at the point $t = \tau/2$. Similarly at the trailing edge of the pulse, the tangent for $t = \infty$ meets Q_{t1} at a point $\tau/2$ before its asymptotic value. This means that the time interval between the two tangent intersections is equal to T even though τ may be large.

B. Electric field profiles

Since the relationship between the electric field in a solid and the carrier drift velocity can be determined by the transient EBC drift velocity technique, a unique relationship of the form $u = f(\mathscr{E})$ can be obtained. If a drift velocity proportional to field is found, the relation is simply $u = \mu \mathscr{E}$ where μ is the drift mobility, but this simple law of proportionality, and a constant value of μ, cannot always be assumed. In the general case it is necessary either to know the form of the function or to measure it by independent drift measurements.

If a narrow sheet of mobile carriers is generated by a short pulse of high velocity electrons near one face of a depleted or intrinsic semiconductor (such as a silicon p–i–n diode), the shape of the resulting induced current pulse will yield information about the field distribution inside the specimen (Zulliger et al., 1969).

If the diode under study has a fairly wide depletion region consisting of a distribution of only partly compensated impurities, $i(t)$ will not follow a simple law as would have been the case for a uniformly doped depleted diode. The externally measured current pulse will thus be of the form

$$i(t) = Qf(\mathscr{E})/L \tag{5.24}$$

where L is the width of the depletion region and Q is the charge of mobile carriers taking part in the conduction process.

The variation of the field in the presence of the fixed space charge of the ionized donors or acceptors is given by Poisson's equation

$$\Delta \mathscr{E} = LqN_{A,D}/\varepsilon\varepsilon_0 \tag{5.25}$$

where $\Delta \mathscr{E}$ is the difference between the electric fields on either side of a region of width L containing a total of N_D ionized donors or N_A ionized acceptors, q is the charge of an electron, ε is the dielectric constant of the material, and ε_0 is the permittivity of free space.

The method of determining $\mu = f(\mathscr{E})$ has already been outlined in Section I.F. Once the function f is known, it is a straightforward matter to reconstruct the field $\mathscr{E}(x)$ as a function of distance from the generation region, and by the method of finite differences to determine the distribution of N_A or N_D from equation (5.25).

C. Carrier recombination

1. Bulk recombination

In an insulator, electron–hole recombination can clearly only take place in the bulk if charge carriers of both signs are generated or injected. An elegant experiment was undertaken by Dolezalek and Spear (1975) by which the recombination kinetics could be determined by causing a cloud of holes to drift through generated electrons in a single crystal of sulphur under the influence of an applied field. Sulphur is a highly resistive insulator and, without carriers generated by some form of ionizing radiation, carrier recombination is virtually absent. If carriers of opposite sign are made to drift through each other in their passage from opposite sides of a specimen of an insulator, this results in a discontinuity in the measured current flowing in the external circuit when they meet. This ingenious method has so far only been used with ionizing pulses consisting of short triggerable light flashes from xenon discharge tubes, but it is included here because, in principle, it could be adapted to electron pulse excitation for wider band-gap insulators, or materials where accurate jitter-free pulse positions and amplitudes are of paramount importance.

Orthorhombic sulphur is characterized by a lattice hole mobility of 10 cm^2 V^{-1} s^{-1} and an electron mobility of 6×10^{-4} cm^2 V^{-1} s^{-1} at room temperature; it thus lent itself well to first trials of the method. The experimental arrangement used for investigating volume recombination in this material is shown in Fig. 5.18. The sulphur specimen in the form of a thin parallel-sided platelet was provided with thin conducting semitransparent electrodes on either side. To reduce the magnitude of unwanted current transients, each electrode was divided into two equal areas. The electrode (E) joined to the pulsed field unit supply was electrically continuous but only half of it was

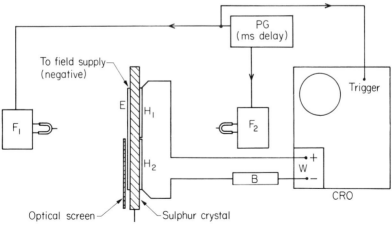

Figure 5.18. Experimental arrangement used for investigating volume recombination of generated electrons and holes in orthorhombic sulphur. F_1 and F_2, xenon flash tubes triggered from the pulse generator PG; E, H_1, and H_2, thin gold electrodes; B, balance network; W, differential oscilloscope amplifier. (Dolezalek and Spear, 1975)

transparent to the exciting radiation. Each half of the opposite electrode (H_1 and H_2) was connected to the input of a differential oscilloscope via a balancing network (B). The two sides were excited by light flashes from two similar xenon discharge tubes (F_1 and F_2). The excitation pulses were longer than the specimen transit time for holes but shorter than that for electrons. The experiment was therefore conducted as follows. The entire volume between the faces of the crystal was filled with photogenerated drifting holes when a photogenerated sheet of electrons had travelled about half way across the specimen. The externally measured electron current pulse then showed a fall in magnitude at the instant the number of drifting electrons was reduced by volume recombination with the holes. The observed transient electron current pulse in the presence and in the absence of photogenerated holes will be described in Section VI.B of Chapter 6.

2. Surface recombination

Clearly, even in a highly resistive insulator, surface recombination is possible where the ionizing radiation generates electron–hole pairs, i.e. geminate surface recombination. The method is based on applying the drift field at a known time interval after the excitation pulse (Dolezalek and Spear, 1975).

In the time interval between the excitation pulse and the field pulse, the rate of loss of carriers by recombination is given by

$$-dn/dt = -dp/dt = \alpha np \qquad (5.26)$$

where n and p are the densities of electrons and holes respectively, and α is the bimolecular recombination coefficient. When $n = p$, this equation becomes

$$1/n = 1/n_0 + \alpha t \qquad (5.27)$$

where t is the time delay between the excitation pulse and the field pulse. An approximate value for the density of electrons (or holes) in the generation region can be obtained from the charge liberated and an effective depth x of recombination beneath the surface. A reasonable estimate for x is about half the diffusion length $(Dt')^{1/2}$ for the most mobile carrier. Thus we can write

$$n_0 = Q_e/Aqx \qquad (5.28)$$

where Q_e is the charge of generated electrons, q is the electronic charge, and A is the surface area under the illuminated electrode.

D. Mean free carrier lifetime

If an excess carrier is generated in an insulator or a semiconductor by ionizing radiation (such as fast electrons) or by injection from an electrode, its free life will be terminated by recombination with a carrier or an ion of the opposite sign, or by escape via an electrode, or by falling into a deep trap. We can define the mean free carrier lifetime (t') as the mean time spent by an excess carrier in an infinite crystal. The value of t' at various distances from the surface of a solid specimen can be determined by a very direct method, provided that carrier release from trapping centres already just filled elsewhere within the crystal does not take place at the same time. This method using "interrupted transits" is explained as follows.

If a sheet of charge carriers is generated near one face of an insulator by a short pulse of electrons, and this is caused to drift across the specimen under the influence of an applied field, it is possible to reduce the applied field to zero when the carriers have moved into the specimen to a predetermined position. Upon resumption of the field, any carriers from the sheet that have diffused into traps will no longer contribute to the observed current, and the decrease in current so observed after this time interval will yield information about the carrier lifetime. Conversely, if detrapping is the dominant mechanism, the observed current will be larger on resumption of the field, and the method of interrupted transits can be used to provide information about detrapping times (Lemke and Müller, 1970).

Interrupted transits have been employed to measure carrier lifetimes with respect to deep traps as well as release times. Since the trapping process in the absence of an applied electric field is basically one of diffusion, the method of interrupted transits provides a very direct means whereby the lifetime of carriers with respect to deep traps can be measured from a semi-logarithmic plot of the current ratio as a function of interruption time. It can yield information about the temperature dependence of the carrier diffusion coefficient if the specimen temperature is varied. It can also provide information about the carrier free lifetime as a function of depth beneath the generation surface if the carrier cloud is interrupted after travelling various distances.

If i_0 is the initial value of conduction current, the current will decay to i_t, where

$$i_t = i_0 \exp(-t_i/t') \tag{5.29}$$

after interrupting the field for a time t_i. Here t' is the lifetime of the drifting carriers with respect to deep traps.

A tracing from an oscillogram of an interrupted electron transit in a single crystal of orthorhombic sulphur is shown in Fig. 5.19. The main features of this pulse are explained as follows. Initially the trace is horizontal and no current flows despite the applied bias field because there are very few free carriers inside the specimen. At the instant when the short electron beam pulse strikes the face of the crystal, the current rises sharply and, apart from

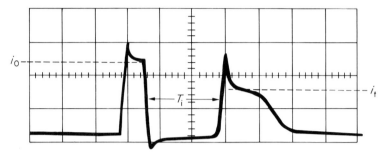

Figure 5.19. Oscilloscope trace of a typical interrupted transit of electrons in a single crystal of orthorhombic sulphur; note the switching transients which must be taken into account.

a short transient "spike", the induced current is substantially constant. The current falls abruptly to zero again when the bias field is interrupted after the drifting cloud of charge carriers has only partly crossed the space between the electrodes. On restoring the field to its original value after a known delay, it can be seen that the amplitude of the current pulse is now smaller. The derived free lifetime results will be described in Section VI.B of Chapter 6.

E. Amorphous semiconductors and the continuous time random walk

The transient conduction pulse shape following short pulsed surface excitation of a lamina of a crystalline semiconductor has already been discussed in a number of situations involving space charges, trapping, detrapping, and other phenomena. These shapes can be interpreted in terms of the excitation conditions and carrier drift mobility and trapping. From analyses of these shapes it is often possible to derive parameters characteristic of the bulk material, such as drift mobility and its field and temperature dependence. However, in a wide variety of amorphous solids, $i(t)$ at the end of a transit falls slowly to zero, rather than abruptly as predicted by the simple theory and equation (5.9). In some cases the tail in the pulse can be explained in terms of carrier diffusion during transit, detrapping, or space charge effects. In the analysis about to be discussed, we will consider only the small signal case, so the influence of space charge can be neglected, but we will discuss long tails.

If carrier diffusion is responsible for the tail, a Gaussian-shaped decay would be expected, i.e. the width of the current decay would exhibit a $T^{1/2}$ dependence. If it can be attributed to delayed generation (excitons) or detrapping, it would be exponential in time, with a width independent of T. But, in the case of phonon assisted hopping, it was shown (Scher and Montroll, 1975; Pfister and Scher, 1978) that, if the transport is considered a stochastic hopping process where the carriers undergo a biased random walk between localized states, the width of the tail is proportional to the transit time of the fastest carriers. Thus, for a given specimen, the width of the tail scales linearly with T if the applied bias is changed. In Fig. 5.20, (a) shows a typical transient current pulse in amorphous As_2Se_3 displaying a long tail due to the wide dispersion of transit times, (b) is from a similar specimen in which the transit time cannot be measured because the pulse shape is completely featureless, and (c) shows (b) plotted on a scale of log i as a function of log t (Scher and Pfister, 1975). The pulse shape can now be resolved into two straight lines tangential to the log i curve for short and long times respectively. Their intersection defines a transit time T which is the time taken for the fastest carriers in the travelling bunch to arrive at the far electrode. This transit time yields information about the distribution of hopping times relative to the actual time for the carriers to move across the thickness L. This highly dispersed motion is known as the continuous time random walk.

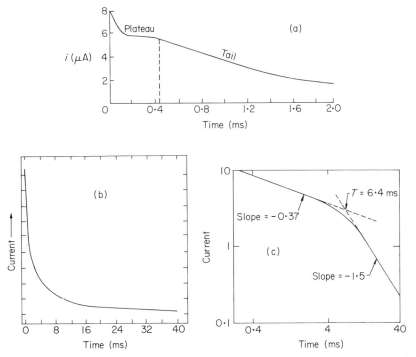

Figure 5.20. (a) Transient current pulse shape in amorphous As_2Se_3, showing the extended tail due to dispersion of transit times. (Scharfe, 1973) (b) Amorphous As_2Se_3 showing an apparently featureless transient current pulse. (Scher and Pfister, 1975) (c) The same data as in (b) but plotted on logarithmic scales.

For strong localization, the hopping time is approximated by

$$\tau_{hop} \approx K\exp(2\rho/\rho_0)\exp(\Delta E/k\theta) \tag{5.30}$$

where ρ is the hopping distance, ρ_0 is the localization radius, and ΔE is the thermal activation energy; K is almost constant with respect to θ and ρ over the range of interest. In variable range hopping, the dominant variable for the different site jumps is ρ rather than ΔE. On the physical model, the maximum of the generated carrier distribution remains close to the point of generation whereas, for sufficiently strong fluctuations in hopping distances ρ, the leading edge of the carrier sheet might have penetrated far into the specimen bulk for the same time. This applies in particular if the distribution of hopping times τ_{hop} extends substantially to times of the order of T or, what amounts to the same thing, times of the same order as the times of observation. The dispersion of the generated charge carrier cloud and the mean displacement of it from its origin increase with time in direct proportion. This property is termed universality, and it implies that, for all amorphous solids obeying this kind of stochastic hopping, graphs of $\log i$ as a function of $\log t$ can be made to coincide, both if $t < T$ and if $t > T$, for a given specimen with different values of applied bias by merely sliding the graphs along the horizontal and vertical scales. This would not be possible if the tails were Gaussian.

Scher and Montroll suggested site jump probabilities of the form $p = kt^{-(1+a)}$, where a $(0 < a < 1)$ depends on the macroscopic parameters ρ, ρ_0, and ΔE. By introducing a boundary at a distance L from the origin, the externally measured current then obeys two simple algebraic laws depending on whether or not carriers are lost by escaping from the specimen entirely via the electrode:

$$i(t) = \begin{cases} i_0 t^{-(1-a)} & \text{for } t < T \\ i_0 t^{-(1+a)} & \text{for } t > T \end{cases} \tag{5.31}$$

It is shown that on a log-log plot this leads to two straight lines which intersect at $t = T$ where $T \propto (L/\mathscr{E})^a$. The theory predicts therefore that the transit time is sublinearly dependent on the specimen thickness. The important implication from this is a "mobility" that is specimen thickness dependent. Further, the sum of the slopes of the tangents of a log-log plot of $i(t)$ as a function of t for short $(t < T)$ and long $(t > T)$ times is equal to -2. The field-dependence of μ is not an unusual feature of many conventional solids, but a particularly important aspect predicted is that μ will also be specimen thickness dependent if it is derived from $u = L/T$ (Pfister, 1974). Thus, if we adopt this definition of μ we shall find that if $\mu \propto \mathscr{E}^{-(1-a)}$ then $\mu \propto L^{-a}$.

Experimentally this has been found to hold for a number of amorphous solids.

If we refer back to equation (5.30) for phonon-assisted hopping, it is seen that τ_{hop} depends exponentially on ρ/ρ_0. Thus, if the concentration of hopping centres can be altered, as might be possible in the case of solid solutions, the mean distance ρ of a hop will change. Hence, a plot of $\ln(L/T\rho^2)$ against ρ should be linear with slope equal to $-2/\rho_0$. This analysis was applied successfully by Mort et al. (1976).

It is clear from this relatively new approach to carrier transport in amorphous solids that unequivocal features of the transient pulse shape are predicted and can be put to experimental test. This has already been done in the case of the amorphous inorganic solids a-Si (Allan et al., 1977), a-Se (Pfister, 1976), and a-As_2Se_3 (Pfister, 1974; Scher and Pfister, 1975), and for organic solids, such as holes in the polymer polyvinylcarbazole (PVK) (Mort and Lakatos, 1970), electrons and holes in the charge transfer complex of PVK with trifluorenone (Gill, 1972; Seki, 1974), and charge transport in an inert organic polymer, polycarbonate doped with a molecule that forms a hopping centre such as N-isopropylcarbazole (Mort et al., 1976). However, a thickness-dependent mobility is not always observed (Tahmasbi and Hirsch, 1980). From particular analyses of the transient pulse shapes, it is possible to derive parameters characterizing the hopping probability, and the field dependence and the time-averaged value of this as a function of specimen thickness. In certain solids it has been possible to derive the localization radius of the carrier. The concept in all these cases of "mobility" as being a property of the bulk material is of limited value.

F. Trap distributions

Although a double-logarithmic plot of $i(t)$ against t is capable of locating the position of a "transit time" if there is considerable dispersion, this cannot be construed as unequivocal evidence for hopping transport. Marshall (1977a,b) pointed out that a similar result can arise if the transit pulse is dispersive owing to almost any mechanism that causes a distribution of transit times. Examples analysed included trap distributions, such as a Gaussian spread of the density of traps about a mean depth, a linear variation of trap density with depth from the band edge (band tail states), a rectangular distribution, and an exponential distribution (Silver and Cohen, 1977). Computer simulations of the expected pulse shapes were performed by a Monte Carlo technique. Using a mathematical model originally explored by Silver et al. (1971) and Marshall and Owen (1971), the trapping time for an initially free carrier was calculated as

$$t_1^* = -t_f \ln X$$

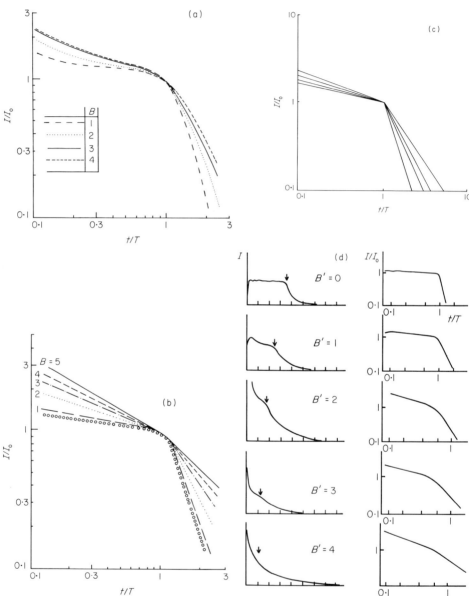

Figure 5.21. Dispersive carrier transport. Transient EBC pulse shapes computed for various trap distributions. (a) Rectangular distributions of different widths, and (b) Gaussian distributions of different widths. (Marshall, 1977b) (c) Exponential trap distributions of different mean depths. (Silver and Cohen, 1977) (d) Amorphous selenium; B' increases as θ is lowered. (Marshall, 1977a) The parameter $B = \delta\varepsilon/k\theta$ describes the extent of the spread of energies within the distribution.

where X is a random number between 0 and 1. In the case of a discrete set of trapping centres, the release time $t_2{}^*$ is expressed in terms of an independent second random number X' by

$$t_2{}^* = -t_r \ln X'$$

A succession of random trapping and release events can be used to compute the progress of a single carrier through the specimen. This is then repeated for about 10^4 carriers and averages are taken to compute the pulse shape. In the situation where the trap depths are distributed over a range, the particular distribution can be incorporated in the calculation of the shape of the carrier pulse expected.

The results of the computer calculations are shown in Fig. 5.21. It is seen that the temperature dependence of the transit time (as experimentally observed with holes in amorphous selenium) is in accord with a distribution of traps, whereas dispersion based on a hopping mechanism with a distribution of hopping distances (as in the Scher–Montroll model described above in Section E) would be expected to be independent of temperature. The break in the double-logarithmic plots marking the transit time depends on the particular trapping parameters, and in some cases Marshall compared his computed pulse shapes with experiment and was able to draw conclusions about the most probable trap distributions. A significant difference from the Scher–Montroll model is that here we again have a field-independent mobility, and it is thus possible to talk about a true fundamental drift mobility property of the solid.

Further analyses of dispersive transport arising from trap distributions were published by Marshall (1978), and as applied to amorphous silicon by Marshall and Allan (1979).

G. Carrier release time from traps

An interesting feature of the shape of the transient induced charge pulse shape was reported by Martini *et al.* (1972). The trap-controlled mobility transit time is given by

$$T' = T[1 + (N_t/N_b)\exp(\Delta E/k\theta)] \tag{5.32}$$

In a certain range of material parameters of t', t'', and T, the 0–72% value of Q is directly proportional to the mean time for a carrier to escape the crystal. This is given by

$$t_e = T + (T/t')t'' \tag{5.33}$$

provided that $T' < T$. This intersection at $Q/Q_\infty \approx 0.7$ occurs for values of $T/t' > 2t'/t'' > 0.1$. The computed values of t/t_e at which $Q/Q_\infty = 0.72$ are

Table 5.2. Values of t/t_e at which $Q/Q_\infty = 0.72$.

T/t'	t/t_e		
	$t'/t'' = 0.001$	$t'/t'' = 0.01$	$t'/t'' = 0.1$
50	0.70	0.70	0.71
20	0.70	0.70	0.70
10	0.70	0.70	0.70
5	0.70	0.70	0.70
2	0.66	0.66	0.65
1	0.38	0.38	0.40

shown in Table 5.2 where the calculated response for t'/t'' is given over a range of values of T/t'.

As shown in Chapter 1, the release time t'' of a carrier from a trap of depth ΔE is given by

$$t'' = f^{-1}\exp(\Delta E/k\theta)$$

where the frequency factor f varies from one material to another. In practice f is found to lie between about 10^8 and 10^{14} Hz but an estimate for a *new or unknown* material is difficult to make. Experimentally, it can be derived from phosphorescence glow curves or conductivity glow curve measurements. In terms of the trap cross-section σ and the effective density of states N_b in the appropriate band, it is given by

$$f = N_b u_{th} \sigma$$

where u_{th} is the thermal velocity of the free carriers. The mean values of t'' for various values of ΔE when f has a value of 10^{11} Hz are given in Table 5.3.

Table 5.3. Release time of carriers at 300 K from traps of different depths.

Depth ΔE (eV)	Release time t'' (s)	Depth ΔE (eV)	Release time t'' (s)
0.1	5.5×10^{-10}	0.4	8.9×10^{-4}
0.15	4.0×10^{-9}	0.5	4.9×10^{-3}
0.2	3.0×10^{-8}	0.6	2.6×10^{-1}
0.25	2.2×10^{-7}	0.7	15
0.3	1.6×10^{-6}	0.8	790
0.35	1.2×10^{-5}	0.9	4.9×10^{-3}

III. Carrier velocity

A. Mobility of either sign of carrier

As a typical example which should provide a picture describing the order of magnitudes involved in the transient EBC time-of-flight experiment, we shall discuss the transport properties of a single-crystalline specimen, 3 mm square and 300 μm thick, bombarded by an electron pulse lasting 100 ns and giving 1 μA current at 30 keV energy. The transient EBC pulse might last a total time of 0·5 μs, and the integrated current pulse (charge transported from face to face) gives rise to a voltage signal passed on to the measuring equipment of about 0·1 V at an applied bias field of 10^3 V cm^{-1}. Obviously all these quantities cited may vary for different materials. From measurements such as this, it is possible to make a series of determinations of the transit time T as a function of applied bias field on the same specimen and to repeat the series at different temperatures θ. Usually, if possible, a number of specimens of the same material are prepared with different thicknesses L. An accurate measurement of this parameter for each specimen is important because it is probably the most critical factor determining the accuracy of the derived drift mobility value. An elegant optical lever thickness probe for these purposes was described by Spear et al. (1963).

If a fall in the transient EBC can be seen, this will yield the carrier transit time T. With the specimen thickness known accurately, this immediately leads to the drift velocity, as already given in equation (5.1), $u = L/T$, where L is the specimen thickness; SCF conditions are assumed. If the velocity is proportional to the applied field, and v is the applied bias, the carrier mobility, already given in equation (5.2), is $\mu = L^2/vT$. The mobility of either sign of carrier can be measured by choosing the polarity of v appropriately.

In insulating solids the effect of trapped space charges is very prominent. Any space charge due to trapped carriers can be cleared by one or a number of neutralizing pulses applied by bombarding the specimen under zero applied bias. This is explained by Fig. 5.22 in which (a) shows the regular sequence of trigger pulses, usually synchronized with the 50 or 60 Hz mains supply; these are used to provide trains of about eight pulses to turn the high-voltage electron beam on for a sequence of eight sub-microsecond bursts, (b). The applied voltage bias is supplied to the specimen also in the form of pulses as shown in (c). They are phased and timed to apply the bias a millisecond or so before the excitation pulse, and to short-circuit the front and back (zero bias) electrodes after the transit pulse is complete. The repetition frequency of the bias pulses is the same as that of the electron beam pulse trains, but the phasing is so arranged that seven out of the eight

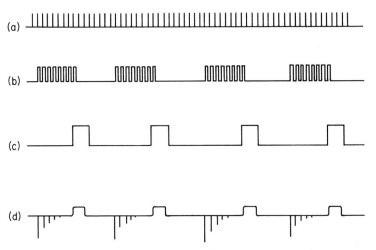

Figure 5.22. Relationship between field and excitation pulses in normal transient EBC time-of-flight measurements: (a) 50 Hz mains frequency reference; (b) electron excitation; (c) electric field; (d) observed conduction pulses.

pulses bombard the specimen under zero bias conditions. These neutralize the internal space charge by generating electron–hole pairs which move under the influence of the space charge field alone; this continues until the latter is reduced to zero. The sixth and seventh excitation pulses act as test pulses to ensure that there is no carrier movement inside the specimen in the absence of an applied field. The induced current pulses displayed on the oscilloscope are as shown in (d). The decaying space charge may be inspected by observing the reverse-direction induced current pulses just before the next forward transit pulse.

The value of μ is determined from the slope of a linear plot of $1/T$ against v, after applying corrections (now to be discussed) for small magnitudes of trapped space charges or a short but measurable carrier free lifetime if necessary. If the line cuts the $1/T$ axis this is probably a sign of limited carrier lifetime t' with respect to deep traps, and if it cuts the v axis this is probably a sign of a trapped surface space charge. The first of these situations was analysed by Spear and Mort (1963) and the second by Spear (1957).

The general expression for the point of intersection of tangents at $t = \infty$ with that of Q_{t1} when t is not too large (i.e. $t \ll T$) was shown to be

$$1/T'' = (1/t')[1 - \exp(-L/\mu \mathscr{E} t')]^{-1}$$

where T'' is the apparent value of the transit time obtained by drawing tangents as usual. Figure 5.23 shows how the graph of $1/T''$ meets the axis at

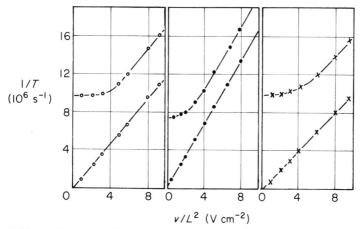

Figure 5.23. Influence of lifetime on a plot of $1/T$ against bias for three different specimens of CdS. (Spear and Mort, 1963) Note how the intercept on the vertical axis yields $1/t'$ directly where t' is the carrier free lifetime; after correction of the experimental curves as described in the text, the points all lie on a line passing through the origin and having a slope for large v the same as that before correction.

a value of $1/T'' = 1/t'$ and, when a correction to T is made by using this equation, the plot of $1/T$ against v then passes through the origin. It is noted from Fig. 5.23 that, when $1/T''$ is plotted against v, the slope of the linear part of the curve is exactly the same as the slope of the straight-line plot of $1/T$ against v. Thus, in either case, the slope is given by L/μ except near $v = 0$.

In the case where a plot of $1/T$ or $1/T''$ against v cuts the abscissa this is usually a sign of a trapped static surface charge. This was observed in transient EBC measurements on amorphous selenium by Spear (1957) shown in Fig. 5.24. The thin charged surface layer gave rise to a small zero error in determining the effective value of v. In the measurements described, the zero offset was substantially constant and amounted to a few volts. In cases such as this the carrier mobility is still determined by the slope of the graph relating $1/T$ to applied bias.

After taking account of the expected precision with which L, v, and L/T can be determined, an overall accuracy in a single determination of μ will be about 10%. However, in many insulating solids, μ can vary considerably from one sample to another. If carrier trapping plays a dominant role, the variation from specimen to specimen provides information about trap density, and the temperature dependence of μ gives information about trap depth. Carrier trapping is more important in insulating solids than in semiconductors because there is a wider range of vacant energies in the band gap

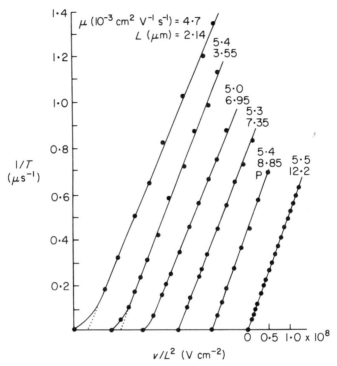

Figure 5.24. Influence of trapped surface space charge on a plot of $1/T$ as a function of bias in layers of amorphous selenium having different thicknesses; the curve marked P is for photoexcitation. The graphs cut the horizontal axis at a point where the change in potential near the electrode is caused by the static space charge near it. Note the change in origin for each curve. (Spear, 1957)

which may happen to accommodate unspecified localized levels arising from defects or neutral impurities.

B. Trap-controlled mobility

If an insulator or high resistivity semiconductor has carrier trapping centres in the bulk, the drift velocity of a cloud of injected charge will be reduced by the trapping–detrapping processes. The model we visualize is this. A carrier drifts towards the appropriate electrode under the influence of an applied electric field for a time t' until it falls into a trapping centre of depth ΔE. If the trap is not too deep or the temperature θ too low, the carrier will be released thermally in a time t'' which is not long compared with the duration of the observation. The effective mobility (μ_t) is thus reduced by the ratio of the time the carrier is free to move to the total time spent in undergoing this cycle. Thus the trap-controlled mobility is given by

$$\mu_t = \mu[t'/(t'' + t')] \tag{5.34}$$

and by using an argument based on the thermodynamic equilibrium of carriers between "free" and "trapped" states (Mott and Twose, 1961), we get

$$\mu_t = \mu[1 + (N_t/N_b)e^{\Delta E/k\theta}]^{-1} \tag{5.35}$$

where N_t and N_b are the density of traps and the effective density of states in the band respectively. A simplification arises if the mobility is strongly trap-controlled; equation (5.35) then becomes

$$\mu_t = \mu(N_b/N_t)e^{-\Delta E/k\theta} \tag{5.36}$$

The mobility is thus an activated process and, by measurements of μ_t at various temperatures, a semi-logarithmic plot of $1/\theta$ against $\ln \mu_t$ yields a straight line (Arrhenius plot) over the range for which equation (5.36) is valid. The slope of this line gives ΔE directly.

If the specimen under investigation contains more than one type of trap, as is usually the case, then as the temperature is lowered the carrier lifetime with respect to the deeper traps eventually dominates, and at very low temperatures complete carrier transits can no longer be seen. This limitation arises because t'' is then much longer than the carrier pseudo-free lifetime.

C. Trap density

The density of shallow trapping centres, N_t in equations (5.35) and (5.36), can be determined if μ is known. As an example, consider Fig. 5.25 which shows the hole drift mobility in a number of specimens of orthorhombic sulphur as a function of temperature. For a number of specimens it can be seen that the mobility is an activated process, and the value of μ_t rises exponentially with the absolute temperature. However, a few specimens display a distinct departure from linearity on this Arrhenius plot, especially at high temperatures (e.g. $10^3/\theta < 5$). This implies that the pre-exponential factor in equation (5.35) is starting to dominate, and by careful fitting of the measured points to this equation, using a least-squares method, it enables μ, N_b, and N_t to be derived for the various specimens.

The value of N_b is about 10^{18} for most high-mobility semiconductors such as Si, but it is approximately proportional to $m_{\text{eff}}^{3/2}$ where m_{eff} is the effective mass of the carriers. N_b may be somewhat higher if the band is very narrow; if the bandwidth is of the order of a few $k\theta$ wide, N_b might be almost as high as the density of molecules (10^{22} cm^{-3}).

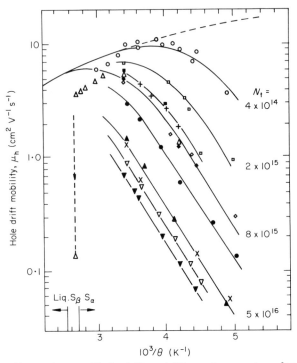

Figure 5.25. Dependence of hole drift mobility on temperature for a number of specimens of orthorhombic sulphur. This graph clearly shows how the trap-controlled mobility gives way to a lattice-dominated transport mechanism at high temperatures, especially if the density of shallow trapping centres N_t is small. The slope of the linear part of the curves gives the activation energy directly. Four of the curves were calculated using the values of N_t indicated. (Adams and Spear, 1964)

D. The motile trap model

If the material happens to be fairly soft, or measurements are being made at temperatures approaching its melting point, the density of defects may change as the temperature is altered. In this case, a number of drift measurements by the transient EBC technique at different temperatures may not give reliable values for the derived trap depth (M. Silver, private communication).

The conduction process is envisaged in terms of a band model in which the electron interacts strongly with a dominant level of traps associated with point defects. If the density of defects is in thermal equilibrium at all temperatures, a modified form of the expression for activated trap-controlled mobility will result. If we now have a low-mobility solid, we can put $N_m = N_b$.

Consider a thermal equilibrium density N_t of defects at a potential energy ε_f higher than that of a completely ordered lattice; ε_f is thus equal to the heat of formation of a defect. Thus, at any temperature θ, if N_m is the density of lattice sites in the ordered state,

$$N_t = (N_m - N_t)\exp(-\varepsilon_f/k\theta) \qquad (5.37)$$

Provided that the density of defects is not too high, this simplifies to

$$N_t = N_m\exp(-\varepsilon_f/k\theta) \qquad (5.38)$$

Now, if each point defect gives rise to a shallow electron trap of depth ε_t, the value of N_t from equation (5.38) can be used in the expression obtained for trap-controlled drift mobility, equation (5.36). Thus

$$\mu_d = \mu(N_c/N_m)\exp[(\varepsilon_t - \varepsilon_f)/k\theta] \qquad (5.39)$$

It may be noted that equation (5.39) predicts that the drift mobility may either be independent of temperature or rise or fall with temperature according to whether $(\varepsilon_t - \varepsilon_f)$ is zero, positive, or negative. If $\varepsilon_f > \varepsilon_t$, the mobility will increase with temperature with an activation energy equal to the difference between the heat of formation and the trap depth.

For this model to be applicable to a series of mobility measurements taken at different temperatures, it is important that the time interval between measurements should be large compared with the time constant for the process by which the defects are formed.

IV. Apparatus

A. Medium and low mobility insulating solids and liquids

This particular class of materials does not lend itself easily to methods of carrier mobility determination such as those involving a combination of conductivity and Hall effect measurements. However, time-of-flight techniques–and especially transient EBC methods–have proved highly successful. By "medium and low mobility" in the present context we refer mainly to those crystalline, amorphous, or liquid insulators for which T lies in the range about 70 ns to 5 ms.

Suitable apparatus is equipment based on modifications of the design described in Chapter 2 for steady EBC of thin films. The specimen is now much thicker and the primary electrons penetrate only a small fraction of its thickness; typical values are $L = 300\,\mu\text{m}$ and $V_p = 35\,\text{kV}$. A schematic diagram of apparatus for transient EBC measurements is shown in Fig. 5.26.

In operation, the electron beam is biased off by applying a voltage of about $-30\,\text{V}$ to the gun modulator grid from the battery (B). A fast-rising

184   ELECTRON BOMBARDMENT INDUCED CONDUCTIVITY

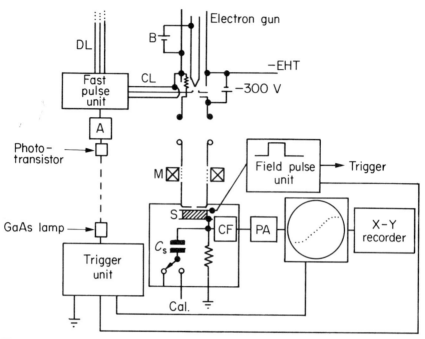

Figure 5.26. Experimental arrangement used in transient EBC drift mobility measurements: DL, delay line; CL, connecting line; M, magnetic lens; S, specimen; CF, cathode follower; PA, preamplifier; A, amplifier. (Spear, 1969)

positive voltage from the delay line pulser is fed via the terminated coaxial line (CL) to the modulator which switches the beam on for the duration (τ) of the voltage pulse. The beam current is simply adjusted by the charging potential applied to the delay line (DL) or by the filament temperature.

The fast pulse unit consists of a coaxial delay line (coaxial cable) which is charged to a set potential via a large resistor and is discharged into a terminated cable of the same characteristic impedance by a mercury-wetted contact high-speed relay. This in turn is driven from a circuit triggered by a phototransistor and an optical link. A suitable circuit was shown in Fig. 2.8. An avalanche transistor can also be used to discharge the delay line. A magnetic lens is used to focus the beam to a spot between 1 and 2 mm in diameter on the specimen (S). A separate magnetic lens can direct the beam into a Faraday cup, if required.

The output from the cathode follower or f.e.t. probe is fed to a wide-band oscilloscope. Sometimes an ordinary c.r.t. display is quite adequate, in which case measurements can be made directly from the screen or via a Polaroid photograph of it. Alternatively, a sampling oscilloscope or a storage oscillo-

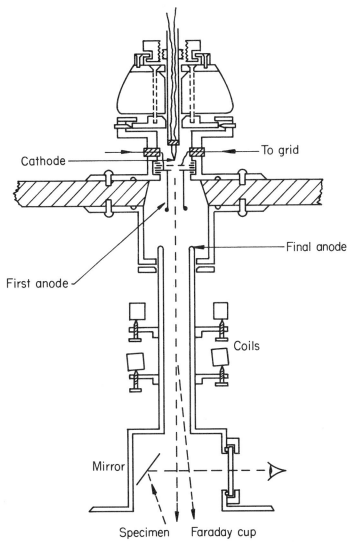

Figure 5.27. Details of electron gun used in EBC drift mobility measurements. (Miller, 1967)

scope may be used; in the case of a sampled display, the output can be traced by a pen-recorder.

Details of an electron gun are shown in Fig. 5.27. The specimen holder is similar to that already described for steady EBC measurements, but care is taken to make the electrical connections short, and the vacuum leadthroughs are of the coaxial type. A typical crystal specimen mounted ready

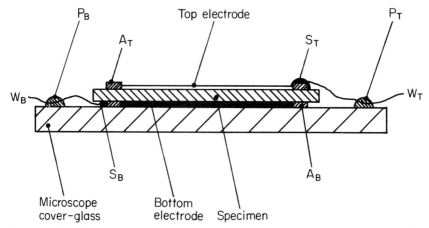

Figure 5.28. Side view of a typical specimen mounted for EBC drift mobility measurements: W_T and W_B, top and bottom contact connecting wires; A_T and A_B, top and bottom contacts made from Aquadag; S_T and S_B, silver paste; P_T and P_B, polystyrene cement. The specimen is held firmly down on the glass with silver paste under the bottom electrode.

for measurements is shown in Fig. 5.28. The glass cover-slip is held in intimate thermal contact with the copper support rod by a film of silicone oil, and a thermocouple made from Pt and PtIr alloy wires about 0·1 mm in diameter is embedded in the oil or as near the specimen as possible. Accurate temperature readings are obtained if the vacuum thermocouple leadthroughs are continuous through the seal, and from then onwards the connections to the voltmeter and reference junction are of the same metals or alloys as those making contact to the other side. A cathode follower circuit enclosed in a small screened box (Eddystone box) is mounted directly beneath the specimen holder. The power supplies for this are provided from batteries in an earthed metal box or from a stabilized power supply driven from the mains. Alternatively an impedance-matching probe using an operational amplifier with a f.e.t. input has been used.

B. Solid gases

Electron and hole transport in crystalline solid gases is possible only when they are grown inside the apparatus. This was effected conveniently by Spear, Miller, and their collaborators using a version of the apparatus just described but modified particularly for growing crystals of gases *in situ*.

To ensure good electrical contact to the side of the crystal where it started to grow, it was necessary to mount the specimen holder above the electron gun, which now fired electron pulses upwards. The specimen holder shown

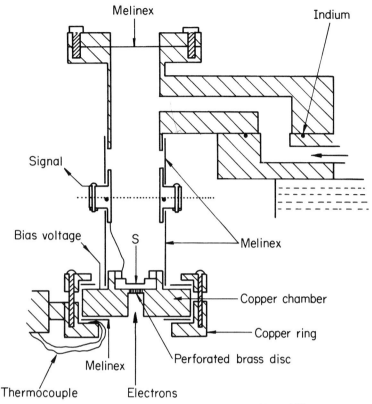

Figure 5.29. Diagram of specimen chamber used in drift mobility measurements on solid and liquid noble gases. This enabled the pre-purified gas to be introduced into the cooling chamber where single crystals of the solid were grown between the electrodes. The electron-permeable gold-plated Melinex bottom electrode separated the gas from the vacuum space, and the electron gun was positioned below the chamber so that contact with the front electrode was ensured. Contact to the reverse face was effected by a spring-loaded electrode. (Miller, 1967)

in Fig. 5.29 was cooled by a miniature Joule–Kelvin cryostat (Air Products, Cryotip), and the gas under study was isolated during growth process from the rest of the vacuum system by a thin film of Melinex fixed on a thin brass supporting disc perforated by a large number of very small holes. One side of the Melinex was made conducting by a thin evaporated layer of gold, and adhesion between the plastics film and the brass was made easier by roughening its surface by lightly etching in a solution of caustic potash. A suitable adhesive is potassium silicate solution.

A diagram of the apparatus is shown in Fig. 5.30. The method used by Miller *et al.* (1968) for single crystals of gases was as follows. The gas under

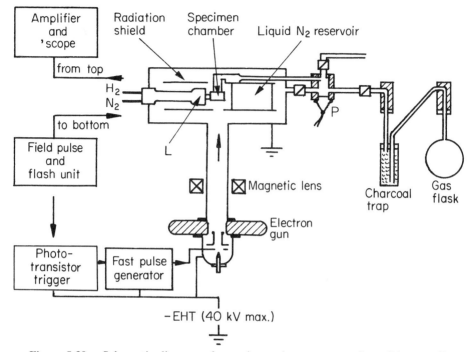

Figure 5.30. Schematic diagram of experimental arrangement for solid gases: P, pressure transducer; L, miniature Joule–Kelvin cryostat. (Miller, 1967)

investigation was condensed into the specimen chamber at temperatures and pressures not far from its triple point. When sufficient liquid had been condensed to fill the space between the two electrodes, the gas supply was shut off, and the system allowed to reach equilibrium. By careful adjustment of the cooling power of the cryostat, it was possible to grow the solid at a rate of less than 5 mm h^{-1}. In a number of subsidiary experiments, crystals were grown by the same method and their surface was examined in reflected light; grain boundaries could be clearly seen, the average grain area being about 20 mm^2 which is about twice the effective area used in the transport measurements, so the effect of grain boundaries was not considered significant. The specimen was bombarded by 30–40 keV electron pulses of 5–100 ns duration.

C. High mobility solids

For measurements on high mobility semiconductors, work by two independent groups yielded modifications to accommodate transient EBC measurements at higher speeds. The first material in this class was silicon, partly because of the technological advances made by about 1964 in the methods for growing large ingots of high purity low conductivity material. These high-speed techniques have now been used for studies of the electron and hole drift velocities in a number of pure semiconducting solids as well as field-profiles in lithium-drifted silicon or germanium diodes.

A description of time-of-flight carrier drift velocity measurements on silicon was published by Alberigi Quaranta *et al.* (1965). Transient charge pulses were induced by single α-particles of 5·5 MeV energy which impinged on a totally depleted silicon surface barrier nuclear particle detector. The output charge was amplified by a 2 ns rise-time amplifier and displayed on a sampling oscilloscope. From these initial trials, the same research group at Modena in Italy advanced to short electron pulse excitation and faster pulse detecting systems which are to be described shortly. It is noted here that the range of a 5·5 MeV α-particle is only about 30 μm in silicon, and therefore the requirement of the technique for surface excitation was well fulfilled. As mentioned, however, there are objections and errors involved when α-particle excitation is used for some measurements. But these were not very clear at the time; these objections were circumvented when electron pulses were used for a similar purpose.

Norris and Gibbons (1967) undertook measurements on silicon specimens with thin alloyed p$^+$ and n$^+$ contacts using an electron beam derived from the grid-cathode structure of a commercially available microwave triode. The contacts were made from gold–antimony alloy or from aluminium. The sample mount was in the form of a coaxial structure. It had a ring contact on the face of the specimen which was exposed to the beam to prevent carrier generation near the edges and consequent induced surface currents. As shown in Fig. 5.31, high speed pulses were applied to the grid of

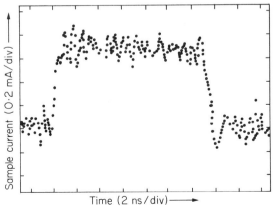

Figure 5.31. Transient EBC sampled pulse shape for a high resistivity silicon specimen with low-temperature diffused contacts. (Norris and Gibbons, 1967)

the gun from a mercury-wetted contact high speed relay which discharged a delay line, and the output current pulses were measured on a sampling oscilloscope. To prevent sample heating due to surface leakage, the specimen bias was applied in the form of short square-topped pulses about 100 µs wide, and generated from a bias pulse unit triggered from the same relay that was used for generating the gun modulator pulse.

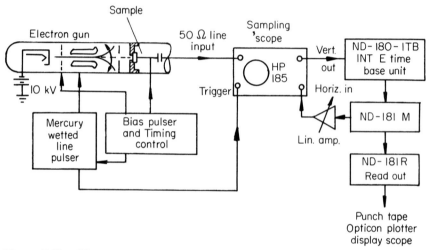

Figure 5.32. Picosecond electron gun and signal averaging system; block diagram of the experimental set-up. (Zulliger et al., 1969)

Details of a similar experimental arrangement are to be found in an article by Zulliger et al. (1969). A block diagram is shown in Fig. 5.32. The principal advances described in this system are those designed for improving the signal/noise ratio of the sampled gain. This was achieved by a signal averaging technique described above in Section I.G.

Wertheim and Augustyniak (1956) pointed out that very short electron pulses from a Van de Graaff generator can be produced by scanning a steady electron beam across a small aperture. This method was used by Zulliger and extensively by the group working at the university of Modena in Italy. Their apparatus was described by Alberigi Quaranta et al. (1970). The cathode of the gun described was an ordinary commercial cathode ray tube oxide-coated cathode. This was arranged so that the cathode could not only be positioned and centred while the gun was working but be replaced fairly easily when it needed renewing, which was normally after not less than several hundred working hours.

A diagram of the apparatus is shown in Fig. 5.33. To lengthen the working life, the cathode space within the vacuum enclosure was isolated from the atmosphere by a vacuum-tight slide-valve (N) when the specimen was changed. In this way the cathode was maintained at a vacuum of about 10^{-4} torr or better throughout its life. High speed deflection of the electron beam was achieved by applying suitable voltages to an electrostatic distributed deflection plate assembly (G) taken from an oscilloscope cathode ray tube.

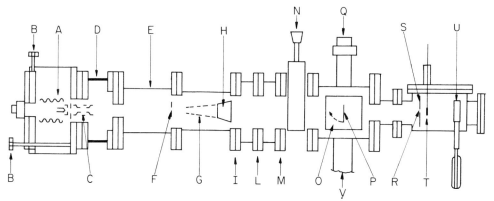

Figure 5.33. Schematic diagram of electron accelerator: A, flexible shaft coupling; B, micrometer screws; C, electron source with accelerating and focusing electrodes; D, insulated flange; E, transparent flange; F, fluorescent circular screen; G, vertical deflection plates (Y plates); H, horizontal deflection plates (X plates); I, L, and M, flanges insulated by segments of vacuum-proof ceramic tubes; N, fullway slide valve; O, transparent flange; P, fluorescent screen; Q, high vacuum gauge; S, circular screen with hole R; T, pickup electrode for observing the electron beam both in dc and pulsed condition; U, coolable sample holder; V, vacuum system connection. (Alberigi Quaranta et al., 1970)

A sinusoidal voltage at a frequency of 48 MHz was applied to a simple pair of plates (H) at right-angles to the plane of the first set, and a square-topped trapezoidal-shaped voltage impulse was applied to the distributed deflection plates. This deflection pulse was 160 ns long. The system of focus and accelerating electrodes (I, L, M) was so arranged that the electron beam was accelerated from an initial potential of about 10–15 kV to 30–40 kV, at which latter potential the beam passed through an aperture (R) in a plate (S). Frequency divider circuits driven from a quartz crystal controlled master oscillator were used to synchronize the deflection voltages applied to the two sets of plates. The phasing of the two deflection voltages was so arranged that the beam was swept across the centre of the aperture plate when it was travelling at its highest sweep speed by virtue of the deflection voltages. A pre-trigger pulse was obtained from the circuits just described, for applying to a pulsed specimen bias supply. This could apply bias voltages of up to 1600 V at a chosen time interval between 1 μs and 10 ms before the electron burst with a time jitter of about 100 ns between them. The duration of the bias pulse was selected in the range between 1 and 200 μs.

The resulting high-speed transient EBC time-of-flight apparatus produced an average electron current of 20 μA at 30–40 kV with a minimum duration of each burst about 70 ps. The repetition frequency could be selected within the range 50–140 Hz, and the time-jitter of each electron pulse with respect to a trigger pulse was less than 80 ps. The number of electrons per pulse for a typical pulse width of 0·2 ns and a beam current of 5 μA was $6·3 \times 10^3$. The measured pulse height jitter was about 1%, which is

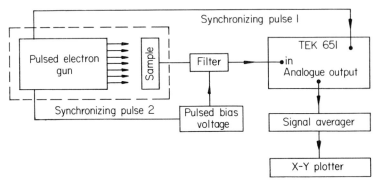

Figure 5.34. Block diagram of experimental arrangement. (Ottaviani *et al.*, 1975)

near the expected variance in a bunch of this size. With this equipment the minimum density of pairs created in a semiconductor was of the order of 10^{14} cm^{-3}.

This experimental arrangement has been used by the Modena group for studies of semiconducting solids. An advanced form of this equipment was described by Ottaviani *et al.* (1975) and the additional peripheral apparatus included a variable temperature liquid helium cryostat which could cool the specimen near 6 K if desired (Adonian MDG-7L-30V). A block diagram of this system is shown in Fig. 5.34.

D. Microwave time-of-flight technique

The method for generating sub-nanosecond electron pulses by means of a swept electron beam and the use of a distributed delay line deflection system leads naturally to microwave techniques for undertaking drift measurements at even higher speeds. If the doping concentration in a semi-

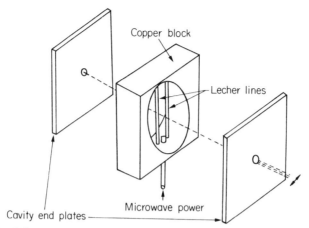

Figure 5.35. Microwave time-of-flight deflection system. (Evans and Robson, 1974)

conductor is fairly high, it is possible only to deplete fairly thin samples before the onset of avalanche breakdown. An example is epitaxially grown n-type GaAs where the sample thickness cannot exceed about $10\,\mu m$ and the transit times to be measured lie in the region of 0·1 ns. Evans and Robson (1974) described a microwave time-of-flight measuring system for semiconducting samples of the kind described.

A very simple electron beam deflection system used a pair of lecher wires tuned to resonate at 5 GHz which was the deflection frequency used. The electron beam passed between the lines as shown in Fig. 5.35, and it was then deflected to and fro past an aperture in a similar way to that employed in the equipment shown in Figs 5.32 and 5.33 and described above in Section C. The particular version used by Evans and Robson is shown in Fig. 5.36. The transit time of generated carriers through the specimen

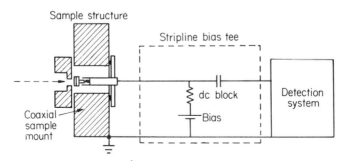

Figure 5.36. Sample mount and detection circuit. (Evans and Robson, 1974)

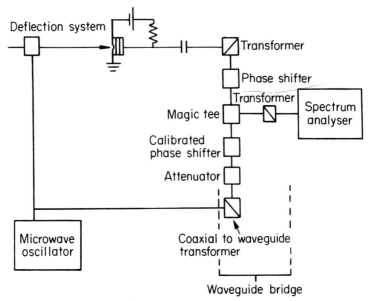

Figure 5.37. Microwave bridge circuit for phase measurements. (Evans and Robson, 1974)

caused a phase delay between the externally measured ac conduction current and the excitation, $\phi = fL/u$ (see Section I.C.5 above). Figure 5.37 shows a block schematic circuit diagram including the microwave bridge used for measuring ϕ.

6
Specific materials properties determined by transient EBC techniques

One of the most important applications of EBC is in the transient EBC method for materials characterization. The principal solids in this context are amorphous and crystalline insulators and high resistivity semiconductors because these do not lend themselves readily to alternative methods of hole and electron transport measurement. This chapter is devoted to a review of the results obtained for the transport properties of such materials and a brief description of their interpretation. In some cases, parallel measurements made on the same materials using light-flash or α-particle excitation to generate excess carriers are described, and the results compared. However, there are difficulties associated with using these methods alone, and a discussion of them has been included in Chapter 5. Furthermore, an interpretation of drift mobility in amorphous semiconductors is difficult if a combination of conductivity and Hall effect is used, since the sign of the Hall coefficient is not unambiguous if carrier transport is via a small polaron hopping mechanism or is characterized by a short mean free path as in amorphous semiconductors. Such ambiguity does not arise when transient EBC methods are used.

The method is sometimes called the time-of-flight technique (ToF) or the transient charge technique (TCT) but neither name emphasizes electron pulse excitation, so we use a more specific name—transient EBC methods. As about to be shown, many different properties such as carrier diffusion coefficients, mean free lifetimes and mobilities, modes of charge transport, trap depths and trap distributions, carrier recombination kinetics, band repopulation times, and numerous other fundamental carrier properties of materials can be deduced from the transient EBC pulse shape, and many more materials parameters can be measured as well as drift velocity from a simple time-of-flight determination.

I. Group IV elemental semiconductors

A. Silicon

The most intensively studied semiconducting solid is silicon owing to its dominance in the manufacture of solid-state electron devices. In particular, the transport properties of charge carriers, such as mobility (μ), diffusivity (D), and drift velocity (u), and their dependence on temperature (θ), electric field ($\mathscr{E}$), "hot" electron effects, crystal orientation, and impurity concentration, are necessary data for a proper design of such devices. In this section we shall summarize the experimental results, many of which were obtained by the transient EBC techniques described in Chapter 5. A review of some of the charge transport properties in crystalline silicon by Jacoboni et al. (1977) also included data obtained by the earlier conductivity techniques on extrinsic samples (Rodriguez and Nicolet, 1969). The diffusivity results were obtained from the transient EBC current pulse rise and fall times, t_r and t_f respectively, according to

$$D = (t_f^2 - t_r^2)u^3/21 \cdot 6L \tag{6.1}$$

where L is the specimen thickness, and the numerical factor in the denominator is related to the 5–95% values taken for the rise and fall times (Nava et al., 1979b).

1. Diffusivity

At low fields within the ohmic region, the diffusivity is related to the mobility by the Einstein relation,

$$D = \mu k\theta/q \tag{6.2}$$

However, at high applied fields, equation (6.2) does not hold without some modification (Canali et al., 1975b), and D becomes a field-dependent tensor (Fawcett, 1973). Thus we have $D_\perp$ and $D_\parallel$ respectively for the diffusivity of electrons or holes in directions perpendicular and parallel to the applied electric field. Experimentally $D_\perp$ is obtained in a transient EBC measurement by using a number of collecting electrodes, appropriately spaced, on the face of the specimen to which the generated charge cloud travels (Persky and Bartelink, 1971). $D_\parallel$ is obtained from equation (6.1).

The experimental data for the diffusivity at 300 K of electrons are shown in Fig. 6.1 and of holes in Fig. 6.2. It is seen, for electrons at least, that experimental data for $D_\perp$ fit the Einstein relation as modified by Canali et al. for fields up to 10^4 V cm^{-1}. This modified relation takes the form $D(\mathscr{E}) = \mu(\mathscr{E})\langle\varepsilon\rangle/q$ where $\langle\varepsilon\rangle$ is the mean "hot" carrier energy.

6. DETERMINATION OF SPECIFIC MATERIALS PROPERTIES   197

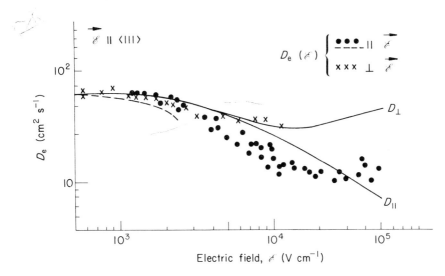

Figure 6.1. Diffusion coefficients of electrons in silicon at room temperature (300 K) as a function of field applied parallel to a $\langle 111 \rangle$ crystallographic direction: $D_\parallel$ from Canali *et al.* (1975a); $D_\perp$ from Persky and Bartelink (1971). Both theoretical curves were obtained with a non-parabolic model.

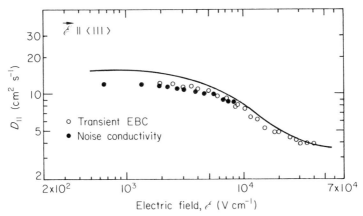

Figure 6.2. Longitudinal diffusion coefficient of holes in silicon at 300 K as a function of field strength applied parallel to a $\langle 111 \rangle$ crystallographic direction. The drawn curve refers to theoretical calculations (Nava *et al.*, 1979b,c)

2. Drift mobility

The electron and hole drift mobilities have been extensively measured by transient EBC methods ever since high purity silicon single crystals first became available (Norris and Gibbons, 1967). Similar measurements were considerably developed and refined by the Modena University group in Italy during the 1970's, and a number of significant papers on this subject have been published (e.g. Canali *et al.*, 1971c, 1973, 1975b; Jacoboni *et al.*, 1977; Nava *et al.*, 1979a; Ottaviani *et al.*, 1975; Reggiani, 1979; Reggiani *et al.*, 1975).

Figures 6.3 and 6.4 show the temperature dependence of the low-field electron and hole mobilities in silicon over the range 8–400 K. At around

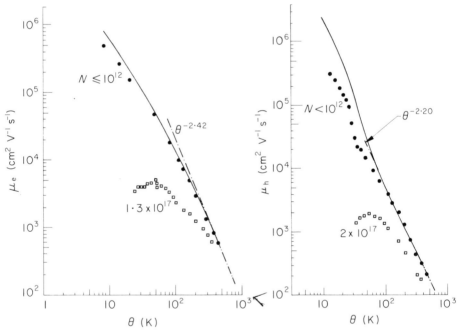

Figure 6.3. (LEFT) Ohmic mobility of electrons in silicon as a function of temperature. The drawn curve indicates theoretical results for pure lattice mobility; the dot-dash line gives the $\theta^{-2.42}$ dependence of electron mobility around room temperature; ● refer to measurements by the transient EBC method (Canali *et al.*, 1975a); □ refer to doped samples with 1.3×10^{17} donors cm^{-3} (Morin and Maita, 1954).

Figure 6.4. (RIGHT) Ohmic mobility of holes in silicon as a function of temperature. The measurements (●) were made by the transient EBC method on high purity silicon; the drawn curve indicates the $\theta^{-2.20}$ dependence of the hole mobility around room temperature (Ottaviani *et al.*, 1975); □ are for a doped sample with 2×10^{17} acceptors cm^{-3}. (Morin and Maita, 1954)

room temperature $\mu \propto \theta^{-\beta}$, where $\beta = 2\cdot 42$ for electrons and $\beta = 2\cdot 20$ for holes. Apart from measurements obtained by the transient EBC technique, others for doped samples measured some years ago by a combination of Hall effect and conductivity are included (Morin and Maita, 1954). These results show dramatically how the influence of impurities affects the drift mobility both of electrons and of holes especially at low temperatures.

3. High-field drift velocity

At high fields, the drift velocity is no longer a linear function of the applied field and it also exhibits anisotropic behaviour with respect to the electric field in the crystal. In the case of electrons, if the field is parallel to the $\langle 100 \rangle$ direction, two of the conduction band valleys exhibit the longitudinal effective mass (m^*_l) in the direction of the field whereas the remaining four show the transverse (m^*_t). The electrons with a small value of m^* will tend to repopulate the valleys containing electrons with a high m^*. The result of this is to reduce the drift velocity of electrons when the electric field is applied along the $\langle 100 \rangle$ direction as shown in Fig. 6.5. A less pronounced anisotropy for holes is due to the presence simultaneously of "light" and "heavy" holes. The drift velocity of holes as a function of applied field for two different orientations of the field is shown in Fig. 6.6.

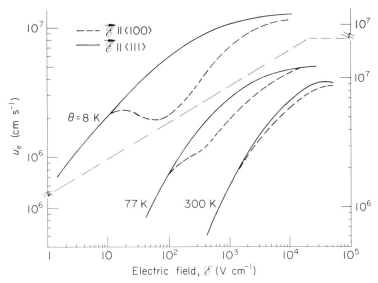

Figure 6.5. Experimental measurement of electron drift velocity in high purity silicon as a function of the field applied parallel to the $\langle 111 \rangle$ and $\langle 100 \rangle$ directions at different temperatures. (Canali *et al.*, 1975a)

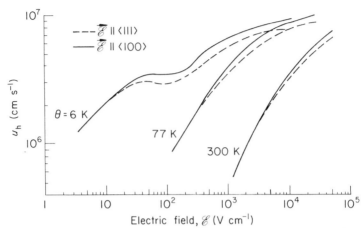

Figure 6.6. Experimental measurement of hole drift velocity in high purity silicon as a function of electric field applied parallel to $\langle 111 \rangle$ and $\langle 100 \rangle$ crystallographic directions at several temperatures. (Reggiani et al., 1977)

The experimental drift velocity results for electrons along $\langle 111 \rangle$ and for holes along $\langle 100 \rangle$ directions are shown in Figs 6.7 and 6.8 respectively.

Measurements of the surface drift velocity of carriers in silicon are incomplete but there is evidence that they are about an order of magnitude smaller than the bulk values and that the high-field saturation drift velocity on an etched surface is about 5×10^6 cm s^{-1} (Fang and Fowler, 1970; Sato et al., 1971).

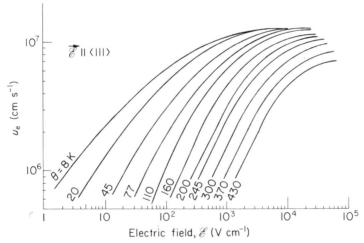

Figure 6.7. Experimental electron drift velocity in pure silicon as a function of electric field applied parallel to a $\langle 111 \rangle$ crystallographic direction at different temperatures. (Canali et al., 1975a)

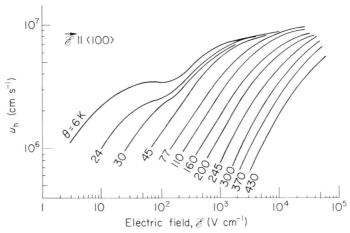

Figure 6.8. Experimental hole drift velocity in pure silicon as a function of electric field applied parallel to the ⟨100⟩ crystallographic direction. (Ottaviani *et al.*, 1975)

B. Germanium

The strong superficial similarities of the semiconductor properties of germanium to those of silicon tempt one to assume that a close parallel may be drawn between their detailed electrical behaviour. Caution must be exercised before doing so, however, because the band structure of the two is not similar and it becomes so only at high hydrostatic pressures (Ahmad *et al.*, 1979). Accordingly, the higher hole drift mobility at 300 K and the saturation drift velocity are affected to the extent of about 20% by an energy-dependent effective mass. Our knowledge of the detailed behaviour of holes in germanium is more complete than that of electrons, and agreement between theory and experiment for silicon is less satisfactory especially in the region of negative differential mobility of electrons.

1. Diffusivity

The hot hole diffusivity in ultrapure germanium at various temperatures has been measured from the rise and fall times of the transient EBC current pulse shape as already described for silicon. Figure 6.9 shows how $D_\parallel$ depends on $\mathscr{E}$ for three temperatures in the range 77–190 K; the lines correspond to theoretical expectations based on the Einstein relationship. Reasonably satisfactory agreement between theory and experiment is obtained especially when consideration is given to the complexity of the measurements. The values of $D_\parallel$ for $\mathscr{E}$ parallel to the ⟨100⟩ and ⟨111⟩ crystallographic directions at 77 K are shown in Fig. 6.10. Agreement between theory and experiment is within a factor of about 2 at high fields and

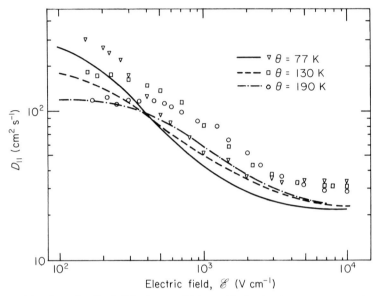

Figure 6.9. Longitudinal diffusion coefficient of holes in ultrapure germanium as a function of field strength at different temperatures. (Reggiani *et al.*, 1978)

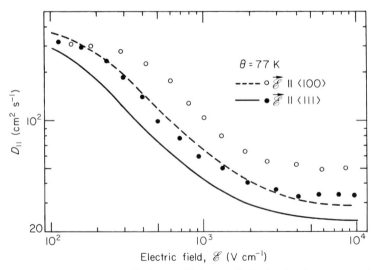

Figure 6.10. Longitudinal diffusion coefficients of holes in ultrapure germanium as a function of field strength applied in two directions at 77 K. (Reggiani *et al.*, 1978)

somewhat better at fields below about 200 V cm^{-1}. It is shown that in the temperature range examined, as the electric field is increased, the values of $D_\parallel$ become almost independent of temperature, in agreement with theoretical analyses of high-field transport in semiconductors.

2. Drift mobility and high-field drift velocity

Experimental results for the hole drift velocity as a function of field applied along the $\langle 111 \rangle$ direction in ultrapure germanium were obtained using the transient EBC technique by Reggiani *et al.* (1977). The results for a number of temperatures in the range 8–220 K are shown in Fig. 6.11. It was found that the hole drift velocities were not significantly different from those measured along different crystallographic directions.

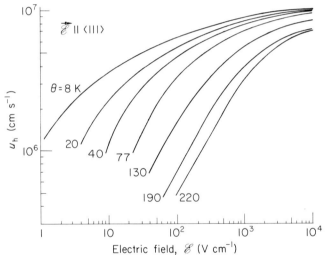

Figure 6.11. Experimental hole drift velocity in ultrapure germanium as a function of electric field applied parallel to a $\langle 111 \rangle$ direction at different temperatures. (Reggiani *et al.*, 1977)

The low-field drift mobility of holes as a function of temperature is shown in Fig. 6.12. The experimental values agree with previous unpublished data (de Laet *et al.*, 1971) and are in good agreement with a Monte Carlo calculation shown by the continuous line in the figure.

Equivalent data for the drift velocity and mobility of electrons in germanium are shown in Figs 6.13 and 6.14 respectively. From the low-field velocity versus field characteristics shown in Fig. 6.13, the drift mobility data shown in Fig. 6.14 were obtained. Again good agreement with previous results (de Laet *et al.*, 1971) was observed and also agreement with a

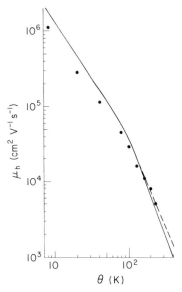

Figure 6.12. Mobility of holes in ultrapure germanium as a function of temperature. (Ottaviani *et al.*, 1973b)

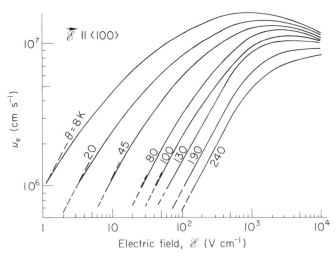

Figure 6.13. Electron drift velocity in germanium as a function of electric field applied parallel to the $\langle 100 \rangle$ direction at several temperatures. (Nava *et al.*, 1976)

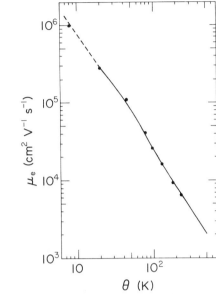

Figure 6.14. Electron drift mobility in high purity germanium as a function of temperature. The continuous line represents a theoretical calculation by Rode (1972), and the experimental points are for transient EBC measurements by Nava *et al.* (1976).

theoretical calculation (Rode, 1972) where a combination of acoustic and optical mode scattering gives a law $\mu \propto \theta^{-1.63}$ above about 90 K and a law $\mu \propto \theta^{-1.5}$ suggesting purely acoustic mode scattering below this temperature.

Below about 100 K and at high fields, a negative differential mobility (NDM) for electrons was observed in agreement with theoretical predictions (Borsari and Jacoboni, 1972). As shown in Table 6.1, the NDM and the drift velocity at the threshold field increased as the temperature was lowered, whereas the threshold electric field decreased.

Table 6.1. Dependence of electron drift velocity at threshold, threshold field, and negative differential mobility on temperature in germanium.

Temp. (K)	Drift velocity at threshold (cm s^{-1})	Threshold field (kV cm^{-1})	NDM (cm^2 V^{-1} s^{-1})
8	1·75 × 10^7	1·0	600
20	1·53 × 10^7	1·5	480
45	1·36 × 10^7	2·0	300
80	1·30 × 10^7	2·7	260
100	1·17 × 10^7	3·5	140

It was pointed out by Nava *et al.* (1976) that there were discrepancies between their results shown in Table 6.1 and similar data for germanium reported earlier (e.g. Baynham, 1969; Chang and Ruch, 1968; Elliott *et al.*, 1967; Fawcett and Paige, 1971; Kino and Neukermans, 1973) and it was felt that the present theoretical model for NDM in germanium should be improved.

C. Diamond

A systematic study of the Group IV semiconducting elements was completed at the end of the 1970's by the inclusion of natural diamond. Unlike the early studies on diamonds for the crystal counter which were undertaken more than 30 years before, the measurements were made on diamonds of relatively high electrical conductivity (known as class IIA diamonds). These contained naturally incorporated impurities of nitrogen and aluminium at concentrations below 10^{19} cm^{-3} which rendered them p-type with a room temperature resistivity greater than about 10^{14} ohm-cm. No neutralizing electron pulses were needed when these were studied by the transient EBC technique because in the measurements the specimens were fitted with different types of contact on their opposite faces. If hole transits were under observation, the build-up of a positive space charge due to hole trapping was prevented by using an injecting contact for electrons as the back electrode (B in Fig. 5.1) and a blocking contact for the front electrode (F). In the case of electron transits, B was made injecting for holes, and F blocking. In practice it was found that, apart from the expected behaviour of an ion implanted layer of P, Li, Al, or B, an injecting contact for either electrons or holes could be made by painted Aquadag followed by vacuum heating for 2–3 hours, and a blocking contact could be made by vacuum deposition of gold, silver, or platinum (Kozlov *et al.*, 1975). No polarization phenomena were observed at temperatures below 500 K.

1. High-field drift velocity

The drift velocity both of electrons and of holes as a function of electric field in the temperature range 85–700 K was measured by the transient EBC technique. Results obtained for specimens cut along the $\langle 100 \rangle$, $\langle 110 \rangle$, and $\langle 111 \rangle$ crystallographic directions and about 300 μm thick are shown in Fig. 6.15. The drift velocities of electrons and of holes approach saturation at about 10^7 and $1\cdot 5 \times 10^7$ cm s^{-1} respectively and they are anisotropic with respect to the crystal axes (Nava *et al.*, 1980; Reggiani *et al.*, 1979). The anisotropy is greater at low temperatures and is more pronounced at lower fields as the temperature is lowered. This behaviour is reminiscent of the behaviour of electrons in silicon. Figure 6.16 shows u_e for electric field parallel to $\langle 110 \rangle$ over a wider range of temperatures.

6. DETERMINATION OF SPECIFIC MATERIALS PROPERTIES   207

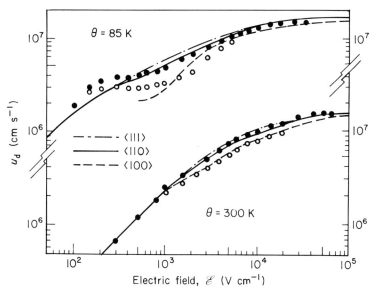

Figure 6.15. Drift velocity of electrons in natural diamond as a function of electric field applied along three different crystallographic directions, for two values of temperature. (Nava *et al.*, 1980)

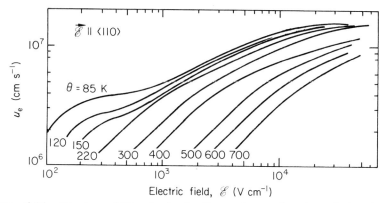

Figure 6.16. Electron drift velocity in diamond as a function of electric field applied in the ⟨110⟩ crystallographic direction, at different temperatures. (Nava *et al.*, 1979b)

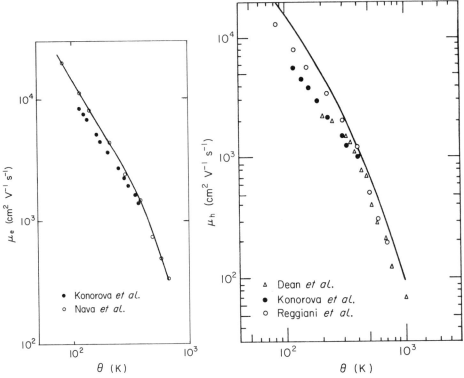

Figure 6.17. (LEFT) The low-field drift mobility of electrons in diamond as a function of temperature. (Konorova and Schevchno, 1977; Nava et al., 1979b)

Figure 6.18. (RIGHT) Hole drift mobility in diamond as a function of temperature. (Dean et al., 1965; Konorova and Schevchno, 1977; Reggiani et al., 1979)

2. Drift mobility

The low-field drift mobility of electrons as a function of temperature is shown in Fig. 6.17, and of holes in Fig. 6.18. Below about 400 K the electron and hole mobilities display a temperature dependence of the form $\mu \propto \theta^{-3/2}$, while above this temperature the mobility of holes at least is controlled by optical phonons. The room temperature values of μ_e and μ_h are 2400 and 2100 $cm^2\,V^{-1}\,s^{-1}$ respectively.

II. Amorphous silicon

Extensive studies of amorphous silicon (a-Si), and especially measurements on material prepared by the glow discharge technique, have contributed greatly to a better understanding of electrical conduction processes in amorphous semiconductors (Spear, 1977). Measurements of the electrical conductivity (σ) and its temperature dependence, field-effect measurements to determine the density of states function, and the measurement of carrier drift mobilities by transient EBC methods have built up a consistent picture well supported by theoretical studies. The apparatus used for drift mobility measurements was similar to that shown in Figs 5.26 and 5.27. It is appropriate here to refer again to a diagrammatic model of the density of states in undoped a-Si (Fig. 6.19) which is important in the discussion that follows. (This may be compared with Fig. 1.11 which gives $g(\varepsilon)$ in greater detail.) The energy ε_A marking the onset of tail states is taken as the fixed reference point in the system. The arrows on the figure indicate that the Fermi energy and the mobility edge both move with respect to ε_A if the temperature is increased (Spear et al., 1980a,b).

Theoretical concepts explain the picture as follows. At energies above ε_C, the electron wavefunctions extend throughout the crystal, and although the Bloch conditions of phase-coherence do not extend this far, the electronic states are extended states. Between ε_C and ε_A the states are localized because the variations in electron energy from one atom to another are comparable to the interatomic exchange energy J. This is because the average value of J becomes smaller as the centre of the mobility gap is approached. The position of ε_C is defined by the Anderson criterion for the rather abrupt transition to a complete lack of electron movement without phonon assistance when the lattice disorder exceeds a given fraction of the bandwidth as explained in Chapter 1. At energies below ε_A and extending to the equivalent energy ε_B near the hole states, the electronic wavefunctions are localized about specific characteristic lattice defects. The separation ε_A to ε_B is known

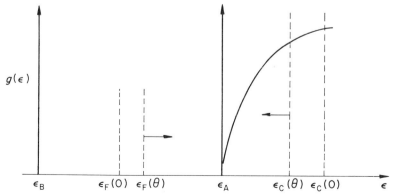

Figure 6.19. Simplified diagram of the density of states distribution $g(\varepsilon)$ in amorphous silicon. (Spear et al., 1980a,b)

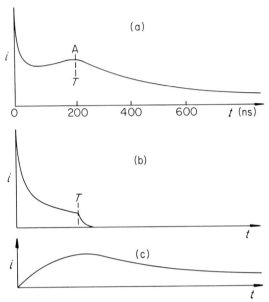

Figure 6.20. (a) Observed transient EBC signal in drift mobility experiments on a-Si films; the cusp at A defined the transit time T of the excess electrons. The signal can be resolved into two components: (b) shows the transit of the coherent group of electrons; (c) shows the incoherent background signal. (Spear and Le Comber, 1972)

as the mobility gap, and the density of gap states depends critically on the method used for preparing the a-Si. This density is particularly low in glow-discharge deposited a-Si, but there can still be variations between different samples prepared under carefully controlled conditions in different laboratories. However, these differences are much smaller than those between specimens produced by this method and samples prepared by evaporation or sputtering. Despite these latter differences, there can be similarities in some respects such as the structure as demonstrated by their radial distribution functions, and dependence of their conductivity activation energy on deposition temperature.

Transient EBC methods were used to study the transport of electrons in undoped a-Si in detail, and also the drift mobility both of holes and of electrons in doped and undoped specimens. In undoped glow-discharge a-Si, the resistivity is found to be strongly dependent on the temperature of the substrate, θ_D (deposition temperature). As noted originally by Chittick et al. (1969), σ could be changed from about 10^{-11} ohm^{-1} cm^{-1} at a deposition temperature of $\theta_D = 0\,°C$ to 10^{-4} ohm^{-1} cm^{-1} at $\theta_D = 500\,°C$. Transient EBC measurements on films about 1 μm thick deposited on Corning 7059 glass

6. DETERMINATION OF SPECIFIC MATERIALS PROPERTIES

fitted with Mo bottom and Al top electrodes were performed using 10–30 ns pulses of electrons of about 5 keV energy. In most specimens the transient current pulse of generated excess electrons had a reasonably well defined discontinuity as shown in Fig. 6.20(a), associated with the transit of a coherent group of carriers [Fig. 6.20(b)]. The transit time (T) as defined by this feature was found to be inversely proportional to the applied bias (v), and a plot of T^{-1} against v yielded a straight line from which μ could be found. The transient current pulse also contained a slower component [Fig. 6.20(c)] attributed to thermal release of electrons from a distribution of trapping centres. The observed pulse shape in (a) arises by a superposition of those shown in (b) and (c) (Spear and Le Comber, 1972).

The drift mobility of excess electrons as a function of temperature was measured on a number of specimens prepared at different deposition temperatures. The results are shown in Fig. 6.21, and it can be seen that, although there is no systematic variation of μ with θ_D, all the mobility values for different specimens can be described by two regions having different activation energies. In region 1, above about 250 K, the activation energy is between 0·17 and 0·19 eV, whereas below this temperature there is a fairly sharp transition to region 2 where the activation energy is between 0·08 and

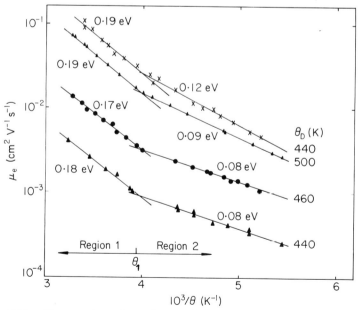

Figure 6.21. Temperature dependence of electron drift mobility for four a-Si specimens prepared by the glow decomposition of SiH_4 using the values of deposition temperature (θ_D) indicated. (Le Comber et al., 1972)

0·12 eV. The modes of electron transport in these two regions will be analysed separately.

In region 1, the clearly defined time T implies that the excess electron packet in transit across the specimen remains fairly coherent. If these carriers move in the extended states above ε_C, this condition can be fulfilled only if the carriers interact through trapping and thermal release with states in a narrow energy range. When there is a distribution of electron states with energy, it is essential that the trapping levels involved correspond to a maximum in the distribution of trapped carriers, at least for the duration T. This is related to the density of states function $g(\varepsilon)$ and the Fermi–Dirac distribution probability appropriate to a quasi-Fermi level ε_F' determined by the excitation provided by the initial pulse of ionizing electrons. With ε_F', corresponding to the transient excess carrier densities used in the experiments, the trapped carrier density is given by

$$n_t(\varepsilon) = g(\varepsilon)\{1 + \exp[(\varepsilon - \varepsilon_F')/k\theta]\}^{-1} \qquad (6.3)$$

where $g(\varepsilon)$ is measured by independent field-effect studies on similar specimens. Figure 6.22 shows the derived trapped carrier density distribution $n_t(\varepsilon)$ using equation (6.3). It can be seen that indeed $n_t(\varepsilon)$ passes through a maximum near ε_A, and its occurrence is possibly linked with the rapid drop in $g(\varepsilon)$ marking the transition between tail states and gap states (Le Comber et al., 1972).

It was shown earlier that the trap-controlled drift mobility for a narrow energy range is given by

$$\mu = \mu_0(N_c/N_t)\exp(\times \Delta E/k\theta) \qquad (6.4)$$

where N_c is the effective density of states in the conduction band, N_t is the density of trapping centres, μ_0 is the electron mobility in the absence of

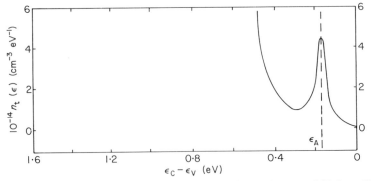

Figure 6.22. Model used in the interpretation of the electron drift in a-Si. The figure shows the distribution $n_t(\varepsilon)$ of electrons trapped in localized states, using equation (6.3). (Le Comber et al., 1972)

trapping, and ΔE is the thermal activation energy for the detrapping process. [This equation is effectively identical with (5.36) explained in Chapter 5, but it is renumbered here for easier reference.]

The value of μ_0 can be taken as the value appropriate to electron motion in the extended states above ε_C, and the activation energy ΔE as $(\varepsilon_C - \varepsilon_A)$. Taking the factor N_c/N_t as equal to $g(\varepsilon_C)/g(\varepsilon_A)$, and with the measured value of $\Delta E = 0.18$ eV, equation (6.4) gives $\mu_0 = 3$ cm^2 V^{-1} s^{-1} in reasonable agreement with expectation. A similar analysis, assuming a linear distribution of trapping states with its origin at ε_A, gives $\mu_0 = 16$ cm^2 V^{-1} s^{-1}. Thus a value of $\mu_0 \approx 10$ cm^2 V^{-1} s^{-1} may be assumed for electron conduction in the extended states at room temperature.

Below 250 K, the generated electrons spend most of their time within the state distribution defined by the region between ε_C and ε_A. Motion of carriers in this region can proceed only by phonon assisted hopping. Here the hopping mobility is given by an expression of the form

$$\mu_H = (\mu_H)_0 \exp(-W/k\theta) \tag{6.5}$$

where $(\mu_H)_0$ is only weakly temperature-dependent. The hopping activation energy W was found to be about 0.1 eV from the experiments. The hopping mobilities of between about 10^{-2} and 10^{-3} cm^2 V^{-1} s^{-1} are in satisfactory agreement with estimated values from the parameters.

Measurements of the electron mobility on doped and compensated glow discharge a-Si provided further insight into the mobility mechanisms involved (Allan et al., 1977). The proper interpretation now requires $(\mu_H)_0$ to be expressed fully as

$$(\mu_H)_0 = (qR^2/6k\theta)\nu \exp(-2\alpha R)\exp(-W/k\theta) \tag{6.6}$$

where R is the average hopping distance, ν is a characteristic frequency of the order of 10^{10} Hz and α^{-1} is the spatial average atomic orbital wavefunction radius in the Slater nodeless approximation. Equation (6.5) is also applicable to donor band hopping in a-Si doped with n-type impurities if R in equation (6.6) is equated to the mean donor separation R_D. The latter is related to the donor density N_D by

$$R_D \approx 0.7 N_D^{-1/3} \tag{6.7}$$

Experimentally, the specimens were prepared using the equipment shown in Fig. 3.1 to provide a variety of samples with different densities of B-acceptors and P-donors. The substrate was held at 550 K and the proportions of PH$_3$ or B$_2$H$_6$ to SiH$_4$ by volume in the starting gas ranged from 57 to 1840 parts per million; the proportion of B$_2$H$_6$ was always very slightly greater than that of PH$_3$. Specimen conductivities were about 10^{-11} ohm^{-1} cm^{-1}, and R_D ranged from 39 to 90 Å.

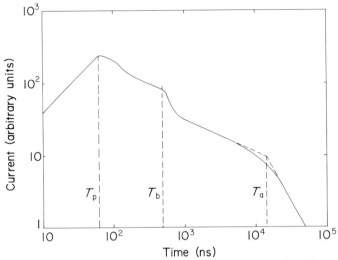

Figure 6.23. Double-logarithmic plot of current transient at 295 K for a lightly doped sample, showing transits at T_a and T_b; T_p marks the end of the excitation pulse. (Allan et al., 1977)

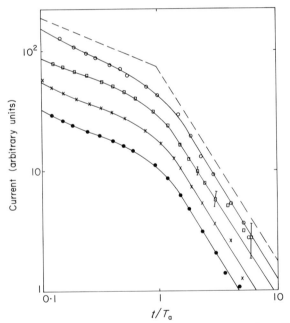

Figure 6.24. Current transients for different temperatures and applied bias plotted logarithmically. Curves are normalized to T_a, the time denoting the intersection of the broken lines; they have slopes of -0.4 and -1.6. (Allan et al., 1977)

Transient EBC methods were used to examine the electron transport by using excitation pulses of 60–100 ns duration and $V_p = 5$–7 kV. Figure 6.23 depicts the transient current pulse on a double-logarithmic scale to show the features observed over an extended time interval. Transient current signals were investigated within the interval 230–370 K, which is the temperature range that had previously been established for undoped specimens where electron transport was shown to occur by a trap-controlled mechanism via extended states above ε_C. Generally, the current transients seen for excess electrons were highly dispersive, and the transit time T_a could best be defined by a plot of log i against log t in accordance with the Scher–Montroll theory. Figure 6.24 consists of a family of such curves, normalized to T_a; it shows that they can be represented by pairs of straight lines which intersect at $t = T_a$, and furthermore that they retain their shape as the applied field or the temperature is varied. By taking T_a as the transit time, the mobility could be derived; the results are summarized in Table 6.2. The table details the

Table 6.2. Summary of specimen data for electron transport; the gaseous doping ratios are given in volume parts per million (vppm).

	PH_3/SiH_4 (vppm)	B_2H_6/SiH_4 (vppm)	N_D (cm^{-3})	R_D (Å)	μ_a (cm^2 V^{-1} s^{-1})	W (eV)	μ_b (cm^2 V^{-1} s^{-1})
1.	57	65	4.7×10^{17}	90	3×10^{-5}	0.1	8×10^{-2}
2.	98	115	8.5×10^{17}	74	9×10^{-5}	0.04	5×10^{-3}
3.	131	150	1.1×10^{18}	67	2×10^{-4}	0.05	—
4.	412	484	3.0×10^{18}	49	5×10^{-4}	0.05	—
5.	1690	1840	5.8×10^{18}	39	8×10^{-4}	0.06	—
U.	—	—	—	—	—	—	1.3×10^{-1}

ratios of PH_3 and B_2H_6 in the SiH_4 gaseous mixture used for preparing six of the samples (U = undoped), the derived values of R_D, and measured values of the room-temperature electron mobility μ_a as defined from T_a and the mobility activation energy W. As a check of equation (6.6), which should be representative of electron hopping between donor sites, $\log(\mu_a/R_D^2)$ is plotted as a function of R_D (Fig. 6.25). It can be seen that this yields a straight line for a range of more than two orders of magnitude of μ_a/R_D^2. The gradient yields a value $\alpha^{-1} = 22$ Å in good agreement with other measurements for phosphorus donors in crystalline silicon, and the intercept $\nu = 10^{11}$ Hz for an average value of W taken as 0.05 eV. There is strong evidence for interpreting the mobility μ_a as donor site hopping where the electron makes phonon-assisted uncorrelated jumps between sites associated with donor impurities. Allan points out that the close similarity between α^{-1} in amorphous and crystalline silicon suggests a similarity in the

216 ELECTRON BOMBARDMENT INDUCED CONDUCTIVITY

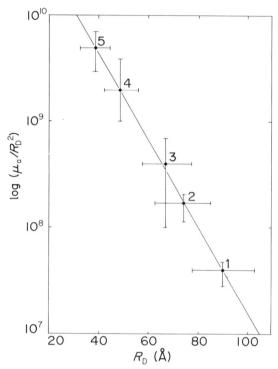

Figure 6.25. Experimental points (the numbers refer to Table 6.2) plotted to show the applicability of equation (6.6) to the observed donor hopping. (Le Comber *et al.*, 1979)

environment of the phosphorus impurities in both phases. The value of $W \approx 0.05$ eV is similar to the value of 0.08 eV deduced for tail state hopping in undoped a-Si below about 250 K, and it suggests a similar disorder potential.

If we now examine the shape of the transient current pulse (Fig. 6.23) at short intervals after excitation, another feature can be seen at $t = T_b$. This was particularly prominent for the lightly doped specimens (samples 1 and 2 in Table 6.2), and by taking T_b as the transit time for carriers moving in an alternative path, a higher mobility value μ_b could be defined. In Fig. 6.26 the donor state hopping mobility μ_a is plotted as a function of $1/R_D$ together with values of μ_b for the two most lightly doped samples and the room-temperature electron mobility in an undoped sample prepared at the same substrate temperature. It can be seen from the figure that the mobility pattern of μ_a follows a smooth path as R_D is varied whereas μ_b fairly clearly belongs to another class. The point in the figure corresponding to the undoped specimen can be grouped fairly well with these others, whose

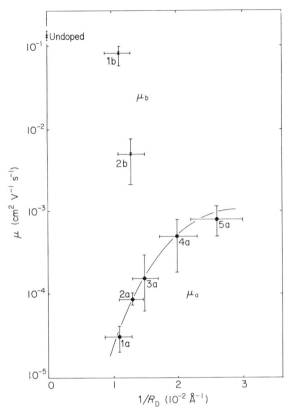

Figure 6.26. The mobilities μ_a and μ_b at 295 K as a function of inverse donor separation R_D. The numbers on the experimental points refer to Table 6.2 and they increase with increasing donor density. (Le Comber et al., 1979)

mobility values approach those appropriate to the trap-controlled mechanism already described for the undoped specimens. To observe transport in two parallel conduction paths it is necessary that complete thermal equilibrium between the electrons in the two groups of states does not occur during transit. From the observed transient EBC pulse shapes, however, it is fairly clear that an appreciable fraction of electrons initially excited into the donor band remain in those states for a time of at least T_a. In the measurements T_a was typically 10–20 μs and this suggests that electron transitions between donor states and extended states above ε_C or to localized states near ε_A are rare.

Although a signal due to the motion of generated excess holes was reported for undoped specimens (Le Comber et al., 1972), there was not a sufficient degree of coherence for the mobility then to be studied by trans-

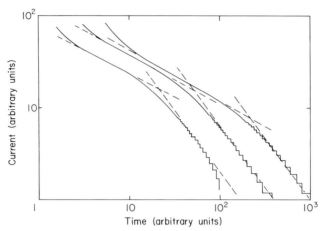

Figure 6.27. Log i against log t plots for a heavily doped sample of a-Si at 335 K; the curves have been repositioned for clarity. The transit times are 38, 105, 220 μs for applied bias $v = 6$, 3, and 1·5 V respectively, and the pairs of tangents to the curves have slopes of $-0\cdot 55$ and $-1\cdot 37$. The structure is due to the transient recorder. (Allan, 1978)

ient EBC methods. In 1972 the Scher–Montroll method of double-logarithmic plots had not been established. A low mobility was deduced and a hopping mechanism suggested but no measurements were made. However, extensive measurements of the mobility of holes in doped compensated a-Si were reported by Allan (1978), and he reported measurements on an undoped specimen which showed that the general pattern of behaviour is the same.

Problems due to rather featureless transient EBC current pulses were overcome by determining T from a Scher–Montroll plot of log i against log t. Experimentally this was achieved by feeding the transient signal into a Datalab/DL905 transient recorder and then passing the stored pulse to a Bryans Southern Instrument X-Y plotter fitted with logarithmic amplifiers. Figure 6.27 shows a family of such curves for a sample prepared from glow discharge in SiH_4 to which PH_3 and B_2H_6 had been added in volume ratios of 1000 and 1150 parts per million respectively. As with electron transits in similar specimens, the log-log plots could be resolved into pairs of straight lines which intersected at the transit time T. All the transit times were found to vary as v^{-1}, in marked disagreement with the Scher–Montroll theory which predicts a field-dependent effective mobility. Further measurements on the same sample, at temperatures spanning the range 250–400 K, all showed the same v^{-1} relationship, so it was possible to define an unambiguous value for the hole drift mobility as a function of temperature. The temperature dependence of μ_p is shown in Fig. 6.28 where it is seen that the

6. DETERMINATION OF SPECIFIC MATERIALS PROPERTIES

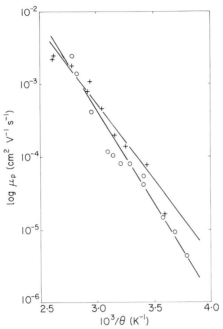

Figure 6.28. Temperature dependence of μ_p in a-Si. Plots are for samples 2 (○) and 5 (+) given in Table 6.3. (Allan, 1978)

mobility is an activated process over several orders of magnitude. Neither μ_p nor its activation energy ε_p shows any systematic dependence on doping level, thus indicating that the observed current path for holes is not linked with the acceptor levels.

The hole transport is interpreted as motion in hole states below ε_V, but trap-controlled by interaction with localized states. The depth of the traps is approximately 0·43 eV. Experimental results for hole transport, and specimen details, are summarized in Table 6.3.

As with electron transport, the presence of this centre can be explained in terms of the measured density of states function. This was determined on similar specimens prepared in the same way by Madan *et al.* (1976) and Madan and Le Comber (1977). The difference ε_C to ε_V is taken as 1·6 eV. With the experimental conditions used by Allan in his transient EBC measurements, a hole conduction current of 5 μA flowed through a specimen with $v = 2$ V, and the density of excess holes was calculated from the measured drift mobility as approximately 5×10^9 cm^{-3}. This leads to a quasi-Fermi level about 0·54 eV above ε_V under the excitation conditions employed. Then, using an analysis similar to that just described for electrons, the density of occupied states can be derived using the appropriate

Table 6.3. Data for the a-Si specimens used in hole drift mobility experiments; L is the thickness of the a-Si film, and the gaseous doping ratios are given in volume parts per million (vppm).

	L (μm)	PH_3/SiH_4 (vppm)	B_2H_6/SiH_4 (vppm)	μ_p (cm^2 V^{-1} s^{-1})	ε_p (eV)
U.	0·82	—	—	$1·5 \times 10^{-5}$	0·3
1.	1·3	28·5	32·5	$1·3 \times 10^{-4}$	0·5
2.	1·18	49·0	57·5	5×10^{-5}	0·54
3.	0·93	65·5	75·0	$1·3 \times 10^{-4}$	0·35
4.	1·13	111	128	2×10^{-4}	—
5.	1·18	500	560	7×10^{-5}	0·45
6.	2·15	1000	1150	2×10^{-5}	0·45

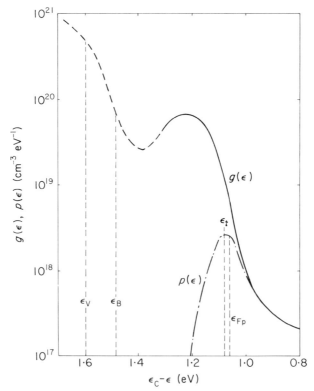

Figure 6.29. Part of the $g(\varepsilon)$ distribution for a-Si. The full curve is deduced from experiment; the dashed curve is an extrapolation beyond the experimental region. The dash-dot curve is the $p(\varepsilon)$ distribution as described in the text. (Allan, 1978)

Fermi–Dirac distribution function. The results are given in Fig. 6.29 where it can be seen that there is a pronounced peak in the trapped hole density $p(\varepsilon)$ at an energy 0·52 eV above ε_V. The trapping centres involved in the hole transport must lie within this state maximum.

No evidence for acceptor band hopping could be observed in boron-doped specimens.

Disagreement with the predictions of the Scher–Montroll theory were most pronounced when it was found that even the most dispersive transient EBC hole pulses showed a linear relationship between T and v^{-1}. The specimen thickness-dependent effective mobility predicted by the theory could not be put to test owing to the rather limited range of L which for the different specimens fell only within the small range 0·82–2·15 μm.

A slightly different approach to the problem of a dispersive transport involving a large number of trapping–detrapping events as related to hole mobility in a-Si was described by Marshall and Allan (1979). This considers that trapping and trap release are stochastic processes whose probabilities can be written as proportional to the logarithm of a pseudo-random number between zero and unity. Thus the free carrier lifetime was written as $t' = -\tau_f \ln X'$, where $0 < X' < 1$. Similarly, the release time was written as $t'' = -\tau_r \ln X'' \exp(\Delta E/k\theta)$, where X'' is another random number such that $0 < X'' < 1$. The logarithmic process tends to emphasize the importance of small values of t' and t'' in relation to large values, and thus the values of t' and t'' defined in this way are not unbiased. By following the progress of a large number of holes in transit across a specimen, and then averaging the detailed passage of any one of them after trapping and detrapping in times defined like this, it was possible to compute the shape of a current pulse. It was shown that a Scher–Montroll plot of the simulated transient current pulse would indeed yield a pair of straight lines, and the intersection of these lines was clearly identified with the arrival of the fastest carriers at the far electrode.

These Monte Carlo simulations also indicated that equilibration of the injected charge carriers is incomplete. It was shown that at $\theta = 666$ K a pulse of holes that has reached equilibrium with trapping centres to a depth of 0·6 eV will be characterized by an effective mobility approximately twice the ultimate value. However, at $\theta = 285$ K the same carrier packet will possess an apparent mobility of about 400 times the equilibrium value. This is useful in emphasizing the marked effect of temperature on the transient current pulse shape for a trap-limited drift mobility involving trapping centres with a broad energy distribution.

The simulation data predict a hole mobility two orders of magnitude smaller than the measured values, but activation energies between 0·43 and 0·74 eV which encompass the measured value of 0·45 eV. The simulation data furthermore predict significantly more field-dependence of mobility than is observed. Also, the slopes of the observed and simulated straight lines on the log-log plots were different; no systematic ratio between the two was found. Thus we can almost certainly rule out the possibility of a range of trapping–detrapping times such as might occur if there was a wide range of hole traps with which the drifting carriers interact, although it is coincidentally constructive to note that this analysis does predict qualitatively many features of the Scher–Montroll theory for the continuous time random walk. However, if the trapping cross-section is independent of energy, the simulation data and the measured values for the density of trapping centres are in good agreement with the values shown in Fig. 6.29 using a much simpler analysis.

III. III–V intermetallic compound semiconductors

Only a few members of this important class of semiconductors have been investigated by transient EBC techniques. In about 1951 these compounds were suggested as suitable alternatives to silicon or germanium in device applications, but in the succeeding years they have displayed unique properties to favour their application in related but not always equivalent devices. Examples may be cited of gallium arsenide avalanche diodes. Schottky barrier or junction diodes, transistors and field-effect transistors which are closely similar to the germanium or silicon counterparts; the very high electron mobility in GaAs makes it more suitable than germanium at high

Table 6.4. Transport properties of some III–V compounds

Material	Low-field carrier mobility ($cm^2\ V^{-1}\ s^{-1}$)	Carrier saturation velocity ($cm\ s^{-1}$)	Threshold field for NDM
GaAs	$\mu_e = 8800$	2×10^7 at 300 K	$3 \cdot 5\ kV\ cm^{-1}$
	$\mu_h = 400$	—	—
GaP	$\mu_e = 300$	—	—
	$\mu_h = 100$	—	—
InP	$\mu_e = 4600$	$1 \cdot 8 – 2 \cdot 8 \times 10^7$ at 300 K	$5–10\ kV\ cm^{-1}$
	$\mu_h = 150$	—	—
GaSb	$\mu_e = 4000$	—	(no NDM)
	$\mu_h = 1400$	—	—
InAs	$\mu_e = 3 \cdot 3 \times 10^4$	—	$1 \cdot 6\ kV\ cm^{-1}$ at 14 kbar
	$\mu_h = 460$	—	—
InSb	$\mu_e = 7 \cdot 8 \times 10^4$	$5 \cdot 6 \times 10^7$ at 77 K	$50\ kV\ cm^{-1}$
	$\mu_h = 750$	—	—

frequencies. However, probably the most significant property of this and other III–V compounds is a region of negative differential mobility (NDM). Also, GaAs can be used for efficient light-emitting diodes in the near infrared. The first of these two properties sparked off a great deal of interest, especially after the announcement (Gunn, 1963) that solid-state diode microwave oscillators could be made from GaAs. However, further studies of related semiconductors showed that NDM is not unique to III–V solids.

Direct measurements of the velocity–field characteristics have been performed by the transient EBC method only in GaAs, InSb, and InP. Alternative measurements of the carrier transport properties of other III–V compounds have been made; the collected results are reviewed in Table 6.4.

A review article by Alberigi Quaranta *et al.* (1971) mentions that NDM has been observed in GaAs, InP, InAs, ZnSe, and CdTe. It has also been observed in silicon at low temperatures and in germanium along the $\langle 100 \rangle$ direction below 130 K.

A. Gallium arsenide

For several years the only reliable measurements of the velocity–field characteristics of electrons in GaAs were those made by conventional transient EBC methods on fairly high resistivity material. The method employed was very similar to that shown schematically in Fig. 5.32 and described in Chapter 5. The specimens were typically of resistivity between 10^6 and 10^8 ohm-cm. An n^+ contact to one face was made by alloying Au and Ge at 500°C, and a blocking contact to the other by first evaporating SiO_2 about 200 Å thick followed by evaporated gold 500 Å thick. The sample thickness was about 300 μm and the area bombarded by energetic electrons was 3 mm in diameter. As shown in Fig. 5.32, the specimen mount was at one end of a parallel coaxial transmission line of 50 ohm impedance. The equivalent capacitance of the specimen and its supports was maintained below 2 pF in order to keep the system rise time less than 100 ps. Figure 6.30 shows typical sample current waveforms, and the derived drift velocity results are given in Fig. 6.31. The measurements were limited to fairly low fields, because of the danger of avalanche breakdown. Ruch and Kino (1967, 1968) discussed these results and concluded that, although they apply to high resistivity GaAs, the velocity–field characteristic should be independent of the sample resistivity provided that they all had the same value of positive mobility at low fields.

Measurements on low resistivity epitaxial layers of GaAs were made by Evans, Robson, *et al.* (1972, 1974) using the microwave transient EBC technique described in Figs 5.35–5.37. Since all the measurements are relative to one for which the drift velocity can be determined independently,

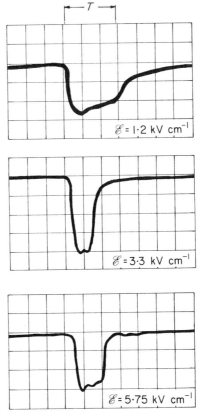

Figure 6.30. Typical sample current waveforms for GaAs excited by a short electron pulse; horizontal scale, 1·3 ns/div. (Ruch and Kino, 1967)

this was undertaken in a similar way to that just described. The 50 ohm coaxial mount fed straight into a sampling oscilloscope having a rise time of 25 ps. The electron excitation pulse was less than 25 ps long. Samples of resistivity 78 ohm-cm and thickness 10 μm were grown on a GaAs substrate. A gold dot 150 Å thick was evaporated on to the epitaxial layer to form a Schottky barrier. (This contrasts with the experience of Ruch and Kino who were unable to make a non-injecting contact without a thin layer of SiO_2 first.) Despite a slight distortion in the shape of the transient EBC pulse due to the capacitance of the sample and its mount, the transit time could still be measured. Concurrent measurements were made using the microwave phase bridge, and also drift velocities where electric field taper within the sample was taken into account. It was found that all their results were in good agreement with each other but their measured saturation drift velocity of 1·4

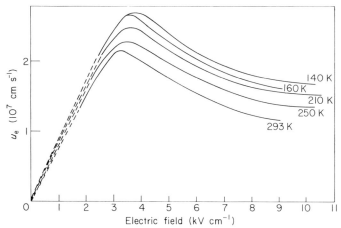

Figure 6.31. Drift velocity of electrons at different temperatures in high resistivity GaAs, derived from the transient EBC results. (Ruch and Kino, 1968)

$\times 10^7$ cm s^{-1} was substantially smaller than Ruch and Kino's value of $2 \cdot 0 \times 10^7$ cm s^{-1}. However, it was noted that the low-field electron mobility was lower in their epitaxial sample than in the former.

Comprehensive measurements of a similar kind but concentrating on the direct time-of-flight were reported by Houston and Evans (1977). The specimens were similar to those just described, and the exciting electron pulse was incident on the GaAs through the gold barrier. Applied fields up to about 10^5 V cm^{-1} were used at room temperature. The electron drift velocity is shown in Fig. 6.32, where it can be seen that the results obtained at lower fields by Ruch and Kino meet the extrapolated values well.

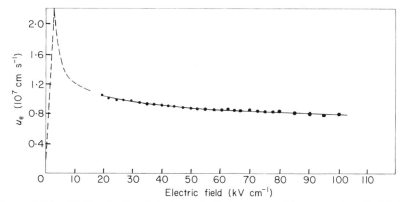

Figure 6.32. Drift velocity of electrons in GaAs over a wide range of applied fields. (Houston and Evans, 1977)

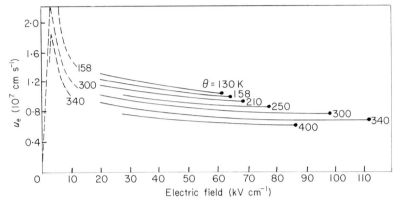

Figure 6.33. Velocity–field characteristics of electrons in GaAs at various temperatures. (Houston and Evans, 1977)

The saturation drift velocity beyond the peak is weakly temperature dependent. The velocity–field characteristics of electrons in GaAs at various temperatures within the range 130–400 K are shown in Fig. 6.33.

B. Indium antimonide

This material has such a short dielectric relaxation time even at 77 K that it is extremely difficult to employ conventional transient EBC methods for measuring the carrier drift velocities. However, Neukermans and Kino (1970) used a method that is really a hybrid of the Haynes–Shockley method and the Spear transient EBC technique. The same short electron pulse excitation as that just described was used. The specimens were 15 ohm-cm p-type InSb with a low-field value of $\mu_h = 1000$ cm^2 V^{-1} s^{-1} at 77 K. The front and back contacts were made by alloying Cd to a depth between 0·5 and 0·7 μm followed by a 500 Å overlay of In. Their size was approximately 1·5 mm × 1·5 mm × 0·6 mm. With these Cd contacts it was found possible to apply fields as high as 650 V cm^{-1} without excessive electron injection.

Excitation pulses consisted of short bursts of electrons of 11 keV energy and about 300 ps long. The observed drift velocity was the ambipolar velocity of the carriers (van Roosbroeck, 1953). Since the mobility of holes is about 60 times smaller than that of electrons at 77 K this was substantially equal to the drift velocity of electrons. Care had to be taken to ensure that the generated electron density was small compared with the background hole density. Figure 6.34 shows the shape of the transient EBC current pulse. In these experiments the bombarded electrode was biased positively with respect to the back but, although the generated excess electrons were

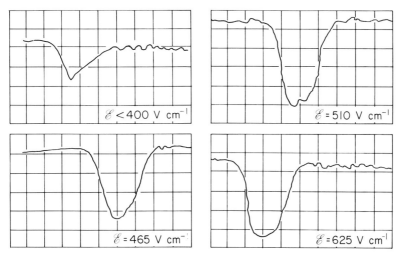

Figure 6.34. Shape of transient EBC pulse in InSb (horizontal scale, 500 ps/div.). Owing to the short dielectric relaxation time, the excess current is mainly carried by holes despite the direction of the applied field. The transit time is equal to that of electrons traversing the sample. (Neukermans and Kino, 1970)

made to travel across the specimen, the current was mainly carried by holes which entered from the back electrode. The excess current $\Delta I = (Q/L)(\mu_e + \mu_h)\mathscr{E}$ flowed as long as the electrons were traversing the sample. Thus the length of the observed pulse yielded the transit time of electrons despite the direction of applied bias. Also, in the region of velocity saturation. Neukermans and Kino showed that the ambipolar velocity was essentially equal to the electron drift velocity.

Under the experimental conditions used, the trapping rates were about 10^9 s^{-1} because the fast traps were unfilled. It was therefore possible to carry out accurate measurements of the velocity–field characteristics which would not otherwise have been possible with longer excitation pulses. It was found that the transient EBC pulse shape became squarer as the field was increased and trapping was overcome. At around 460 V cm^{-1} the pulse shape was sufficiently clear to allow an accurate measurement of the transit time of electrons. Above about 650 V cm^{-1}, electron injection from the negative contact in the absence of excitation made the measurements impossible despite their using a pulsed bias field only 200 ns long. The experimental results are shown in Fig. 6.35; sketched on the same diagram are the results of a single-band model calculation by Persky and Bartelink (1969) and an interband transfer model by Fawcett and Ruch (1969). Good agreement with the latter calculation lends support to an interpretation of NDM based on transfer of electrons from the (000) band minima to the (111) subsidiary

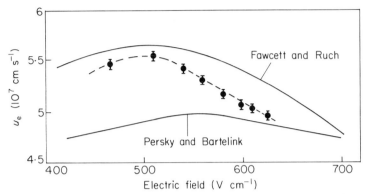

Figure 6.35. Velocity-field characteristics of electrons in InSb at 77 K. The theoretical calculations by Persky and Bartelink (one band NDM) and by Fawcett and Ruch (one band NDM and transfer) are also indicated. (Neukermans and Kino, 1970)

minima, in agreement with the conclusions by Smith *et al.* (1969) on the pressure dependence of the threshold field.

IV. II–VI compounds

A. Cadmium sulphide

The first member of this class of semiconducting insulators to be studied to an intensive degree was CdS, mainly owing to the successful technique pioneered earlier by Frerichs (1947, 1949) for growing good quality single-crystal platelets from the vapour phase in an atmosphere of H_2S and H_2. The starting materials were of carefully controlled purity and it was subsequently found that sulphur vacancies could be compensated by adding halide dopants such as chlorine or iodine. Very closely self-compensated specimens of high resistivity could be produced in this way. The halogen could either be added after crystal growth by low temperature diffusion, or be added to the gaseous atmosphere during growth and thereby assist also in the vapour-phase transport of the reactants. It has since been discovered that the property of self-compensation is common to all the II–VI compounds, and it probably also occurs to some extent in the III–V semiconductors. The mechanism of self-compensation can be explained in the case of CdS as follows. Normally, an uncompensated CdS crystal as grown will contain a number of sulphur vacancies in the lattice due to a stoicheiometric excess of Cd; this results from the kinetics of the vapour-phase reaction. These S vacancies act as donors, and thus the CdS so produced is an n-type extrinsic semiconductor. If now halogen atoms are allowed to fill the vacancies in substitutional S sites, an agglomeration of two such centres will restore charge neutrality. This process of compensation (i.e. two halogen atoms substitutionally per sulphur vacancy) continues until the negative charge of halogen impurities exactly balances the charge due to the stoicheiometric excess of cadmium. The process is self-regulating in so far as it is virtually impossible to make the CdS p-type by addition of too much halogen. However, halogen atoms in interstitial lattice positions act as donors.

The electrical transport of generated electrons and holes in CdS platelets was studied by the transient EBC technique by Spear and Mort (1962, 1963). In order to avoid entry of carriers via the bombarded electrode after the end of the excitation pulse, these workers found it necessary to use a blocking top contact and a short pulse electrical bias field. The blocking contact consisted of a thin insulating Pyrex glass blown film about 0·5 μm thick, and the back contact was a thin conducting evaporated metal layer of Ag or Au.

The drift mobility of electrons at fairly low values of applied field is shown in Fig. 6.36 for three specimens. The curves were obtained by fitting the experimental mobility results at different temperatures to the trap-controlled mobility formula

$$\mu_e = \mu_0[1 + (N_t/N_c)\exp(\Delta E/k\theta)]^{-1} \quad (6.8)$$

(see Chapter 5). Here ΔE is a dominant level of electron trapping centres at a distance of ΔE below the conduction band edge, N_t is their density, and μ_0 is

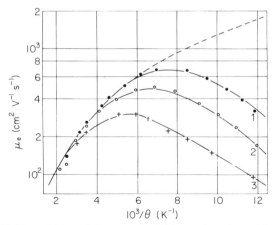

Figure 6.36. Drift mobility of electrons in CdS as a function of inverse temperature. The three experimental curves are for crystals from different sources; the dashed line (μ_0 = lattice mobility) is an extrapolation using equation (6.8). (Spear and Mort, 1963)

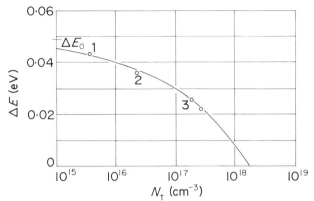

Figure 6.37. Dependence of trap depth ΔE on trap density N_t; the numbered points refer to the corresponding curves in Fig. 6.36. The drawn curve represents $\Delta E = \Delta E_0 - a N_t^{1/3}$ with $\Delta E_0 = 0.049$ eV and $a = 4.2 \times 10^{-8}$ cm^{-1} eV. (Spear and Mort, 1963)

the lattice mobility. Consistent results were found when μ_0 was assumed to have a temperature dependence of the form $\mu_0 \propto \theta^{-3/2}$ commonly associated with an acoustic mode phonon scattering mechanism. A least-squares fit of the results to sets of these equations led to a value of $\mu_0 = 265$ cm^2 V^{-1} s^{-1} at room temperature and a temperature dependence of μ_0 given by $\mu_0 = 1.28 \times 10^6 \theta^{-3/2}$ cm^2 V^{-1} s^{-1} where θ is measured in K.

The shallow electron traps which are responsible for the activated electron mobility are situated at about 0·04 eV below the conduction band. The dependence of ΔE on trap density is shown in Fig. 6.37. Following an analysis of a density-dependent activation energy by Pearson and Bardeen (1949), the measured values of ΔE were fitted to a law of the form $\Delta E = \Delta E_0 - a N_t^{1/3}$, where a is a constant. The activation energy of this particular trap at infinite dilution was $\Delta E_0 = 0.049$ eV; although its nature was not ascertained it is conceivably associated with a centre resulting from a self-compensation mechanism occurring during growth.

The low-field hole mobility behaves in a somewhat different way. The transient EBC technique was used to study the mobility in crystals obtained from widely different sources (Fig. 6.38). However, a straightforward trap-controlled mechanism must be ruled out because it would imply that all vapour-phase grown crystals have the same density of defects irrespective of origin. We are thus led to interpret the hole transport in terms of trapping within centres that are a fundamental property of the band structure. The results described here can be attributed to the presence of two closely spaced valence bands. The hole drift mobility was determined under conditions where t' was less than the transit time, by the method of drawing tangents to the $Q(t)$ curve as shown in Chapter 5. The hole drift mobility normalized to

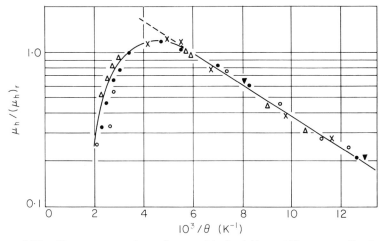

Figure 6.38. Temperature dependence of hole drift mobility normalized to its room temperature value $(\mu_h)_r$, for five specimens from various sources: ● RCA laboratories, $L = 75\,\mu\text{m}$; ▼ AEI laboratories, $L = 100\,\mu\text{m}$; △ RCA laboratories, $L = 85\,\mu\text{m}$; ○ EMI laboratories, $L = 56\,\mu\text{m}$; × RCA laboratories, $L = 67\,\mu\text{m}$. (Spear and Mort, 1963)

its room temperature value is shown in the graph, and in this case the mobility results were fitted to a law of the form.

$$\mu_h = \mu_2[1 + bN\exp(\Delta E/k\theta)][1 + N\exp(\Delta E/k\theta)]^{-1} \quad (6.9)$$

where μ_2 is the hole mobility in one valence band and $\mu_1 = b\mu_2$ is the mobility of holes in an adjacent band. Two closely spaced valence bands separated by 0·02 eV have been deduced from the dichroism of the optical absorption edge (Gobrecht and Bartschat, 1953), and the close agreement with the activation energy of 0·019 eV for the hole mobility is strong evidence to support the interpretation of the fundamental hole transport in terms of motion within two valence bands. This leads to a lattice mobility for holes of about 15 cm² V⁻¹ s⁻¹ at room temperature. This compares with a value of 23 cm² V⁻¹ s⁻¹ reported by Le Comber *et al.* (1966).

Trapping centres in CdS crystals were studied by Bradberry and Spear (1964) who found that crystals (grown either as platelets by vapour phase transport or as boule crystals in a closed tube) could be classified according to the nature of the trapping centres found in them; these were distinguished as type A or type B, and Fig. 6.39 shows the temperature dependence of the electron drift mobility for the two types. The class A crystals are the same as those that yielded the results for Fig. 6.36. The shallow traps influencing the electron mobility in class B crystals were however found to lie at a distance 0·16 eV below the conduction band edge. In particular, it was found that the

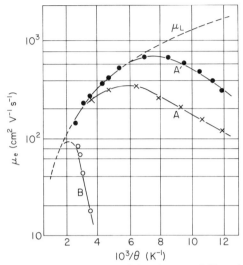

Figure 6.39. Temperature dependence of the electron drift mobility in type A and type B crystals of CdS (μ_L = lattice mobility). Solid curves were calculated from equation (6.8) using the following parameters: curve A, $\Delta E_1 = 0.043$ eV, $N_1 = 3.5 \times 10^{15}$ cm^{-3}; curve A', $\Delta E_1 = 0.030$ eV, $N_1 \approx 10^{17}$ cm^{-3}, $N_2 < 10^{14}$ cm^{-3}; curve B, $\Delta E_2 = 0.16$ eV, $N_2 = 5 \times 10^{16}$ cm^{-3}. (Bradbury and Spear, 1964)

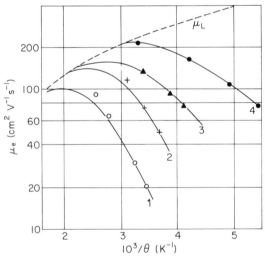

Figure 6.40. Effect of heat treatment on type B crystals of CdS. The solid curves are calculated: curve 1, untreated crystal, $\Delta E_2 = 0.16$ eV, $N_2 = 4 \times 10^{16}$ cm^{-3}; curve 2, short heat treatment, luminescence weakened, $N_2 = 8.7 \times 10^{15}$ cm^{-3}; curve 3, further heat treatment, luminescence quenched, $N_2 = 2.6 \times 10^{15}$ cm^{-3}; curve 4, different crystal, luminescence quenched, $N_2 = 3.4 \times 10^{14}$ cm^{-3}. (Bradberry and Spear, 1964)

density of these centres could be reduced by more than an order of magnitude by moderate heat treatment *in vacuo* at 350°C for 2 hours. This reduction was accompanied by complete quenching of the green luminescence edge-emission at 77 K which always tended to be present in type B crystals before the heat treatment. An analysis was made of the experimental drift mobility in terms of an interaction with two types of centre; in general it was found that $(N_2/N_c)\exp(\Delta E_2/k\theta)$ was much greater than $[1 + (N_1/N_c)\exp(\Delta E_1/k\theta)]$ for type B crystals, and therefore the expression simplified to equation (6.8) by which the results could be analysed in terms of the deeper traps only. Here N_2 and N_1 are the densities of the deeper and shallower electron traps respectively, N_c is the effective density of states at the bottom of the conduction band, and ΔE_1 and ΔE_2 are the thermal activation energies of the two types of trapping centre. Figure 6.40 shows how the heat treatment progressively reduced the density of the 0·16 eV centres and thereby increased the electron drift mobility. The derived values of some of the electron trapping parameters are shown in the caption. It is seen that the density of the deep centres was within the range 4×10^{16} cm^{-3} to $3\cdot 4 \times 10^{14}$ cm^{-3} for the different crystals and for the same crystals at different stages in the annealing process. These and other experiments also showed that, although luminescence is quenched when annealing reduces the density of the 0·16 eV centre, this centre is not directly associated with the luminescence process.

Higher-field measurements of the transient EBC in single crystals of CdS and ZnS by Le Comber *et al.* showed an important velocity saturation effect which occurs when the velocity of the drifting charge packet reaches or exceeds the velocity of sound; this arises principally owing to strong coupling between the carrier and piezoelectric modes of lattice vibration. Coupling both to shear waves and to transverse waves was found, especially if trapping was not too extensive. The limit for saturation effects occurred when the drift velocity had fallen to 0·37 of its trap-free value.

The measured drift velocity of electrons and holes as a function of applied field is shown in Figs 6.41 and 6.42 for some CdS samples. It can be seen that a sharp discontinuity occurs when the drift velocity reaches the velocity of sound. This can be explained in terms of a model that shows how a carrier can be self-trapped by the electric field gradients surrounding it when the lattice suffers a dilatation due to its presence. Thus, if the carrier travels faster than the associated acoustic wave, energy will be transferred from the applied electric field to the acoustic wave. The time taken for the interaction to occur was shown to be less than the resolving time of the apparatus, which was 10 ns.

The acoustoelectric saturation effect in CdS was further studied by Somerford and Spear (1971). These workers used the transient EBC technique to

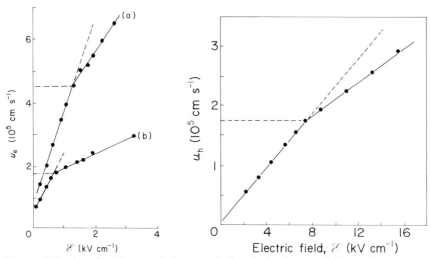

Figure 6.41. (LEFT) Measured electron drift velocity in two CdS specimens as a function of electric field applied (a) parallel and (b) perpendicular to the c-axis. (Le Comber et al., 1966)

Figure 6.42. (RIGHT) Hole drift velocity in CdS, as a function of the applied field, showing interaction of generated holes with piezoelectric modes. (Le Comber et al., 1966)

measure the carrier drift velocities but they also fitted a piezoelectric transducer coupled to the back surface of the specimen to observe the generated acoustic wave. The experimental results confirm the findings by Le Comber et al. Figure 6.43 shows the observed transducer output. The effect of the conditions surrounding the bombarded electrode were also investigated by using steady illumination in conjunction with pulsed electron bombardment. It was possible to deduce that the effect of the barrier field can cause a

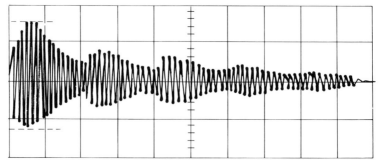

Figure 6.43. Initial part of the observed ultrasonic signal (19 MHz) traced from an oscilloscope photograph. Horizontal scale, 4×10^{-7}s per division; vertical scale, 50 mV per large division. (Somerford and Spear, 1971)

significant increase in local carrier density. This can explain the rapid build-up of the acoustoelectric interaction as predicted by White (1962).

Extensive studies of CdS have been made using direct current EBC as reported in Chapter 3 where the role of X-rays generated in the front electrode was shown to play a significant part in the excitation process. A review of luminescence and trapping in CdS phosphors is to be found in the book by Curie (1963), and of photoconductivity and trapping in the book by Bube (1960).

B. Cadmium telluride

This has probably been studied more than any other II–IV compound apart from CdS. The interest was largely sparked off by the possibility of its use as a nuclear radiation detector (Zanio et al., 1968), an application that will be described in Chapter 7. In common with CdS it also has the property of self-compensation during crystal growth.

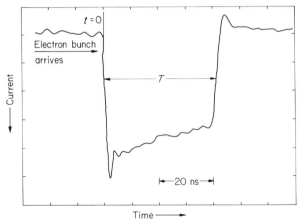

Figure 6.44. Transient EBC pulse shape in a specimen of CdTe observed at room temperature. Applied field 4200 V cm^{-1}; excitation pulse length $\tau = 40$ ns; thickness $L = 1\cdot 9$ mm; drift velocity $u = 4\cdot 75 \times 10^6$ cm s^{-1}; derived mobility $\mu_e = 1130$ cm^2 V^{-1} s^{-1}. (Bell et al., 1974)

The waveform of the transient EBC is shown in Fig. 6.44. The transit time was used to derive the carrier drift mobilities in the range of applied fields below 2 kV cm^{-1} after making sure that the drift velocity was proportional to the field. The drift mobility curve for electrons was fitted to a trap-controlled mobility law. The electron mobility as a function of temperature is shown in Fig. 6.45 and that of holes in Fig. 6.46. A summary of experimental data obtained in halogen compensated samples is given in Table 6.5.

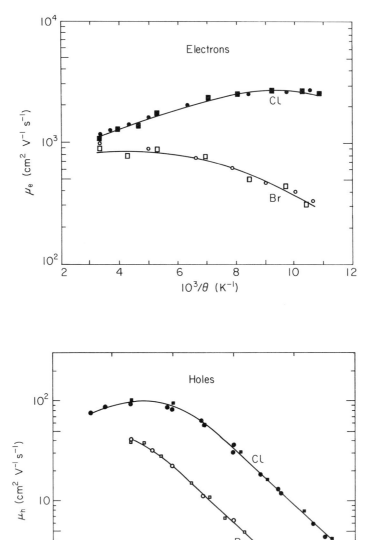

Figures 6.45 and 6.46. Drift mobility of electrons and holes as a function of inverse temperature in two samples of semi-insulating CdTe compensated with Cl or Br; the points represent the experimental transient EBC data and the drawn curves are theoretical. (Bell et al., 1974)

Table 6.5. Summary of experimental results for the activation energy and impurity concentration determined by the transient EBC technique on four CdTe samples compensated with halogen ions.

	Electron traps			Hole traps		
Dopant	depth (meV)	density (10^{16} cm^{-3})	source	depth (meV)	density (10^{15} cm^{-3})	source
Br	25 ± 2	20 ± 2	Br	135 ± 5	18 ± 2	$(V_{Cd}X_{Te})$*
Br	26 ± 2	18 ± 2	Br	142 ± 5	15 ± 2	,,
Cl	28 ± 2	2·4 ± 0·5	Cl	137 ± 5	1·9 ± 0·5	,,
Cl	27 ± 2	2·5 ± 0·5	Cl	140 ± 5	2 ± 0·5	,,
Br	—	—		—	—	Cu?
Br	—	—		—	—	Cu?
Cl	50 ± 5	1 ± 0·5	?	350 ± 10	6·3 ± 0·5	Cu?
Cl	48 ± 5	2 ± 0·5	?	370 ± 10	6·4 ± 0·5	Cu?

* Cd vacancy; halogen on Te site.

Generally it is extremely difficult to assign any given trapping level to a specific impurity, but the deep hole trap at about 0·35 eV above the valence band was tentatively attributed to an unintentional impurity such as copper, at concentrations less than 1 in 10^6, from the growth crucible (Bell et al., 1974).

The self-compensation process in II–VI compounds in general was discussed by Canali et al. (1974). Their explanation of the process is substantially the same as that already discussed for CdS. However, they were able to identify the sources of some trapping centres associated with the compensation. Another way of looking at the same thing is as a solid solution of CdX_2 in CdTe (X = halogen). It can be shown that the hole trap at about 0·14 eV is probably due to the nearest neighbour associate between a cadmium vacancy and a halogen atom. The shallow electron trap is due to a halogen atom. Such an atom, when introduced substitutionally on a Te site, acts as a donor with an activation energy between 25 and 50 meV. The deeper electron trap at about 50 meV in the Cl-doped samples could not be assigned, although it is significant that its presence in the Br-doped specimens could not be detected, possibly because the other traps had an order of magnitude higher concentration and therefore a greater chance of masking their presence.

The density of traps was calculated from the 0·72% value of the transient EBC integrated current pulse rise time T' as described in Chapter 5 (Table 5.2). The density of traps was calculated from equation (5.32) in the form

$$T' = T[1 + (N_t/N_c)\exp(\Delta E/k\theta)]$$

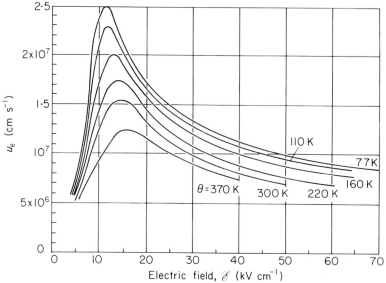

Figure 6.47. Electron drift velocities in CdTe as a function of electric field in the temperature range between 77 and 370 K. The values shown here are the highest measured at every temperature and correspond to samples with a relatively long free lifetime. (Canali *et al.*, 1971b)

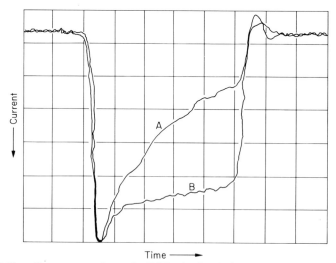

Figure 6.48. Current waveforms in CdTe obtained after short pulse electron excitation showing the electric field dependence of mean free lifetime t', at a field of (A) 7·8 kV cm^{-1} and (B) 32 kV cm^{-1}. The transit times are identical but waveform B shows that the mean free lifetime increases substantially. The horizontal sensitivity is 1·43 ns/div. (Canali *et al.*, 1971b)

assuming that, by the time these measurements were undertaken, T and ΔE were known. The experimental equations and a discussion of equilibria during growth of these specimens were published by Bell et al. (1974).

At high fields, the transport properties in CdTe are more complicated and it is no longer possible to describe their motion in terms of a field-independent mobility. The drift velocity of electrons in the temperature range 77–370 K and at electric fields up to 70 kV cm^{-1} was measured by the transient EBC technique by Canali et al. (1971b). The measured drift velocity of electrons in high resistivity CdTe as a function of field and for various temperatures is shown in Fig. 6.47. The sample resistivity was in the range 10^8–10^9 ohm-cm, and the EBC measurements were performed by 40 keV electrons, using about 10^4 electrons per pulse lasting 100 ps. Specimen contacts were of evaporated gold and the crystal thickness ranged from 0·45 to 1·20 mm. The bombarded area was about 10 mm^2.

The results given in Fig. 6.47 show that the electron drift velocity rises to a peak at an applied field of about 14 kV cm^{-1} before falling with rising applied fields beyond this value. The peak electron velocity increases from $1\cdot25 \times 10^7$ cm s^{-1} at 370 K to $2\cdot5 \times 10^7$ at 77 K; the corresponding value of the electric field decreases from 16 to 11 kV cm^{-1}. Canali et al. point out that all the curves except the one for 370 K correspond to superohmic behaviour below the peak. This is possibly due to scattering by both polar optical phonons and ionized centres. It was found that specimens for which the mean free lifetime t' was large had a high mobility and presumably a lower concentration of ionized scattering centres. When the electric field is increased beyond a certain value, the electrons are not able to dissipate their energy to optical phonons because the scattering mechanism is elastic and their energy increases markedly. This continues until inter-valley scattering between the central conduction band valley and secondary minima occurs. For very high electric fields the electron mobility is probably determined by both inter-valley scattering and polar optical mode scattering.

The transient EBC pulse observed in a specimen for which the applied field was chosen to lie on either side of the maximum, so that the drift velocity was the same, is shown in Fig. 6.48. This trace clearly shows that the mean free lifetime of electrons increases dramatically as the field is increased. This is probably a hot electron effect together with the effect on lifetime due to positively charged attractive trapping centres.

The same techniques were used by Ottaviani et al. (1973a) to study the hole mobility and the Poole–Frenkel effect (field assisted detrapping). The temperature range covered was 130–430 K and the electric fields were between 5 and 50 kV cm^{-1}. The specimens used in the particular series were high resistivity samples produced by compensation with indium. The room temperature resistivity lay within the range 10^8–10^9 ohm-cm. The mean free lifetime of holes in these specimens was between 10 and 40 ns, and the trapping responsible for limiting the free lifetime was due to very deep traps. Indeed, trapped carriers in these centres had such a long thermal release time-constant that Ottaviani and his colleagues were obliged to use subsidiary methods to prevent internal polarization; in this case, a suitable light flash was sent between two successive measurements or by using a pulsed bias voltage. Also, the density of carriers generated by the primary electron excitation was maintained at a low value. Other methods for destroying polarization charges by additional electron excitation pulses are described in connection with the technique for transient EBC measurements on high resistivity materials devised by Spear (see Chapter 5).

In Fig. 6.49 the hole drift mobility is plotted as a function of $\mathscr{E}^{1/2}/k\theta$ for several temperatures. The experimental results cannot be interpreted in terms of ionized impurity scattering (which should show a $\theta^{3/2}$ temperature dependence as well as possible sub- or super-ohmic behaviour). However, at the low temperatures for which these results apply, the data can be fitted to a set of parallel straight lines. The slope of these lines gives the Poole–Frenkel constant, $\beta = q(q/\pi\varepsilon\varepsilon_0)^{1/2}$, where the field-lowering of the potential barrier due to the electric field is given by the law $\Delta E_t = \beta \mathscr{E}^{1/2}$, q is the electronic charge, and ε and ε_0 are the dielectric constant and the permittivity of free space respectively (Hartke, 1968). The experimental value of β was $2\cdot 6 \times 10^{-4}$ eV V$^{-1/2}$ cm$^{1/2}$ which compares well with the theoretical value of

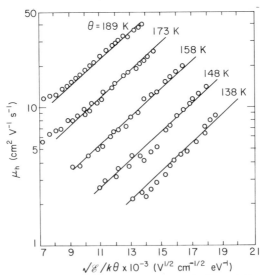

Figure 6.49. Hole mobility in CdTe as a function of $\sqrt{(\text{field})}/k\theta$ at five different temperatures in a sample 205 μm thick. The slope of the drawn lines gives the Poole–Frenkel coefficient (see text). (Ottaviani *et al.*, 1972)

$2\cdot 8 \times 10^{-4}$ eV V$^{-1/2}$ cm$^{1/2}$. The activation energy of the hole trap in question was found to be 150 meV and their density about 5×10^{16} cm^{-3} which was comparable to the density of electron traps in the same specimens. Ottaviani *et al.* tentatively assigned this trap to a cadmium vacancy.

C. Other II–VI solids

1. Cadmium selenide

Drift mobility measurements using transient EBC techniques were undertaken by Canali *et al.* (1972) on single-crystal material grown from the vapour phase by the forced convection method (Corsini-Mena *et al.*, 1971). Good quality very pure semi-insulating platelet crystals about 10 mm × 5 mm × 0·2mm were grown, and contacts were produced by evaporating gold. Pulsed bias was applied in the usual way to avoid heating effects, and the sample was mounted at one end of a 50 ohm coaxial line so that the induced conduction might be displayed directly on a sampling oscilloscope. Measurements were performed both at low excitation levels and at high levels where space-charge-perturbed transits (SCP) could be observed. In both cases the same value of carrier mobility was obtained from the appropriate carrier transit times.

Figure 6.50 shows the drift velocity as a function of applied field for both holes and electrons measured on the same sample. These measurements show that the velocity is directly proportional to the applied field over the range $1\cdot 5 \times 10^3$ to 10^4 V cm^{-1}. The samples yielded the same values for room-temperature mobility, 720 and 75 cm^2 V^{-1} s^{-1} for electrons and holes respectively. The mean free lifetime of carriers with respect to deep traps as measured from the current waveforms was 20 ns for holes and 30 ns for electrons. These results are only in fair agreement with those of Petravicius *et al.* (1977) who, using similar transient EBC techniques, obtained 180 and 20 cm^2 V^{-1} s^{-1} respectively for electron and hole mobility at room temperature, with specimens prepared by static sublimation.

The experimental results were interpreted by Canali in terms of hole polar optical scattering, which plays an important role in the conductivity of holes in II–VI compounds. The ratio of the effective hole masses in the light and heavy hole bands (which are separated in energy by about 0·04 eV) is greater than a factor of 2. It is apparent that the coupling between holes and polar optical mode scattering is dominant; these results also indicate that, at electric fields below about 10^4 V cm^{-1}, the mean energy of heavy holes is very close to the thermal one. This interpretation may be compared with an explanation of the hole transport in CdS in which the split valence band is dealt with as a trapping–detrapping phenomenon, and consistent results can

242 ELECTRON BOMBARDMENT INDUCED CONDUCTIVITY

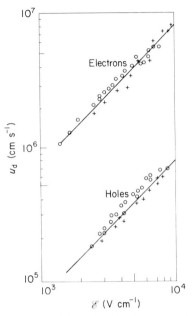

Figure 6.50. Hole and electron drift velocities in two different samples (○,+) of CdSe at room temperature as a function of applied field. The drift mobilities are $\mu_h = 75$ and $\mu_e = 720$ cm^2 V^{-1} s^{-1} (Canali et al., 1972)

be obtained provided also that the density of trapping centres involved is assumed to be a fundamental property of the perfect crystal.

2. Zinc sulphide

Similar determinations to those reported for CdS and CdTe were made on both cubic and hexagonal allotropes of ZnS (Le Comber et al., 1966). These difficult measurements were made on very pure single crystals ranging in thickness between 85 and 1150 μm, and the transient EBC method was employed to measure the electron drift mobility, which ranged from about 280 cm^2 V^{-1} s^{-1} at room temperature to 80 cm^2 V^{-1} s^{-1}; the lower value could be due to shallow trapping, probably by native defects. This is in fair agreement with a value of $\mu_e = 120$ cm^2 V^{-1} s^{-1} deduced by Kröger (1954). The hole mobility could not be measured with any certainty, but it was found to be at least an order of magnitude less than μ_e. The most significant result of the measurements was the observation of an electron drift velocity saturation (or at least a partial saturation effect) at about the velocity of sound. It was well known that ZnS is piezoelectrically active, and these saturation effects were very similar to those found in CdS.

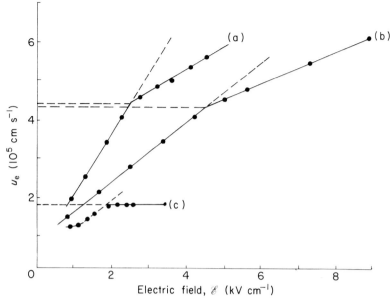

Figure 6.51. Measured drift velocity of electrons in three specimens of ZnS as a function of electric field applied (a and b) along a $\langle 111 \rangle$ direction in the cubic form, and (c) perpendicular to the c-axis in the hexagonal form. (Le Comber *et al.*, 1966)

Figure 6.51 shows the results obtained for both cubic and hexagonal ZnS. It can be seen that in sample (c) the drift velocity of electrons abruptly ceased to rise with increasing applied field, but this was not typical; more often, the velocity–field dependence showed a discontinuity of slope, as in (a) and (b), depending on orientation. As with CdS, it was found that the interaction with acoustic modes stopped when the specimen was cooled sufficiently for the electrons to spend about 0·3 or more of their pseudo-free lifetime in traps.

3. Zinc oxide

As remarked in Chapter 3, ZnO is unstable in a vacuum and tends to lose oxygen by decomposition under the action of electron bombardment. For this reason, all reliable drift mobility measurements reported have been undertaken in air using photoexcitation.

High resistivity single crystals can be produced by compensation with either lithium or sodium, using a high temperature diffusion technique. These atoms enter the lattice substitutionally in zinc sites and act as acceptors to compensate for native defects due to oxygen vacancies (which act as donors). Sodium-doped ZnO was measured by Jakubowski and Whitmore

(1971) using short pulse photoexcitation. By taking the microscopic electron mobility as equal to the Hall mobility, 133 cm^2 V^{-1} s^{-1} at 300 K, they found that the drift mobility was an activated process over the range 240–370 K, with an activation energy of 0·24 eV, and a density of trapping centres between about 5×10^{15} and 2×10^{16} cm^{-3}. These figures compare with an average trap depth of 0·27 eV for lithium-compensated specimens measured by the same technique (Seitz and Whitmore, 1968). The results on both the Li- and Na-compensated specimens were not obtained with material sufficiently pure to permit derivation of any fundamental properties of the carrier transport process.

V. Solid and liquid gases

A. Noble gases (neon, argon, krypton, and xenon)

Electron and hole drift mobility measurements have been made on all the noble gas solids (and some liquids) with the exception of radon. The most extensive and direct measurements of the transport properties of generated excess carriers were made using the transient EBC technique with the apparatus designed for growing the crystals *in situ* in a specimen chamber cooled by a miniature Joule–Kelvin cryostat. The details of this equipment and some comments about crystal growth were given in Chapter 5 (Section IV.B) and in Figs 5.29 and 5.30. The same apparatus was used to measure other solid gases, such as CO, N_2, O_2, H_2, and CH_4; the experimental results for the latter group of solid gases will be described below in Section B.

The first transient EBC results were published by Miller *et al.* (1968), and a comprehensive review of all the results for the transport properties of noble gases was compiled by Spear and Le Comber (1977). Figure 6.52

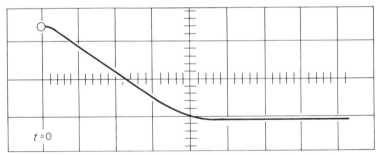

Figure 6.52. Transient EBC integrated current pulse in liquid argon due to the motion of generated electrons (horizontal scale, 50 ns/div.); specimen thickness 360 μm, applied field 1·18 kV cm^{-1}. (Miller, 1967)

shows the observed transient EBC charge pulse in liquid argon after excitation using short electron pulses of 40 keV energy. At low electric fields a mobility could be defined by the slope of the tangent at the origin of a plot of drift velocity versus field. It is clearly shown from the shape of the charge pulses that a transit time could be measured unambiguously. The electron free lifetime (t') was found to depend strongly on the purity of the starting gas and, although commercially available "ultrapure" grade in which the total proportion of impurities was of the order $1 : 10^6$ was used, further purification was essential. Of several methods, an activated charcoal trap at liquid-nitrogen temperature, sodium–potassium alloy at 300°C, and a molecular sieve were all tried. The activated charcoal trap and the molecular sieve were found to be the most successful at removing unwanted impurities, especially oxygen. In the case of xenon, which condenses at liquid-nitrogen temperature, purification was achieved by using a non-evaporable, physically adsorptive, metal alloy getter (CTAM/440; Getters GB Ltd) which operated very successfully to remove oxygen at 400°C despite being already almost saturated by the noble gas. Alternatively, a molecular sieve (Oxysorb; Greisheim GmbH) could be used just as effectively for any of the gases. Crystals were grown from the liquid at pressures and temperatures near the triple point. The rate of crystal growth was as low as 5 mm h^{-1}; the crystal thicknesses were within the range 100–600 μm and the useful face area was about 10 mm^2. The experimental electron drift velocity results for solid argon, krypton, and xenon as a function of applied field are shown in Fig. 6.53. The graph includes drift velocity measurements by Pruett and

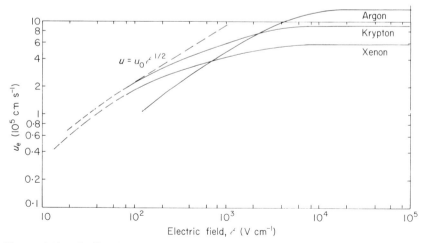

Figure 6.53. Drift velocity of electrons in solid noble gases as a function of applied field. (Miller *et al.*, 1968)

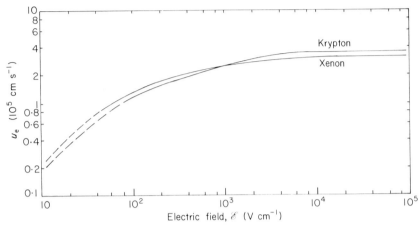

Figure 6.54. Drift velocity of electrons in liquid noble gases as a function of applied field. (Miller et al., 1968)

Broida (1967) who used α-particle excitation, but they were only able to make measurements at high applied fields owing to the low generation efficiency; their drift velocity measurements fit very well into the upper part of the transient EBC curves. Measurements by Miller et al. for the liquids were also undertaken in the same apparatus; the results for krypton and xenon are shown in Fig. 6.54. It can be seen that high values of electron drift velocity at moderate applied fields occur for all the solids and liquids, and the velocities saturate in the region of 10^5–10^6 cm s^{-1}. Theoretically μ_e in the solid would be expected to be twice that in the liquid, and this relationship is found to hold reasonably well.

A physical interpretation of the almost complete velocity saturation was suggested by Spear and Le Comber (1969). The low-field electron mobility gave experimental evidence for control predominantly by acoustic mode scattering; thus $\mu_0 \propto \theta^{-3/2}$ where θ is the absolute temperature. At higher fields, the drift velocity of electrons could be fitted to the hot-electron theory of Shockley (1951), and as more recently developed by Cohen and Lekner (1967). This predicts a velocity proportional to the square root of the applied field. Spear and Le Comber suggested a field-dependent effective mass of the electrons to explain the almost complete velocity saturation that occurs at high applied fields. The modified Shockley theory by Cohen and Lekner predicts a velocity peak which was not observed.

The hole transport in the noble gas solids was reported by Le Comber et al. (1975). Unlike the transport of generated electrons, the hole drift velocity was found to be proportional to applied field up to the highest fields used (50 kV cm^{-1}). This enabled a field-independent drift mobility μ_h to be deduced.

By restricting the thickness of the single-crystal specimens of Ne, Ar, Kr, and Xe to 100 μm or less, it was possible to obtain clear transient EBC pulses whose shape was not distorted by short hole lifetimes; the hole drift mobilities close to their triple-point temperatures are shown in Table 6.6 which also summarizes the low-field electron drift mobilities in the solid and in the liquid.

Table 6.6. Summary of electron and hole drift mobilities in noble gases.

Material	Triple point (K)	Drift mobility (cm^2 V^{-1} s^{-1})		
		electrons (solid)	electrons (liquid)	holes (solid)
Ne	25	600	$1 \cdot 6 \times 10^{-3}$	$1 \cdot 05 \times 10^{-2}$
Ar	84	1000	475	$2 \cdot 3 \times 10^{-2}$
Kr	116	3600	1800	4×10^{-2}
Xe	161	4000	1900	$1 \cdot 8 \times 10^{-2}$

An experimental difficulty occurred in the hole mobility measurements due to a slow build-up of charges in deep traps. Although the usual precaution of using one or more electron pulses under zero bias between forward conduction pulses was used to reduce space charge accumulation, as described in Chapter 5, it was still often necessary to use a pulse repetition frequency as low as 2 Hz or even less. In some instances it was found necessary to make single-shot measurements and to display the transit current on a fast storage oscilloscope (Tektronix 7623). The specimen thickness (which in the case of the solid could not be easily predetermined owing to the flexible top electrode) was deduced from measurements of the transit time of positive ions in the liquid; the mobility of these ions was already known from transient EBC measurements using a rigid specimen chamber. Unavoidably, the flexible electrodes tended to become distorted slightly as the liquid solidified, so these measurements of the absolute values of mobility may be up to 50% in error, except for Kr where the error may be up to a factor of 2. However, the relative errors in a particular temperature run were probably less than 20% for Kr and less than 10% for the other noble gases. The values of μ_h were of the order of 10^{-2} cm^2 V^{-1} s^{-1}, almost five orders of magnitude less than the corresponding electron mobilities. The temperature dependence of deduced mobility values is shown in Figs 6.55 and 6.56.

Le Comber *et al.* (1975) showed that the most probable explanation of the low hole mobilities is carrier localization through formation of the R_2^+

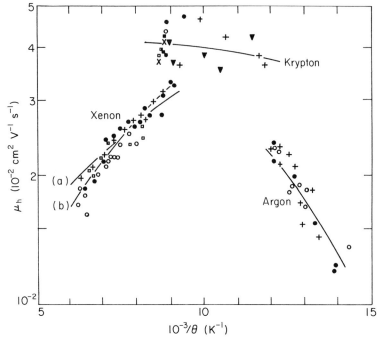

Figure 6.55. Temperature dependence of the hole drift mobility in solid noble gases (Le Comber et al., 1975). The points were measured in a number of experiments using samples ranging in thickness as follows: Ar 50–85 μm; Kr 60–130 μm; Xe 250–385 μm. The curves were calculated from the non-adiabatic theory of Holstein (1959) using equation (6.10) and the following values of the parameters:

Ar:	E_b = 70 meV	J = 12 meV	$\hbar\omega_0$ = 7 meV	γ = 10	a = 3·90 Å
Kr:	40	4	15	2·7	4·12
Xe(a):	20	1·5	25	0·8	4·47
(b):	5	0·76	30	0·17	4·47

molecular ion. This results in small polaron formation involving strong carrier–lattice vibrational coupling. The experimental hole mobility results could be fitted to the small polaron non-adiabatic hopping theory of Holstein (1959), and it was found that reasonable values of the parameters for a good fit were obtained. The expression for the mobility in the non-adiabatic case is

$$\mu = (qa^2/k\theta)(J^2/\hbar^2\omega_0)[\pi/\gamma\mathrm{csch}(\hbar\omega_0/2k\theta)]^{1/2}\exp[-2\gamma\tanh(\hbar\omega_0/4k\theta)] \quad (6.10)$$

where we assume that the carriers interact predominantly with phonons of vibrational frequency ω_0. Here $\gamma = E_b/\hbar\omega_0$ is the dimensionless interaction parameter which gives the strength of the carrier coupling to vibrational

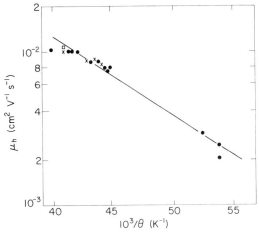

Figure 6.56. Temperature dependence of the drift mobility of holes in solid neon. (Loveland *et al.*, 1972)

modes; E_b is the small polaron binding energy [see Section III.C in Chapter 1 and the approximation to equation (6.10) given by equation (1.32)]. The deduced parameters for Ne, Ar, Kr, and Xe are shown in Table 6.7.

The non-adiabatic hopping theory of Holstein was used to fit the results to equation (6.10), except for neon where the applicability of this approach is in doubt; argon is a marginal case but it seems that the neon results can best be fitted to the adiabatic hopping theory (Emin and Holstein, 1969) which gives

$$\mu = (qa^2/k\theta)(3\omega_0/4\pi)\exp[-(E_b - 2J)/2k\theta] \quad (6.11)$$

where the symbols have the same meaning as in equation (6.10). The limits of applicability was laid down by Holstein can be met for neon but the large values of $\hbar\omega_0$ for xenon suggest that adiabatic hopping should be excluded, although this does not apply to Ar. However, the values of J, E_b, and $\hbar\omega_0$ were obtained for neon by fitting the experimental results for holes to

Table 6.7. Polaron interaction parameters for holes in solid noble gases.

Solid	a(Å)	E_b(meV)	J(meV)	$\hbar\omega_0$(meV)	γ
Ne	3·12	~150	63	8–25	*
Ar	3·90	60–90	8–18	5–9	~10
Kr	4·12	~40	~4	15	~3
Xe	4·47	1–50	0·5–2·5	10–50	0·1–1

* Adiabatic transport.

equation (6.11) and obtaining a value of $\hbar\omega_0$ and of $(E_b - 2J)$; E_b was then estimated by extrapolating values derived for the higher atomic weight noble gases. The type of phonon interaction was discussed to a certain extent by Le Comber *et al.*, and two alternatives were put forward; these were vibrational modes of the R_2^+ molecule or zone boundary acoustic phonons which are known to lie within the energy range 5–8 meV (Shukla and Salzberg, 1973). The results for holes in Table 6.7 show that the general trend in progressing from Ne to Xe is a decrease in polaron binding energy, a decrease in the exchange integral J, and a lack of any systematic variation of the phonon energy involved.

Electron drift in liquid neon is especially interesting in this class of elements studied by transient EBC techniques (Loveland *et al.*, 1972); the electron drift velocity is considerably smaller than that in other noble gas liquids (see Table 6.6). The possibility of Ne^- ion drift being responsible for the observed EBC pulses was questioned by Allen (1976), but ionic motion was excluded on the following grounds. The product of the ionic mobility and the fluid viscosity at all temperatures should be constant (Walden's rule), but in the case of the observed transient EBC pulses for negatively charged carriers this was not obeyed. Thus it is most probable that the electron transport in liquid neon is by an electronic "bubble state". Such an entity consists of an electron surrounded by a void due to fairly strong repulsion of the surrounding gas through its pseudopotential not being counterbalanced by the comparatively weak polarization potential (see Chapter 1, Section III.C.5), so the excess electron in the liquid travels within a spherical void of radius about 6·4 Å.

B. Molecular gases (nitrogen, oxygen, carbon monoxide, hydrogen, and methane)

The solid noble gases can be considered to form molecular crystals in so far as the components of the crystal unit cells are entities in which the valence orbitals are completely filled. However, the crystalline solids of the diatomic gases are of particular interest since the basic theories of small polaron transport (Holstein, 1959; Emin and Holstein, 1969) are based on a model consisting of a linear array of diatomic molecules. A more generalized approach to the problem of small polaron transport in a three-dimensional array was adopted by Emin (1975). Qualitatively the results of the calculation are not very different in the two cases. Transient EBC measurements have been made on the solid phases and on the liquid for four out of the five gases mentioned. Results for CO, O_2, and N_2 were reported by Loveland *et al.* (1972), for H_2 by Le Comber *et al.* (1976), and for CH_4 by Wilson and Le Comber (1977).

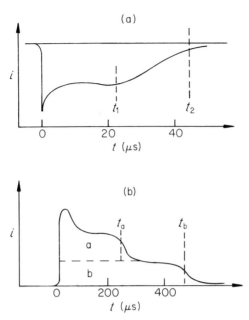

Figure 6.57. Observed pulse shapes in solid and liquid nitrogen. (a) The shape of the transient current pulse due to the transit of excess electrons in solid N_2; the transit time was defined as $T = \frac{1}{2}(t_1 + t_2)$. (b) A positive ion transit in liquid nitrogen ($L \approx 30$ μm), suggesting the drift of two species a and b. (Loveland et al., 1972)

The observed transient pulse shape due to the motion of excess electrons in solid N_2, and the conduction pulse associated with the transit of positive ions in the liquid phase are shown in Fig. 6.57. The deduced carrier mobilities of electrons in both the α-phase and the β-phase are shown in Fig. 6.58. There was no indication of hot electron transport at high applied fields of the kind found in the noble gas solids at fields up to 10^5 V cm^{-1}. The drift velocity was always strictly proportional to the applied field. It is seen that the value of μ_e is within the range 10^{-3} to 8×10^{-4} cm^2 V^{-1} s^{-1} at the various phase transition temperatures. These mobility results were interpreted in terms of Holstein's non-adiabatic hopping theory, discussed already in connection with the hole transport in solid noble gases. The results are summarized in Table 6.8.

It is surprising that intramolecular vibrational stretching modes are not responsible for the molecular distortions that result in the predominant carrier interaction in solid N_2 or solid O_2. It was suggested by Siebrand (1964) that, because the intermolecular orbital and vibrational wavefunction overlaps were small in comparison with those between the component atoms of the diatomic pair, intramolecular vibrational modes would be

252 ELECTRON BOMBARDMENT INDUCED CONDUCTIVITY

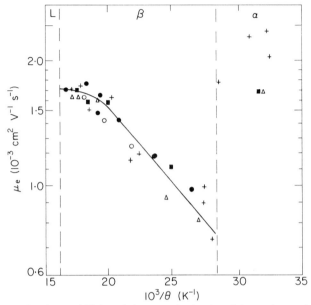

Figure 6.58. Carrier mobilities of electrons in the β and the α phase of solid N_2; the symbols refer to different specimens between 30 and 100 μm thick. (Loveland et al., 1972)

the dominating interaction if any carrier localization occurred. However, these modes correspond to a phonon energy of about 200 meV and the predictions by Siebrand cannot be upheld. The predictions do not fit the experimental values even though an extra electron is localized on an N_2 molecule occupying a molecular π^* molecular orbital which lies principally

Table 6.8. Electron and hole mobilities ($cm^2\ V^{-1}s^{-1}$) for solid nitrogen, oxygen, carbon monoxide, hydrogen, and methane at the indicated phase transition temperatures; also shown are ranges of fitting parameters to the non-adiabatic hopping formula [equation (6.10)] for the γO_2 and βN_2 results. Here c is the coordination number of the molecules in the solid.

	Phase	θ(K)	μ_e	μ_h	E_b (meV)	$\hbar\omega_0$ (meV)	$Jc^{-1/2}$ (meV)
N_2	β	63	1.7×10^{-3}		22–32	3.5–7.5	0.7–1.0
N_2	α	35	$\sim 2 \times 10^{-3}$				
O_2	γ	54		2.3×10^{-3}	40–65	3.5–8.5	3–6
O_2	β	43	$\sim 8 \times 10^{-3}$	$\sim 10^{-3}$			
CO	β	68	3×10^{-3}				
H_2	—	13.5	7×10^{-6}				
CH_4	—	87	980				

in the median plane. Loveland *et al.* suggested that the vibrational modes involved are probably associated with precessional modes which are known from Raman spectra to occur at energies less than 12 meV. The cause of carrier localization in the first instance is not known, but the possibility of dimer formation similar to $(O_2-O_2)^+$ was suggested.

The electron mobilities in the molecular gas solids (Table 6.8) differ markedly from those in the solid noble gases; in the latter they are around 1000 cm^2 V^{-1} s^{-1} whereas here they are many orders of magnitude smaller. However, there are two exceptions where similarities can be seen: the electron mobility in solid CH$_4$ is comparable to that in solid noble gases, and the hole mobility in solid oxygen is similar to that in solid noble gases. The transport of electrons, and probably of both signs of charge carrier in the diatomic molecular gas solids, seems to be by a non-adiabatic intermolecular hopping mechanism. The carrier is most likely localized by dimer formation, and the phonon frequencies involved are associated with precessional modes. In the corresponding liquids, both positive and negative ions are mobile with a mobility in the region of 10^{-3} cm^2 V^{-1} s^{-1} for N$_2$, O$_2$, and CO. The ionic mobility increases linearly with viscosity as predicted by Walden's rule.

Measurements of electron drift mobility in liquid and in solid hydrogen were reported by Le Comber *et al.* (1976). These measurements were undertaken by the transient EBC technique, and despite the favourable excitation method they proved extremely difficult owing to poor extraction efficiency from the generation region and also the build-up of internal space charges. To overcome these problems, the excitation pulse length was made 100 μs, and measurements were of the single-shot kind (i.e. allowing at least 10 s between successive field pulses and with many space-charge neutralizing electron pulses under zero applied field). The results showed that electrons are the mobile carrier in solid H$_2$, with a drift mobility of 6×10^{-6} cm^2 V^{-1} s^{-1} at 13 K. The ionic mobilities in the liquid were found by the same technique to be $\mu_- = 5 \times 10^{-3}$ cm^2 V^{-1} s^{-1} and $\mu_+ = 4 \times 10^{-3}$ cm^2 V^{-1} s^{-1} just above the triple point at 15 K.

Although the carrier drift mobilities in liquid methane had been known for some time (see e.g. Schmidt *et al.* 1973), the temperature dependence of this parameter and the drift of excess electrons in solid methane were not known until Wilson and Le Comber (1977) reported measurements by the transient EBC technique. It was found that the low-field drift mobility of excess electrons in liquid methane was about 450 cm^2 V^{-1} s^{-1}, in agreement with earlier published values. The electron drift velocity in the liquid increased linearly with the applied field up to about four times the velocity of sound c; above about $4c$ the drift velocity increased sublinearly with applied field. Measurements on the solid proved more difficult, and the field-dependence of the drift velocity could not be measured in any detail. However, just above the triple point temperature (87 K), $\mu_e = 980 \text{ cm}^2 \text{ V}^{-1} \text{ s}^{-1}$, approximately twice the value in the liquid. The temperature dependence of μ was found to be similar to that in the liquid.

VI. Selenium and sulphur

A. Selenium

Selenium has a special place in solid-state history, as well as being a particularly important semiconducting and photoconductive solid. The amorphous form has important applications in such equipment as xerographic photocopying machines. The three allotropic forms which have been studied by EBC techniques are the monoclinic allotrope (commonly called Se_α), the amorphous form, and the semimetallic grey form. The latter was mentioned in Chapter 1 where the dc EBC under steady bombardment was described; the steady EBC of amorphous selenium (a-Se) was described in Chapter 3, and the EVE of a-Se p–n junctions in Chapter 4. Methods whereby the amorphous allotrope has been employed in devices using EBC will be discussed in Chapter 7. The charge transport properties of monoclinic and amorphous forms have been studied by the transient EBC technique and these experimental results are discussed here.

Amorphous selenium is produced by evaporating selenium pellets on to a cool substrate in a high vacuum or on to a substrate certainly not much hotter than about 50°C. There might be a slight tendency for the glassy transparent deep-red films so produced to crystallize spontaneously within small areas with time, especially if the prepared layers are stored in a warm atmosphere. This tendency is reduced considerably if small concentrations of arsenic impurities are present; the results about to be described refer to pure selenium. Related measurements on selenium alloys have been reported by Gill and Street (1973) (selenium–sulphur), Perron (1967) (selenium–tellurium), Owen and Robertson (1970) (selenium–arsenic), and Takahashi (1979) (As-Se-Te ternaries).

Amorphous selenium consists of long-chain molecular spirals of Se_n in a solid solution with Se_8 ring molecules. The simultaneous presence of both entities was not appreciated for many years, but careful measurements of the electron drift mobility by Spear (1957), and that in monoclinic selenium (1961), by the transient EBC method showed clearly that in both materials electron transport is an activated process having an activation energy of about 0·25 eV in both Se_α and a-Se. These results are therefore consistent with the simultaneous presence of the two molecular units in a-Se (Mott and Davis, 1971, p. 371).

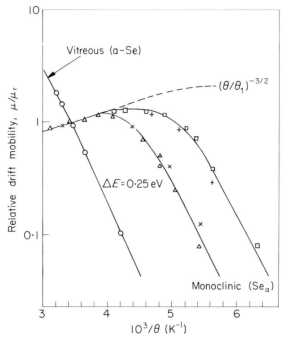

Figure 6.59. Electron mobilities in a-Se and in Se_α plotted as a function of temperature. (Spear, 1961)

Figure 6.59 shows the electron mobilities in amorphous selenium and in Se_α on the same graph plotted as a function of temperature. At room temperature the value found for the mobility in Se_α was between 1·7 and 2·3 cm² V⁻¹ s⁻¹ and in a-Se it was between 4·7 and 5·5 × 10⁻³ cm² V⁻¹ s⁻¹. An interpretation of these results about a decade later showed that it is very likely that the room temperature lattice mobility of electrons in Se_α is about 10 cm² V⁻¹ s⁻¹ and that similar trapping both in the monoclinic form and in the amorphous variety is responsible for the activated mobility. These traps are probably associated with the Se_8 rings.

The hole mobility could not be measured in Se_α on account of the short free lifetime, but the mobility of holes in amorphous Se was reported by Spear (1957). The room temperature value of about 0·15 cm² V⁻¹ s⁻¹ increased with rising temperature as $\exp(-0·16/k\theta)$, so providing evidence for trapping centres lying 0·16 eV above the valence band. The temperature dependence of the electron and hole mobilities in amorphous selenium is shown in Fig. 6.60.

As the temperature is lowered the shape of the transient conduction pulse becomes progressively more diffuse, until eventually the discontinuity at

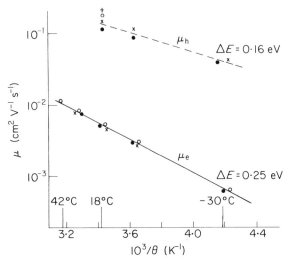

Figure 6.60. Temperature dependence of electron and hole mobilities in a-Se. (Spear, 1957)

$t = T$ marking the transit time of the generated carrier pulse can no longer be seen by the methods originally used in the mid-50s to mid-70s. However, by using a log-log plot to locate T according to the Scher–Montroll theory, measurements on a-Se have been extended to much lower temperatures, using photoexcitation (Pfister, 1976). A Monte Carlo analysis of the expected dispersive pulse shapes, for different pulse height distributions, has shown that the most likely distribution is Gaussian. The characteristic temperature describing the width of the spread was about equal to the glass transition temperature of a-Se (Marshall, 1977a,b). For electrons the peak of the Gaussian distribution occurred at 0·25 eV below the conduction band.

B. Sulphur

Sulphur is an elemental molecular solid which can occur in several allotropic modifications. The form studied most extensively is the orthorhombic, commonly called S_α, which is built from ring molecules of S_8, sixteen of which make up the crystal unit cell. The other form studied is the liquid within the temperature range below that at which the rings start to break up. The room temperature bulk resistivity of S_α is about 10^{18} ohm-cm, which renders this material unsuitable for such measurements as Hall effect and conductivity which might enable carrier drift mobilities or concentrations to be deduced. For this reason all measurements of carrier transport have been undertaken on generated excess charges.

The hole transport properties in solution-grown single crystals were studied by Adams and Spear (1964). The transient EBC method was used as described in Chapter 5, and a large number of crystals grown from solution in CS_2 were measured. The hole mobility varied from specimen to specimen but it was found from measurements at different temperatures that a consistent picture could be drawn in which transport occurred within a fairly narrow band and, at lower temperatures, was dominated by shallow trapping in a level of centres situated about 0·19 eV above the band edge. A computer was used to obtain a least-squares fit of the experimental points to the expression for trap-controlled mobility [equation (6.8)]. From the parameters of this fit, taking $m^*_h \approx m$ and assuming that μ_0 was of the form $\mu_0 = b\theta^{-3/2}$, the values of N_t were deduced as between 4×10^{14} and 10^{17} trapping centres cm^{-3}. There is reason to believe that this activation energy can be attributed almost entirely to vibrational interaction with S_8 molecular modes when the hole is localized at a lattice defect (Gibbons, 1970). The computer results led to a room temperature lattice mobility of holes of about $10\ cm^2\ V^{-1} s^{-1}$. The free lifetime (t') of the holes was found to be about $20\ \mu s$ for solvent crystals grown at 20°C. Figure 5.25 showed the experimental hole drift mobility as a function of temperature.

The transport of generated electrons is by quite a different process. Quantum mechanical analysis of the band structure by Chen (1970) and a much simplified approach by Gibbons (1966, 1970) have shown that it is quite possible for an excess electron to be localized on a molecular site in the solid for many periods of lattice vibration. This being the case, the molecule on which the electron is localized has time to become distorted significantly owing to the presence of the electron. In S_8 it appears likely that the eight-membered puckered ring molecule may flatten, or at least partly flatten. The electron may then move through the lattice by a series of uncorrelated phonon-assisted site jumps (non-adiabatic small polaron hopping transport). It appears that the symmetry of the molecular vibration corresponding to a flattening of the ring is A1 whose fundamental has an energy of 0·0265 eV, and the symmetry of the molecular orbitals whose weak overlap gives rise to the localized states is also A1. This identical space-group symmetry might be coincidental. The infrared absorption spectrum of sulphur shows that the A1 mode is weakly dispersed, which means that the intermolecular vibrational coupling to other modes is very small, and therefore the polaron will be strongly localized.

According to the non-adiabatic hopping theory of Holstein (1959) the hopping mobility is given by

$$\mu = (qa^2/k\theta)(1/\hbar)(\pi/2E_b k\theta)^{1/2} J^2 \exp(-E_b/2k\theta)$$

where $E_b = \gamma \hbar \omega_0$ is the binding energy of the polaron (ω_0 is the vibrational

258 ELECTRON BOMBARDMENT INDUCED CONDUCTIVITY

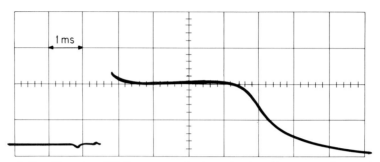

Figure 6.61. Transient EBC current pulse in orthorhombic sulphur for low intensity excitation. (Gibbons, 1965, unpublished)

frequency with which the electron predominantly interacts, γ is a dimensionless interaction parameter) and J is the molecular orbital exchange energy.

Electron mobility measurements for orthorhombic sulphur crystals have been made on solution grown crystals using the transient EBC technique (Adams et al., 1964; Gibbons and Spear, 1966), on natural crystals by the transient space-charge-limited current method (Many et al., 1965), and on solution and vapour phase crystals using photoexcitation (Gill et al., 1967). The transient EBC current pulse due to electron transits is shown in Fig. 6.61, and the space-charge-perturbed transient EBC in a similar specimen in Fig. 6.62. All these results are substantially in agreement despite the different purities and conditions of crystal growth. The electron mobility is shown in Fig. 6.63, from which it can be seen that the room temperature value is about 6×10^{-4} cm^2 V^{-1} s^{-1}; it rises with increasing temperature θ according to $\exp(-0.17/k\theta)$. Making use of the vibrational assignments of the infrared absorption peaks by Scott et al. (1964) and the molecular orbital calculation of Chen (1970), a fit of these results to Holstein's non-adiabatic hopping

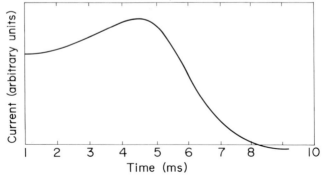

Figure 6.62. Transient EBC electron current pulse under SCP conditions resulting from high excitation in a single crystal of S_α. (Gibbons and Papadakis, 1968)

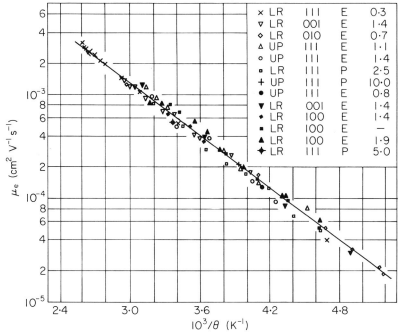

Figure 6.63. Temperature dependence of electron drift mobility in orthorhombic sulphur; all specimens, irrespective of source, fall on this common curve. The measured hole mobilities in the specimens and their crystal orientations are tabulated in the inset (LR = laboratory reagent; UP = ultrapure; P = photoexcitation; E = electron excitation). (Gibbons and Spear, 1966)

theory leads to $E_b = 0.48$ eV, $\hbar\omega_0 = 0.0265$ eV, and $\gamma = 18$. This lends support to the conclusion that mainly the A1 modes are involved, both in the intermolecular overlap integrals and in the vibrational spectrum.

In liquid sulphur up to about 200°C (at which temperature the S_8 rings start to break up) both electrons and holes "hop" (Ghosh and Spear, 1968). The hole mobility falls by nearly five orders of magnitude on melting; this can be attributed to the reduction in the mean value of the intermolecular overlap integral J resulting from free molecular rotation. The experimental results for the drift mobility of carriers in liquid sulphur are shown in Fig. 6.64. It can be seen that the magnitudes of the mobilities and activation energies of both signs of carrier are similar, being about 10^{-4} cm^2 V^{-1} s^{-1} at 199°C and rising with increasing temperature approximately as $\exp(-0.17/k\theta)$. Evidence to support the suggestion that this may be attributed to flattening of the S_8 molecule in the presence of an excess hole, almost identical with that arising as a result of an excess electron (the A1 assignment), may be gained by a strict comparison of the S–S bonding in S_8^+

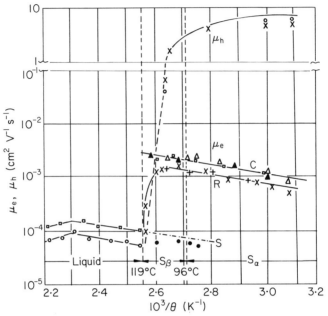

Figure 6.64. Temperature dependence of electron and hole mobilities, μ_e and μ_h, measured during melting of single-crystal platelets of sulphur (the discontinuity in the mobility scale should be noted): C, electron mobility in single crystal; R, electron mobility obtained from re-solidified specimen; S, electron mobility in the super-cooled liquid. (Ghosh and Spear, 1968)

with the isoelectronic ion $N_2H_4^+$ and of S–S in S_8^- with the isoelectronic Cl_2^-. In these cases ionization is accompanied by a change in the molecular orbitals from sp^3 hybrids to sp^2 hybrids (Symons, 1963; Brivati et al., 1965). In S_8 this would occur through the bonding sp^3 changing to non-equivalent sp^2 molecular orbitals and the ring approaching or attaining planarity.

A comparison of the electronic structure and properties of orthorhombic sulphur and monoclinic selenium was made by Dalrymple and Spear (1972). The two elemental molecular solids are similar in so far as they are both built from molecular rings of eight atoms in a puckered configuration. These authors have taken all the available drift mobility measurements on both crystalline solids; they were used with other measurements of the pressure dependence of the electron drift mobility (Dolezalek and Spear, 1970) and the optical properties in the vacuum ultraviolet (Cook and Spear, 1969) in conjunction with the band structure analysis by the authors previously mentioned. They show that in all probability the band structure of the two solids can be represented by the simplified diagram in Fig. 6.65. The molecular orbital levels of the isolated molecule are broadened into energy

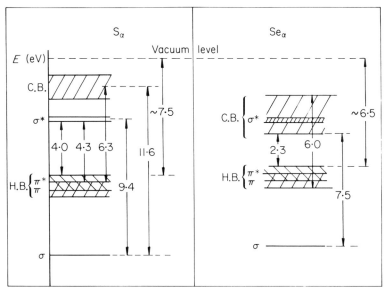

Figure 6.65. Suggested energy level scheme for S_α (left) and Se_α (right). (Dalrymple and Spear, 1972)

bands in the solid due to neighbour interactions which are mainly weak van der Waals forces. The nomenclature adopted here (and used in many of the cited references) is mainly for convenience, and the use of the term "π-orbital" does not imply its strict identity in the quantum mechanical sense to the conjugated molecular orbitals that occur in planar rings such as benzene or in long chains such as polyacetylene. Pairs of $3sp^3$ hybrids on adjacent atoms in S_8 are not parallel, and hence the π and π^* nomenclature adopted here does not apply exactly to the precise chemical molecular orbital conventional terminology. These molecular orbitals are better called quasi-π orbitals. They arise from the significant overlap of the non-parallel bonding and antibonding $3sp^3$ atomic orbitals on adjacent atoms in the ring. However, as just explained, the S_8^- or S_8^+ ions are probably planar, and these would therefore have a formal "π system" of molecular orbitals. Thus, if the crystal is flooded by light (such as in a spectrographic measurement) many of the generated electrons and holes will be localized on planar S_8 ions. Conclusions based on allowed and forbidden transitions between electronic states of a given symmetry are therefore justifiable on the basis of transitions between molecules of which at least one was a true π-system. The lowest empty state of the isolated molecule is $\sigma^* \, 3sp^3$ in the case of S_8 but there are also empty states arising from 4s and 3d orbitals lying above in close proximity. In the solid the intermolecular overlaps between σ^* $3sp^3$ states are insufficient to lead to a broad band of conducting states, and thus the principal conduction band arises from the overlaps of the 4s and 3d orbitals. However, since the electron transport in the solid is via a small polaron intermolecular hopping mechanism, we may deduce that if an electron is excited into the conduction band it will immediately fall back into the electron band that lies just below it in energy; then it will propagate by the suggested mechanism. In the case of selenium, however, the equivalent band to the extremely incoherent electron "band" in sulphur is that arising from σ^* $4sp^3$ molecular orbitals. In this case these states in the solid lie within the conduction band which arises from 5s and 4d atomic orbitals of the isolated Se atom. Thus there is no localized state equivalent to the σ^* $3sp^3$ state in sulphur, and both electrons and holes can propagate by a narrow band mechanism. This interpretation is borne out by measurements of the spectral dependence of the complex dielectric constant obtained from reflectivity data.

The mechanism by which generated charge carriers recombine in orthorhombic sulphur was studied by Dolezalek and Spear (1975). The experimental method, described in Chapter 5, depends on generating carriers by a short burst of ionizing radiation on both sides of a thin laminar specimen. In the experiments described, light from triggered xenon flashlamps was used and a time delay was incorporated so that a cloud of drifting

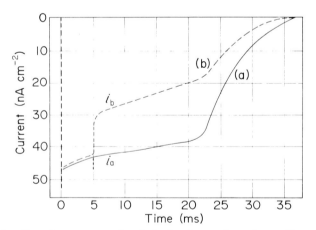

Figure 6.66. Transient current signals in S_α observed on the oscilloscope screen when a drifting charge cloud of holes passes through a cloud of electrons travelling in the opposite direction. (a) Curve corresponding to the transit of generated electrons when these are excited on one face of the laminar specimen. (b) Current pulse trace when a cloud of holes is generated at the opposite face 5 ms later. As recombination between the electrons and holes takes place, the conduction current falls from i_a to i_b. (Dolezalek and Spear, 1975)

holes could be made to pass through a drifting charge packet of electrons during its transit from one side of the specimen to the other. The observed current waveform is shown in Fig. 6.66. The decrease in current from i_a to i_b, due to recombination as the two oppositely charged clouds of carriers passed through each other, is clearly visible. From the deduced value of recombination coefficient it was found that the process could be fitted extremely well to the Langevin model of recombination (Helfrich and Schneider, 1966). This model is one in which the recombination is a diffusion-controlled process and the predicted bimolecular recombination coefficient is given by

$$\alpha = (q/\varepsilon\varepsilon_0)(\mu_e + \mu_h)$$

The measured values of α were between 0·7 and 1·2 cm³ s⁻¹ for different specimens where μ_h varied from 1·4 to 4·6 cm² V⁻¹ s⁻¹. In the same series of measurements it was found that a Poole–Frenkel model can account for the field dependence of generation efficiency even at applied fields as small as 10^2 V cm⁻¹.

7
Applications of EBC and EVE in electron devices

The first practical application of the EBC effect is still in use today in a typical cathode ray tube. The phosphor screen of such a tube is composed of a thin layer of fluorescent material on the glass faceplate, and the phosphor is invariably a good insulator. Its free surface would become charged by the electron beam unless some means existed for completing the circuit containing it. At voltages up to about 10 kV this is achieved by the process of secondary electron emission. So long as the secondary emission coefficient σ exceeds (or is equal to) unity, the number of electrons leaving the phosphor will be the same as the number arriving, and the free surface will stabilize at the anode potential of the tube. A more detailed explanation of this mechanism was given in Chapter 3. However, brighter and sharper images can only be obtained by using higher anode voltages. In the range above about 10 kV all insulators exhibit a falling characteristic of σ against V_p, and in the range of interest, above 20 kV, all phosphors have values of σ less than unity. If, during tube manufacture, the phosphor is deposited on a transparent conducting film on the faceplate, or if the side of the screen facing the gun is metallized, the EBC is sufficient to stabilize it at the potential of the metallization. Both methods have been used for projection TV tubes at anode voltages up to 100 kV, and the second of the two methods is frequently still used in modern cathode ray display tubes. An EBC gain of approximately unity is needed for this to work but, as shown in Chapter 3, values of g up to several hundred can be obtained in a variety of insulators under appropriate conditions.

It is not altogether surprising that the EBC current gain of an insulating film was the first electron bombardment induced current (EBIC) phenomenon to be exploited in an electron device as an amplifying process. Both of the industrial laboratories where the EBC effect had first been used for cathode ray tube screens came up with inventions relating to amplification of TV images. More than 30 years ago, at the RCA laboratories at Princeton, a number of charge storage tubes were devised whereby a two-dimensional charge pattern corresponding to a TV image could be amplified and stored on a thin film of SiO_2 or amorphous Se, and at the EMI laboratories in London, a low-light-level television camera tube using an EBC amplifying target of As_2S_3 was invented.

Twenty years then passed, and technology associated with silicon planar processing advanced, before use of the EVE came into its own; applications of the EVE in silicon had been suggested by Sclar and Kim (1957). In the years between the initial measurements on thin insulating films and measurements on electron bombarded semiconductor junction targets, it was found that the latter opened up a field where very large values of substantially noise-free current amplification were possible in a wide variety of target formats. The physical phenomena of the EVE have been dealt with in Chapter 4. Values of g up to 5000 or so can be fairly readily obtained in silicon, and the target response time is determined almost entirely by the transit time of the generated carriers across the space-charge region associated with the junction; this time might be only a few nanoseconds.

The decade from about 1970 saw the emergence of a number of hybrid vacuum tube–semiconductor devices. These and other applications of EBIC (such as in scanning electron microscopy) that are described in this chapter are only a selection from all the possible uses.

I. Devices using semiconductor junctions

A. Photosil EVE hybrid multiplier photocell

The large steady-state EVE gain that is attained when a high-energy electron strikes the region of a semiconductor near a p–n junction has been applied to achieve virtually noise-free charge amplification in a hybrid vacuum tube–semiconductor multiplier photocell. Attempts to build such a device were made in the 1960s, but difficulties were encountered with degradation in the semiconductor diode as a result of contamination by alkali metal vapours during tube processing, and by increases in diode back-bias leakage currents. However, a successful version (Fig. 7.1) was built using remote photocathode activation techniques. The device consists of an evacuated glass envelope containing a semitransparent photoemissive surface at one end (A), a series of electron focusing and accelerating electrodes (B and C), and a structured silicon p–n diode target (D) at the other end. The target shown in the figure is for a quadrant photodetector; it consists of four square diodes in close proximity on the same silicon slice.

In operation, D is held at a potential of about +15 kV with respect to A. Light from the object being observed is focused on A, and the emitted photoelectrons are accelerated and focused on the target where EVE signal

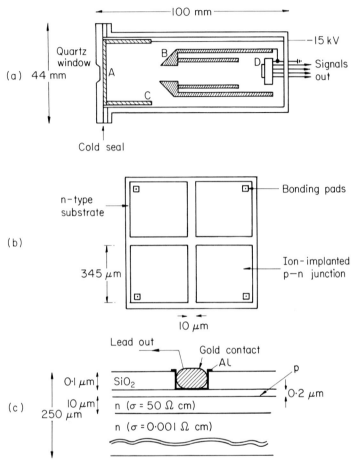

Figure 7.1. Photosil hybrid multiplier photocell: (a) cross-section of tube; (b) plan view of quadrant photosil target; (c) section through part of diode target and bonding pad. (Fegan and Craven, 1977)

pulses corresponding to the object are produced. The four diodes have separate connections to their p-type (ion-implanted) regions and a common connection to their n-type substrate. In use, the diodes are back-biased to about 10 V, so providing a depletion region of depth about 4 μm in the 50 ohm-cm epitaxially grown surface layer of the target. At this value of bias, the shunt capacity of each diode is about 4 pF per quadrant and this is presented in shunt to the input stage of the amplifier. The signal/noise ratio of the output is determined by the noise generated in the first stage of the amplifier, the system bandwidth, and the diode current in the absence of photocathode illumination (dark current). The bandwidth needed for a

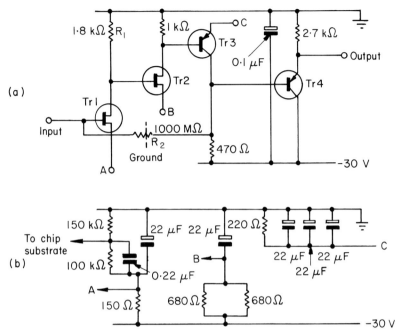

Figure 7.2. (a) Charge sensitive preamplifier; (b) associated biasing components, in which Tr1,2 are BF815 and Tr3,4 are BC214. (Fegan and Craven, 1977)

given output pulse rise time, or repetition rate, is set by the Nyquist criterion which states that the minimum bandwidth Δf needed for an output of N pulses in time t seconds is given by $\Delta f = \frac{1}{2}(N/t)$. Also, the amplifier noise in its early stages is set by the voltage noise generated by the fet input. A suitable circuit is shown in Fig. 7.2.

Such a device was described by Fegan and Craven (1977) who used it in a system for photon counting. This involved pulse shaping circuits and pulse height discriminators to provide an output corresponding to the arrival of single photoelectrons from the photocathode. The pulse height distribution is shown in Fig. 7.3; the dark current is also shown, and the final output was obtained by subtracting the output in the presence and in the absence of illumination. The ratio of signal amplitude to rms noise amplitude is then

$$r = N_s/(N_s + 2N_d)^{1/2} \qquad (7.1)$$

where, for some predetermined pulse-height window, N_s is the expected count due to the signal alone and N_d is the expected dark count for the same time interval. A single large peak can be clearly distinguished corresponding to single photoelectrons, and also a smaller peak which was interpreted as being due to inelastically scattered (back-scattered) electrons from the

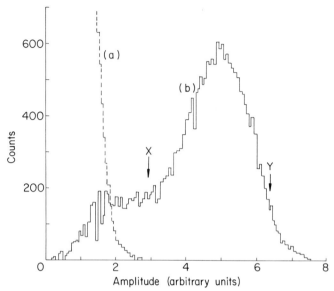

Figure 7.3. Pulse height distributions from one quadrant of the photosil target at $V_p = 12$ kV: (a) trace obtained in dark; (b) count with lamp on after subtracting (a). (Fegan and Craven, 1977)

target. The system delivered output pulses about 6×10^{-7} s wide and it was therefore capable of a high count rate before "pile-up" and also a reduced sensitivity to current noise. Pile-up is the effect that occurs when the pulse rate is so high that the amplifier output zero level progressively rises and eventually it is incapable of amplifying further inputs because of saturation effects associated with a large dc pedestal.

The pulses due to arrival of single electrons at the target were separated from the back-scattered peak by adjusting the discriminators to accept only pulses having an amplitude higher than X (Fig. 7.3) which is the back-scattered plateau. The window was set to accept pulses with a height less than Y which was arbitrarily chosen to lie symmetrically on the other side of the peak maximum. This eliminated counts that occasionally occurred in the dark or under illumination which were up to 30 times greater in amplitude and might be attributable to spurious ion bombardment. This emphasizes one of the advantages of photon counting methods over analogue techniques where such discrimination is virtually impossible.

This quadrant photosil was suggested in an autoguiding system for an astronomical telescope (Jelley, 1973). A subsidiary telescope, linked to the main guidance system of the principal telescope, is trained on a bright star and the image is focused on the photosil which is fixed to the telescope. The output from the four quadrants of the photosil can sense the relative movements of the field of view and the image of the star; the field is then kept steady by a simple servosystem.

The results with such an autoguider were described by Jelley (1980) and by Argue et al. (1977). It was tried on the 15 cm F/15 telescope at Cambridge (England); calculations indicated that the dimmest star suitable for guiding would have magnitude $B = 11 \cdot 5$ at an elevation of $50°$, assuming an integrating time constant of 3 s, and the rms fluctuations on the differential signals from opposing quadrants should be equal to or less than 14%. In practice it was found that a star of brightness $B = 11 \cdot 5$ gave a total of about 62 counts s^{-1} to which the sky background contributed about 22 counts s^{-1}. The dark rate from the photosil was variable but at the time of the reported measurements it corresponded to about 12 counts s^{-1}. This dark rate for their p^+–n diode target tended to rise with increasing bombardment dose.

Choisser (1976) published results of an investigation into this effect. The creation of fast surface states at the Si–SiO$_2$ interface by irradiation with energetic electrons is irreversible and it began to appear after an absorbed electron dose of about 10^4 rads in his diodes which had a SiO$_2$ passivating layer $1 \cdot 0$ μm thick. Beyond this dose (which corresponds to 15 h operation at 25 kV and an average diode flux of 100 photoelectrons s^{-1} per diode) it increased with dose until eventually settling down, after about 10^7 rads, to a value 15 times greater than its initial value. This is consistent with the maximum density of F-centres in alkali halides that had been estimated by Seitz (1946), about 10^{19} cm^{-3}. However, before the establishment of this condition, which did not occur until after thousands of hours of continuous target bombardment in a telescope, the system was able to hold the rms fluctuations in RA to $\pm 0 \cdot 1''$ and $\pm 0 \cdot 2''$ for stars of magnitude $B = 5$ and 9 respectively. Under good conditions the guider worked usefully with guidance stars down to $B = 11$, as calculated.

B. Digicon multichannel photocell

The Digicon electron tube (Fig. 7.4) consists of an evacuated glass envelope containing a transparent photoemissive cathode on the window at one end, and a diode array consisting of 40 independent junction diodes on the same slice of silicon at the other end. The operation is very similar to that of the photosil (see Section A above), and its application is mainly for examination of weak star spectra. The spectrum is focused on the photocathode so that the direction of dispersion is parallel to the line of diodes. Photoelectrons emitted from the photocathode are accelerated to a potential of 15–30 kV and focused on the diode array by a solenoidal magnetic field provided by coils outside the tube. The image electrons are accelerated gradually by means of a number of internal ring electrodes evenly spaced along the length

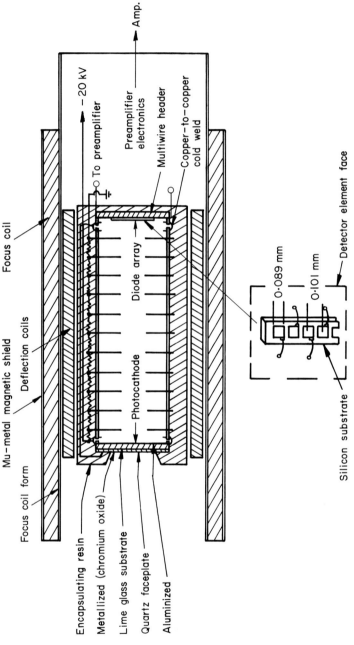

Figure 7.4. The Digicon multichannel photodetector; details of the multielement diode target are also shown. (Beaver et al., 1972)

of the envelope. These are supplied with appropriate voltages, using a conventional resistive bleeder chain which, with the remainder of the tube, is encapsulated in a resin insulator to prevent corona breakdown.

In the multielement diode target, the n-type silicon substrate acts as a common connection to all the diodes, and each p^+ diffused area is taken out by a separate connection to a ceramic plinth on which the diode array is bonded; these connections are then brought out radially through the tube envelope by means of 0·2 mm platinum wires. The tube is fabricated from a lime-soda glass envelope (Corning 0080 glass) and the accelerating rings and the final tube seal are of OFHC grade copper of the Housekeeper sandwich disc type. Contamination of the diodes by alkali metals used in manufacturing the photocathode is prevented by a remote processing technique on a glass disc during which time the diodes are closed off. After checking its photosensitivity, the photocathode disc is transferred to the end of the tube while still connected to the high-vacuum pumps; the photocathode panel is cold-welded by a high-pressure metal-to-metal diffusion seal using copper. This is the final step, and the tube is now removed from the pump.

The Digicon is operated with the photocathode at a high negative potential, and corona is minimized by cementing a quartz glass panel on the end window. This is supplied with a transparent conducting coating on the outside to provide a connection which is operated at earth potential. When photoelectrons liberated from the photocathode are accelerated and focused on the diode elements of the array, the charge amplification arising through the EVE gain is large enough to ensure that the required signal due to arrival of signal electrons is greater than the equivalent noise generated by any amplifiers associated with the circuit containing the diodes. The remaining sources of extraneous signals (noise) are the photocathode dark current and, when used in astronomy, the night sky air-glow components. The first source is about 2–3 counts h^{-1} per diode, and the second is equivalent to a 4th magnitude star per square degree, or approximately 10^{-13} lumen within a 0·1 mm disc at the prime focus of the Mount Palomar 5 m telescope. With a photocathode quantum efficiency of 10–25%, the air-glow spectrum thus contributes an unwanted signal of about 10 electrons s^{-1} per diode if the star image is only dispersed over the array, but about 400 per hour in high-dispersion spectroscopy where 1% of the starlight might be dispersed over the same area.

When employing such a tube for these purposes, it is essential to make full use of the data arising from the emission of a single electron from the photocathode, to collect enough to make the result statistically significant above spurious signals arising from night-glow components, tube dark emission, and amplifier noise, and to use pulse-height discriminators in each

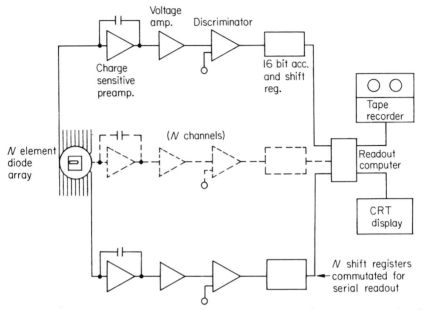

Figure 7.5. Schematic diagram showing the electronic circuitry system associated with a multichannel array. (Beaver *et al.*, 1972)

channel so that single very large or very small pulses can be discounted. A block schematic diagram showing an arrangement of suitable circuits for these purposes for a tube having a number of separate output channels is shown in Fig. 7.5; a suitable low-noise charge sensitive preamplifier has been described in connection with the photosil (Fig. 7.2). To avoid problems associated with the star spectral structure near the size of a diode element, the spectrum in this region is stepped alternately by magnetic deflection half the distance between them. The system has been used successfully for high-dispersion stellar spectroscopy as well as light intensity scans across nebulae and other low-intensity extended astronomical sources.

Clearly the use of multi-diode arrays in a Digicon tube is possible, and a 212 diode version was described by Choisser (1976). These tubes used a linear diode array target (Fig. 7.6) consisting of 200 image diodes each 300×40 μm^2 on 50 μm centres, and 5 alignment diodes at each end 300×40 μm^2 but with their major axes parallel to the length of the array. Two background monitor diodes were included and the total length of the array was 10 mm. With so many individual channels, great care was taken to overcome any possible non-uniformity by stepping the image. In Fig. 7.7 the upper curve shows the response to a uniformly illuminated test pattern consisting of a number of bars whose spacing decreased gradually from left

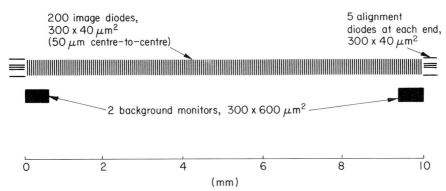

Figure 7.6. Diagram illustrating the target diode configuration in a 212 diode version of a Digicon. (Beaver et al., 1976)

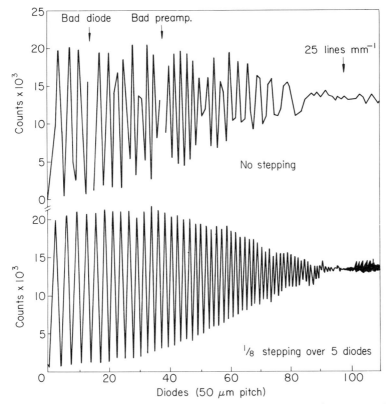

Figure 7.7. Response to resolution test chart with substepping (lower curve) and without substepping (upper curve). (Beaver et al., 1976)

to right; the figure indicates the scale of the bar-pattern. The upper curve shows the Digicon output without stepping, and several blemishes can be detected. The lower curve shows the same pattern imaged by the same tube but with a deflection waveform applied so that each photocathode location is sampled by five adjacent diodes. The slight increase in statistical noise is not noticeable.

As shown in Chapter 4, the avalanche gain of a diode promotes the total EBC gain. The possibility of making a Digicon or photosil by using the additional gain provided by biasing the bombarded diode into avalanching was tried by Choisser (1977); an additional factor of up to 30 at an avalanche diode bias of 300 V was confirmed by his experiments.

C. Intensified charge coupled devices (ICCD) and self-scanned arrays

Even with 200 or so individual diodes in the target array of a Digicon, the problem of large numbers of independent connections is quite a manufacturing nuisance, especially when each lead through the envelope is potentially a cause of a vacuum leak. As already indicated, the number of parallel circuits needed for signal processing eventually reaches the stage where the physical presence of such a large number becomes very uneconomical and inconvenient. In practice the only item of circuitry close to the tube is each channel preamplifier; the remainder of the electronics might be positioned as far as 30 m away. However, larger numbers of separate diodes render such a tube a practical impossibility. Considerations such as these led quite naturally to the development of tubes working on the same general principles as the photosil and Digicon but in which the electron bombarded target was a silicon self-scanned photodiode array or a charge coupled device.

A linear self-scanned 1024-channel tube was built by Choisser *et al.* (1974). This employed a Reticon RL-1024 self-scanned linear array of photodiodes, modified in that the thick scratch-resistant overlayer of SiO_2 was removed. It was mounted on a ceramic header with 13 pins for access to the circuitry from outside the vacuum. The diode array was bombarded by the accelerated photoelectrons through the front face which was passivated by a 1 μm thick layer of SiO_2. The on-chip circuitry was protected from electron bombardment by a metal mask which had only a single 0·5 mm slot to provide electron access to the diodes themselves. The tube was used for astronomical observations which were described by Tull *et al.* (1975). Their system used a small computer to operate the Digicon deflection coils, accept the serial input data from the diodes, add the signals corresponding to each picture point, and subtract the background and fixed pattern noise. The ultimate limit on the integration time possible was set by the diode leakage, and the array could be refrigerated by cooling the tube header; when the header was cooled to $-76°C$ the integration time could be increased to 60 s. Under these conditions it was shown that the readout noise was predominantly photoelectron shot noise. At an overall accelerating voltage of 30 kV the self-scanned Digicon is capable of detecting as few as 1 photoelectron every 4 minutes. It was used in the coudé spectrograph of the 2·7 m telescope at McDonald observatory; in observing part of the spectrum of Zeta Ophiuchi (Fig. 7.8) for 17·2 min, with an equivalent time spent on background subtraction, $1·5 \times 10^5$ photoelectrons per channel were detected with a standard deviation of 0·32%.

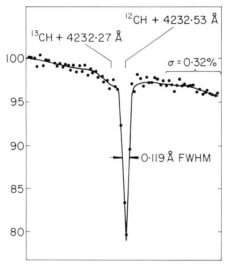

Figure 7.8. Part of spectrum of Zeta Ophiuchi; dispersion 0·46 Å mm^{-1}. (Tull *et al.*, 1975)

A similar tube for astronomy, described by Currie and Choisser (1976), employed a tube envelope and focusing structure similar to those of the photosil (Fig. 7.1) but the target consisted of the Fairchild CCD201/202 buried-channel 100-element 100-line charge coupled device. As with the self-scanned array just described, the on-chip transfer circuitry was protected by a metallic overlayer. It was found that a single photoelectron produced a charge packet of 1770–2040 ionization electrons per element, and the on-chip preamplifier generated a noise equivalent to 300 electrons. Cooling was provided to keep the diode leakage down to a low level in the same way as described for the self-scanned Digicon. Figure 7.9 shows the various sources of noise appearing at the ICCD output, using stellar magnitudes per square arc-second as a measure of brightness (i.e. a logarithmic scale, with larger "magnitudes" corresponding to fainter stars).* With pulse-height discriminators set to distinguish between 1, 2, 3, and 4 . . . photoelectrons, they could normally assume a mean rate of 400 photoelectrons per frame. For a faint star limit at the 5 m Mount Palomar telescope, an estimate of 27·5 stellar magnitudes was calculated for a signal/noise ratio of unity as set by the statistical noise in the night sky background.

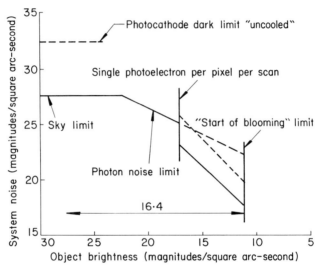

Figure 7.9. Sources of noise when the intensified charge coupled device is used in astronomy. (Currie and Choisser, 1976)

Another version of an ICCD was described for use in measurements of high temperature plasmas in Tokamak fusion reactors (Lowrance *et al.*, 1979). The tube had a 130 mm diagonal S.20 photocathode and a 160 × 100 element CCD. The electron optics of the tube achieved a photocathode–CCD demagnification of 30:1. For Thomson scattering measurements in the plasma diagnostics measurements, the ICCD tube was gated to collect photoelectrons only for 30 ns, which considerably eased the problem of gathering the relatively few photons of interest in the presence of a large background radiation level. This was achieved by applying a gating pulse to the G3 electrode of the zoom electron optical system. The target consisted of a thinned backside-irradiated CCD, and the photoelectron accelerating

---

* The visual magnitude of a star is determined by its illuminance outside the earth's atmosphere. The ratio of illuminances produced by two stars differing by 1 magnitude is given by $(100)^{0.2} = 2.512$. The reference for a zero magnitude star is taken as 2.65×10^{-6} lux. As an example. Venus at its brightest has a magnitude of -4.3.

voltage was 20 kV. At this value of V_p the measured EVE gain was about 2500, but the "dead layer" on the side on which the primary electrons were incident was rather thick, absorbing approximately 11 keV electron energy before the g versus V_p curve began to rise significantly. The dominant readout noise in this tube was the rms noise in the dark current, which was about 10^5 electrons per picture point. The rms value of this being about 300, if the average output due to the incidence of a single photoelectron on the CCD was about 2500 the system would be almost entirely limited by noise in the photoelectron input, even for very weak exposures. The maximum exposure was set by the elemental storage capacity of the CCD, which was about 10^6 electronic charges. The maximum exposure is therefore 360 photoelectrons and the signal/noise ratio would be determined by the quantum noise in the photoelectron flux.

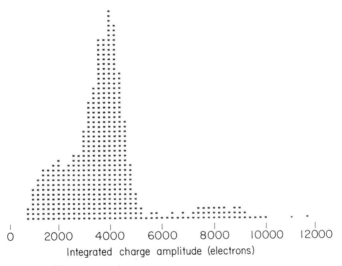

Figure 7.10. Histogram of single electron events. (Lowrance *et al.*, 1979)

Figure 7.10 shows a histogram of single electron events. It was found that the total integrated charge on the CCD from the impact of a single 20 kV photoelectron was 4300 electrons. Depending upon where the photoelectron landed, each event occupied between 3 and 6 picture elements. The diagram shows the clearly resolved peak at 4300 electrons and also a minor peak at twice this value where two photoelectrons struck the same or adjacent picture elements during an integration period. The full width at half maximum (FWHM) of the single photoelectron peak is 1400 electrons, thus enabling pulse height windowing to be used with a high confidence level.

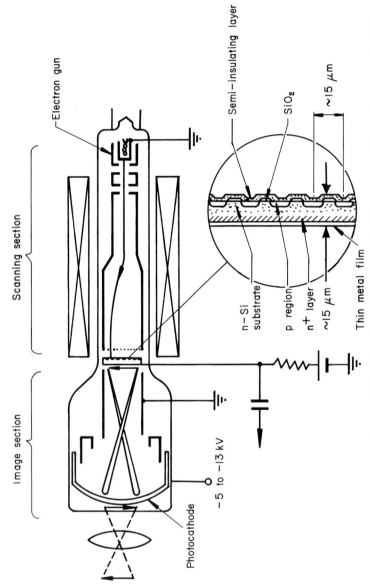

Figure 7.11. Schematic diagram of the silicon intensifier (SIT) television camera tube, and cross-sectional view of the Si target. (Miyashiro and Shirouzu, 1972)

D. Silicon intensifier target (SIT) television pickup tube and SIT scan converter

1. Television pickup tube

A number of versions of a television camera tube for viewing low-light-level scenes have been produced since it was announced that targets of silicon p–n junction photodiode mosaic arrays of the kind used in the silicon vidicon could also be used under electron excitation (Crowell and Gordon, 1967; Gordon and Crowell, 1968).

A television camera tube uses a scanned electron beam to interrogate point-by-point a stored charge pattern on a target corresponding to the scene being televised. The charge pattern may be generated through the primary process of photoemission (CPS Emitron; Gibbons, 1960), or by that of photoconduction (the vidicon; Weimer et al. 1950). A significant increase of sensitivity is attained by using a charge storage target which is capable of integrating the charges corresponding to the photoemitted electrons or the photocurrent for the duration of one TV frame time, t_F. The increase of sensitivity over alternative devices that have no storage (such as an image dissector or photomultiplier) is equal to t_F/t_e where t_e is an element time. Typical values are $t_F = 20$ ms and $t_e = 100$ ns; this factor is thus about 2×10^5. Modern TV pickup tubes almost invariably use a storage target. In these tubes the source of noise in the video signal is determined by that generated in the early stages of the amplifier used in the output connection. It was clear for many years that the sensitivity of such camera tubes to weakly illuminated scenes could only be usefully enhanced by pre-signal amplification. In the SIT version this is achieved by accelerating and focusing electrons emitted from the photocathode at one end of an evacuated envelope on to a silicon target structured to consist of a mosaic of many small isolated p–i–n or p–n diodes. A diagram of the television pickup tube is shown in Fig. 7.11.

Miyashiro and Shirouzu (1971, 1972) described an experimental version of this tube as follows. The figure shows how it consists of an image section and a scanning section. The image section is similar in design to an electrostatically focused image intensifier, and the scanning section is similar to a magnetically focused and deflected vidicon. In place of the conventional image intensifier phosphor screen is a silicon diode mosaic target, and the other side of this target is scanned by an electron beam from the scanning section. A photoemissive cathode is provided on the interior surface of the tube faceplate. The photocathode liberates electrons in proportion to the brightness of the optical image focused on it. Photoelectrons corresponding to the scene being televised are accelerated to a potential of about 5–13 kV and focused on the silicon target by applying suitable potentials to the internal electrodes. The effective image size is 9.5×12.7 mm^2 and the magnification of the electron lens is unity. A charge flows in the circuit containing the target but it is larger in magnitude than the incident charge by a factor equal to the EVE gain; as shown in Chapter 4 this is about 1500 for $V_p = 10$ kV and 2500 for $V_p = 12$ kV. The diodes constituting the target mosaic are reverse biased to 5–15 V by applying a potential between the signal contact and the cathode of the gun. Although the signal contact is effectively held at a fixed potential, contact to the diode islands is made only by virtue of the electron beam while it is scanning across them. The charge flowing in the interval between such scans slightly discharges the diode capacitance. Thus a charge pattern is built-up on the scanned surface of the silicon target corresponding to the scene being televised, and this remains without degradation by virtue of the diode capacitance and the low inter-diode leakage until the surface is scanned again (about 17–20 ms later). Each time the floating surface of the target is scanned it restabilizes at the potential of the gun cathode (cathode potential stabilization), and in so doing it deposits a charge equal and opposite to that which flowed in the interval between scans. This induces in the circuit containing the target a current whose amplitude is instantaneously proportional to the magnitude of the charge deposited on the diode being scanned; it is amplified by a low-noise video amplifier connected to the signal contact which, as shown in Fig. 7.11, is a common n$^+$ connection to all the diodes. The pre-signal amplification provided by the image section yields virtually noise-free intensification, and therefore the ratio of the output signal to amplifier noise is enhanced by a factor of about 2000; this is the factor by which the sensitivity of the tube is enhanced when compared with a similar photoemissive type without pre-signal intensification.

The SIT is essentially similar to the target used in the silicon vidicon; it is made by conventional silicon planar integrated circuit techniques. The starting material is an 18–20 mm diameter disc of 10 ohm-cm n-type silicon having a mean hole lifetime of about 100 μs; its thickness before processing is 10–15 μm. [Some manufacturers start with somewhat thicker material, and etch away a circular area on the back after the mosaic diffusion step, leaving a rim of silicon around the edge for added strength (see Crowell and Labuda, 1969).] An array of small p–n or p–i–n diodes is formed by diffusing boron to a depth of about 2 μm through holes in a grown SiO$_2$ layer. The diodes are about 6 μm in diameter with 15–20 μm spacing, making a total of about 5×10^5 isolated diodes in all. Signal contact (the common diode n$^+$ contact) is made by diffusing phosphorus to a depth of 0.1–0.3 μm on the other side; this also provides a low-S surface in the generation region, and the EVE gain is substantially greater than with an untreated surface. A similar process is used as a gettering step by forming a phosphorus glass over the etched surface and thus reducing the diode leakage. To eliminate the influence of stray light on the diodes, the signal contact is made opaque by covering the n⁺ layer with an aluminium film about 300 Å thick. The scanned surface of the target is covered by a continuous resistive "sea" to prevent repulsion of the scanning electrons by charges in the SiO_2 on the target left between diodes after photolithography. This sea may be a layer of Sb_2S_3 or CdTe with a sheet resistivity of about 10^{13} ohm per square.

The signal output is substantially independent of the reverse bias applied across the diodes, since the induced current is already saturated even under zero bias. However, the target capacity is reduced if the diodes are biased, and carriers liberated near the n⁺ side of the target then have less distance to travel by diffusion before reaching the edges of the depletion region. If the bias exceeds a certain limit (of the order of 5–15 V), the depletion regions of adjacent diodes actually touch, and inter-diode leakage will occur because adjacent p⁺ diode regions will be joined by a labyrinth of fully depleted material and charge leakage will occur in a manner determined by the potentials of adjacent islands and of the resistive sea; thus image degradation occurs if the bias exceeds this limiting value. The effect is analogous to that operating in a CCD. Holes liberated in the silicon target by the fast primary electrons from the photocathode normally depend on diffusion to travel about half the target thickness to the edges of the depletion region. After target processing, the hole lifetime is about 1 μs which is quite adequate for this to occur (the diffusion length $\lambda = (Dt')^{1/2} \approx 50\ \mu m$). The target capacity lies between 2000 and 4000 pF depending on the applied bias. Target leakage is kept down to about 6 nA by the gettering process, thus giving a diode storage time constant of more than 1 s. The typical maximum output at standard television scanning rates is 0·2–0·5 μA, and the output video signal is proportional to the light input over a range of nearly $10^3 : 1$.

An overall tube sensitivity of about 500 mA per lumen can be obtained. This corresponds to an illumination on the face of about 2×10^3 lux for a standard 25 mm diameter photocathode and a signal output of 0·5 μA. The lag (or residual signal) characteristic of a typical tube under different signal conditions is shown in Fig. 7.12. This feature is particularly noteworthy in so far as a relatively high speed of response is attainable for very weakly illuminated scenes. With this tube it is possible actually to see true photon noise on the television monitor when the SIT gain is high and the scene illumination is small (Lubszynski, 1971).

Variations on the SIT have been produced. Engstrom and Rodgers (1971) described a version in which the low-S surface of the target is buffered by a layer that absorbs some of the incident electron energy. In this way a range of target gain more than a factor of 1000 : 1 can be achieved by

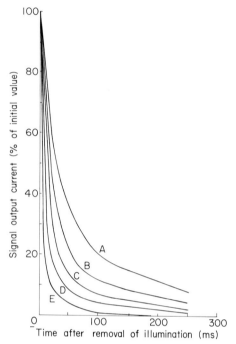

Figure 7.12. Typical persistence characteristics of the SIT tube: A = 25, B = 50, C = 100. D = 200. E = 500 nA. (From RCA data sheet)

varying the image section voltage. The effect is very similar to that obtained with an ordinary etched surface.

The SIT television pickup tube has been used for spectroscopy and direct imaging as an aid to astronomical observation. Kent (1979) described measurements using such a tube cooled with solid carbon dioxide in conjunction with the 1·5 m telescope at Mount Palomar; by using cooling in this way, integration times of more than 100 s were attainable. Absolute photometry and spectrophotometry measurements were taken to an accuracy of 4%, and wavelength calibrations were within 0·1 of a picture point (for the system mentioned this was equal to about 0·2% of the picture width).

As shown in Section III of the Appendix, when a television imaging tube of this kind is used for scientific observations, the highest signal/noise ratio in the video output signal is obtained when the scanning speed is considerably lower than that used in normal domestic television. A typical video bandwidth of about 50 kHz for a 625-line picture may be somewhere in the region of the optimum, which would then need a frame repetition rate of about 4 s, and a line scan rate of about 7 ms per line.

2. Scan converter

Silicon diode array targets of the kind used in the SIT television pickup tube are capable of storing the charge image for at least a TV frame time. The attainable limit of storage time at room temperature is set by diode leakage, the maximum practical time being about 1 s. If such a target is used in a double-ended electron tube containing an electron gun at each end, the input and output signals can both be video waveforms. One gun is operated at a potential of about -10 kV with respect to the target. The other gun is similar to that in a vidicon; it is operated at a low voltage in the cathode potential stabilization mode as in the scanning section of the SIT television tube just described. The high voltage gun is known as the writing gun. Such a double-ended tube using a silicon diode array target with a writing gun at one end and a reading gun at the other (Fig. 7.13) was initially described at a

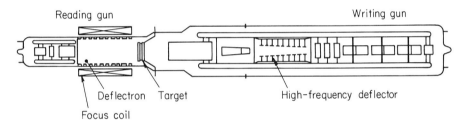

Figure 7.13. Cross section of dual-gun SIT scan converter. (Bates *et al.*, 1977)

conference in 1967 with the main emphasis on using such a device for television scan conversion, random access memories, and scan compression and repeat. Hayes (1975) described such a tube specifically for measuring high speed transients in an oscilloscope. For signals in the range above about 500 MHz it is generally necessary to use photographic methods in conjunction with a high-speed cathode ray tube. Usually such methods need a greatly reduced scan amplitude which is compensated at a later stage by enlarging the photograph. Photographic data cannot be used directly with a computer. Typical commercially available standard wide band oscilloscopes have a photographic writing speed of 6×10^{10} tracewidths per second, or 3 cm ns^{-1}. The tube illustrated in Fig. 7.13 was shown to give a perfectly adequate display at 2·4 GHz; this represented a 6-fold improvement in photographic writing speed and an equivalent speed of 50 cm ns^{-1}. Life tests were briefly described by Hayes, who noted that operating the tube continuously for 2000 hours resulted in a doubling of the diode leakage current in the central area. Degradation due to soft X-rays from the reading beam was negligible at a reading gun anode voltage of 250 V.

E. Evoscope fixed pattern generator

A video signal corresponding to a stationary image or a fixed pattern, such as a test chart or television station identifying logo, is often produced by using a television camera and an illuminated photograph or a drawing. Sometimes this is not convenient or economically desirable, and for static images of this kind a video pattern generator could be used.

A cathode ray tube with a modified target was described by Miyazaki *et al.* (1966). The target consisted of a slice of single-crystal silicon covered with a Schottky barrier contact of gold whose thickness varied according to the image required. In the diagram of the tube shown in Fig. 7.14 the thickness of the gold layer and its variation are exaggerated as an aid to the explanation that follows. In use, a constant electron beam is scanned over the target surface in the form of a television raster. The gold Schottky barrier is reverse biased to about 10 V, and a voltage is produced between the bulk silicon and the gold contact by virtue of the electron voltaic effect (Chapter 4). The magnitude of the target gain (g) depends on the thickness of the gold contact which was chosen to lie within the range 20–500 nm. A video output signal many times larger than the electron beam current could thus be detected and amplified by the circuit containing the target. The output signal was modulated in accordance with the thickness of the gold pattern which could either be evaporated or silk-screen printed. The starting material was 500 ohm-cm silicon, so the depletion layer thickness was 36 μm at 10 V reverse bias. The penetration depth of 10 keV electrons in silicon is only about 1 μm, so all the generated carriers were produced in the high-field depletion region.

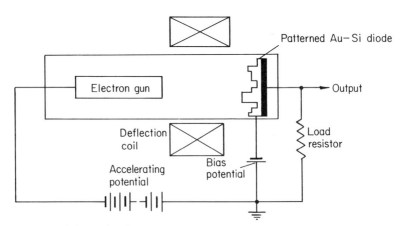

Figure 7.14. Schematic diagram of the Evoscope using an Au–Si patterned target in a cathode ray tube. (Miyazaki *et al.*, 1966)

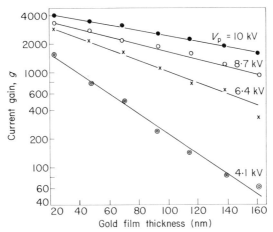

Figure 7.15. Dependence of current gain g in a fixed pattern generator on both the thickness of the gold film covering the silicon target and the accelerating potential V_p. (Miyazaki et al., 1966)

The way in which g depends on the thickness of the gold contact is shown in Fig. 7.15. It is seen that for $V_p = 10$ kV a g value of 4000 was obtained for a gold thickness of only 20 nm, whereas it fell to about 1500 for a thickness of 160 nm. At 500 nm the gain would be zero since this thickness is sufficient to absorb all the incident electrons; this was the thickest gold layer used. The peak white output signal of about 0·25 mA from the tube provided an electron beam current of 0·1 μA, and the resolution was determined almost entirely by the size of the scanning spot.

F. Electron bombarded semiconductor (EBS) microwave devices

So far in this chapter we have discussed electron devices that make use of the EVE gain in a p–n junction target or a Schottky barrier structure. In some cases a charge pattern is built up on an array of small diodes by integrating the induced current and storing this across the reverse-biased junction capacity. A time dependent current (video signal) is then obtained by interrogating the stored charge sequentially by a scanned electron beam of lower energy or by charge transfer in CCDs. An output signal is thus obtained corresponding to an input charge pattern across the target area, and this may be subsequently displayed as in TV by using a cathode ray tube for viewing, or else the output may be passed on to signal processing circuits. Such devices are generally used for data acquisition, and the output power level is usually only sufficient to operate the subsequent circuits without introducing additional noise sources.

In other cases charge integration may be accomplished in a computer store external to the tube, especially when the input rate (e.g. photon flux from a star) is low enough for pulse-height discriminators to be used. In tubes such as the photosil, the Digicon, and ICCDs the sensitivity is high enough for the detection of output pulses due to the arrival of single photoelectrons. Charge integration is used to enhance the signal/noise ratio of the input signal, which is equal to $N^{1/2}$ where N is the number of individual photon counts, provided that the amplification stages add no additional noise components. At high photon flux (e.g. more than about 10^7 photons s^{-1} per image point) analogue circuits are used; the upper limit is then set by the particular design of the tube target, and this maximum video output is typically 100–500 nA. At these high output levels the signal/noise ratio is determined by the first stages of the amplifier; it is typically 10^{-10} A for a bandwidth of 5 MHz and a stray capacity in shunt with the input of 25 pF.

The Evoscope generates a video signal by virtue of the EVE and its variation across the target corresponding to a predetermined pattern within the vacuum enclosure. In this tube there is no charge storage and the output current is determined by the properties of the target being scanned at each instant. Again, the question of the tube providing a source of power does not arise since relatively large output currents can be obtained from reasonably low electron beam currents, and wide video bandwidths can be obtained using fairly conventional low-noise circuits. However, the principle demonstrated by the Evoscope can be adapted for microwave power tubes and amplifiers. A number of such tubes, known as electron bombarded semiconductor (EBS) devices, were developed during the 1970s; the results from the most comprehensive programmes were reviewed by Bates *et al*. (1977). EBS tubes in general are microwave power amplifiers or pulse generators.

The principle of operation of an EBS microwave device is similar to that of the Evoscope except that the electron beam may be amplitude modulated and the target is designed for higher power generation at microwave frequencies. Another version of an EBS microwave tube uses a deflected beam and, as in the Evoscope, the output from the target is determined by the properties of the area being scanned. The main difference here is that in a deflected beam EBS pulse generator the target is split into a relatively small number of fairly large diodes; an output signal is obtained only from those that are excited by the beam, and if the beam only partly covers one of them its output is proportional to the overlap area. There is no vacuum tube equivalent to the deflected beam EBS amplifier which uses a target consisting of only two rectangular p–n junction diodes side-by-side. A diagram of such a tube is shown in Fig. 7.16. The electron gun can provide an electron beam of Gaussian cross-section (typical of a conventional gun with a crossover), or a laminar flow gun can be used; the latter is unusual since there is no

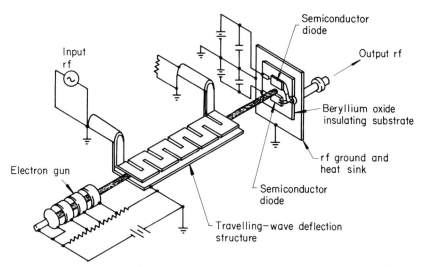

Figure 7.16. Deflected beam microwave power tube. (Bates *et al.*, 1977)

cross-over (i.e. no initial focusing of the beam inside the structure of the electrodes) and the final focus on the target plane is a rectangular area with fairly well defined edges. The tube can be used as a microwave power amplifier by applying the input to the travelling wave electron beam deflection system (shown in the diagram as a meander line structure) and extracting the power output from the two reverse biased target elements in a push–pull circuit configuration. The power gain ultimately rises from the energy supplied to the EBC target from its power supplies. A Gaussian beam profile is used for deflected beam EBS amplifiers but, if such a tube is used principally as a fast pulse generator, a voltage step is applied to the deflection system and the substantially rectangular cross-section beam from a laminar flow gun is used.

In power tube applications, the maximum output current from the device is limited by the target, which is usually n-type single-crystal silicon. The maximum output current is determined by the space-charge-limited current condition, and the maximum voltage is set by the limit of avalanche breakdown. If we assume that the bias voltage is held to a value of 0·6 times the bulk breakdown voltage v_A, the peak output current will be approximately

$$i_P = (\tfrac{1}{2}Z_L)^{-1} v_A L \tag{7.2}$$

where Z_L is the diode output load impedance and v_A is inversely proportional to the diode thickness L.

The generalized space-charge-limited condition for a high-field diode in which the carrier drift velocity u is not proportional to the applied field is given by

$$i_P \approx 2u\varepsilon Av/L^2 \qquad (7.3)$$

where A is the diode area. Putting $v = v_A$ and $u = u_{sat}$ in equation (7.3), with equation (7.2) we get the optimum diode thickness for the highest output power as governed by these two limitations:

$$L_{opt} = 4\cdot5(AZ_L)^{1/2} \qquad (7.4)$$

where we have taken for silicon a dielectric constant of 11·5 and a value of $u_{sat} = 10^7$ cm s^{-1}. Here L_{opt} is measured in μm, A is in mm^2, and the load impedance is in ohms.

In practice the silicon target is made from high conductivity n-type silicon on which an epitaxial layer of the required resistivity is grown. The resistivity of the epitaxial layer can be determined from the relationship between depletion depth and applied voltage for n-type silicon,

$$L_d \approx 0\cdot5(\rho v)^{1/2} \qquad (7.5)$$

where ρ is the silicon resistivity in ohm-cm and L_d is in μm. Thus the target design criteria are governed by choosing the depletion layer thickness for optimum output power as given by equation (7.4) and selecting the appropriate epitaxial silicon resistivity from equation (7.5) so that the depletion thickness L_d is equal to L_{opt} at the chosen voltage of 0·6 times avalanche breakdown voltage. The epitaxial layer thickness is also made equal to L_d so that the semiconductor is fully depleted at the normal operating bias voltage. This provides a target structure that is reasonably rugged, but the bulk silicon does not add significantly to the series electrical resistance of the diode.

If the diode excitation is pulsed, either by using a swept electron beam or by switching the beam rapidly, the output pulse rise time T_r is determined by the target diode capacitance and its load impedance. If the thickness L is optimized according to equation (7.4), the capacitance will depend on the diode area A. It can be shown on this basis that T_r is proportional to $(AZ_L)^{1/2}$. For $Z_L = 50\,\Omega$, $T_r = 100$ ps when $A = 0\cdot05$ mm^2, and $T_r = 10$ ns when $A = 500$ mm^2.

The maximum average output power from an EBS device is determined by the thermal characteristics of the target diode and especially its heat dissipation capabilities. A satisfactory manufacturing technique, which allows acceptable tube baking and outgassing temperatures while on the pump and high average power dissipation when in use, is one that involves soldering the silicon target on to a beryllia plinth with Au–Ge eutectic alloy. The bake-out temperature is kept below 300°C. In addition to these precautions it is necessary to protect the silicon diode edge by a fairly thick layer

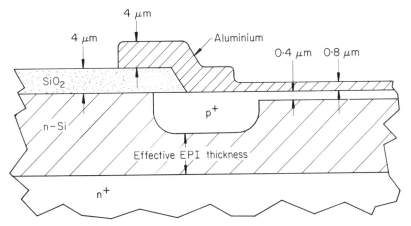

Figure 7.17. Electron bombarded semiconductor (EBS) power tube target structure shown in cross-section near the p–n junction edge. (Bates *et al.*, 1977)

of phosphorus-doped SiO_2 (see Fig. 7.17). Some of the contaminants removed from the tube during normal outgassing and processing are alkali ions; these are positively charged and would readily drift through undoped SiO_2. Phosphorus-doped SiO_2 is effective at immobilizing alkali ions, so this is important in protecting the target both during tube processing and throughout its operating life.

Section G below is concerned with degradation effects which were first recorded by Wolfgang *et al.* (1966) and by Abraham *et al.* (1966). One of the original problems and a solution have just been described. Another problem may be attributed to X-irradiation of the passivating oxide layer at the junction edges. X-rays generated near the diode by the fast electron beam liberate electron–hole pairs in the oxide. The electrons are mobile in SiO_2 and are rapidly extracted, leaving a trapped positive space charge in the oxide. In practice this causes a high electric field at the junction edges and may reduce the breakdown voltage of the diode by as much as a factor of 2. This can be prevented by covering the periphery of the diode with a layer of aluminium 4 μm thick over a 4 μm layer of passivating oxide.

A passivating layer of oxide over the junction edges becomes positively charged under the influence of soft X-rays or stray fast electrons. In a n^+–p silicon diode the electric field induced by the trapped holes weakens the internal field at the Si–SiO_2 interface; thus neither the breakdown field nor the leakage current is affected. However, a p^+–n diode target is nearly always used for EBS devices because of the higher electron mobility, and the previously mentioned precautions need to be taken to prevent a slow rise in reverse bias leakage current. In practice, during the first hour or so, the reverse bias breakdown voltage does fall to about 0·9 of its initial value but it then remains constant for many thousands of hours.

Unlike devices that flood the semiconducting target with electrons, such as occurs unavoidably in the photosil or Digicon, the passivating oxide is not exposed directly to the ionizing beam. The choice of top contact metallization in power tube applications is governed by three factors: absorption of energy from the primary beam, resistivity, and compatibility with tube and target processing. As was shown in Chapter 2, the stopping power for electrons in the energy range around 10 keV is governed by the density of the medium to approximately within a factor of 2 (Fig. 2.11). Hence, a useful figure-of-merit for the choice of top contact metallization is the quotient σ/ρ where σ is the bulk conductivity and ρ is the density; values for a range of contact materials are given in Table 7.1. This table shows that a diffused

Table 7.1. Figure of merit for top contact. (Bates *et al.*, 1977; Siekanowicz *et al.*, 1974)

Material	Density, ρ (g cm^{-3})	Bulk conductivity, σ (10^{-4} Ω^{-1} cm^{-1})	Figure of merit, σ/ρ (10^{-3} cm^2 g^{-1} Ω^{-1})
Aluminium	2·7	38	140
Titanium	4·5	1·2	2·7
Chromium	7·8	7·7	10
Copper	8·9	59	66
Nickel	8·9	15	17
Nickel silicide	5	4	8
Molybdenum	10·2	17	17
Silver	10·5	63	60
Palladium	12	9	7·5
Tungsten	19·3	18	9·3
Gold	19·3	42	21·7
Nichrome	8·25	0·9	1·1
Platinum	21·4	9·4	4·4
Platinum silicide	12	3	2·5
p$^+$ silicon	2·3	0·1	0·45

silicon p$^+$ layer forms a poor contact for high-frequency power applications; in practice such a layer is usually covered with a film of aluminium. For Schottky barriers, Nichrome or chromium is often used, although these metals do not have a low resistivity for high-frequency applications. Where power dissipation is no problem, and where very high frequencies are not involved, gold is usually chosen for Schottky barrier contacts for good diode characteristics as in the Evotron tube.

In non-deflected beam EBS amplifiers, the target structure is the same as that in Fig. 7.17 but the electron gun is not fitted with deflector plates; it is

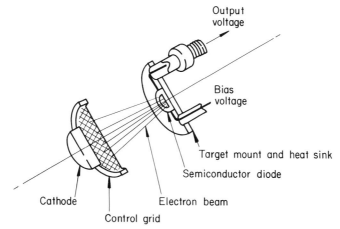

Figure 7.18. Layout of internal elements of a non-deflected beam EBS pulse generator.

almost identical to some conventional microwave power tube guns. This kind of tube (Fig. 7.18) can be used as a high-voltage switch. Such devices have been made that provide up to 400 V output pulse with less than 2 ns rise time into a 100 ohm load; the peak power delivered can be 1·6 kW and the duty cycle is up to 3%. A deflected beam EBS rf and video pulse amplifier typically provides 25–30 dB power gain at frequencies up to 320 MHz and will deliver an output power of 175 W into a load impedance of 25 ohm; such a configuration will provide an output pulse rise time of 1 ns.

G. Degradation phenomena

Some indication of the influence of radiation on the oxide layers associated with EBS or SIT device targets has been given in Sections A–F. Electron–hole pairs generated in an SiO_2 layer covering a silicon p–n junction will give rise to a bulk space charge due to trapped holes, and electric fields due to the proximity of this to a junction edge will cause an increase in reverse bias diode leakage current and a drop in junction breakdown voltage. A second but irreversible effect is the creation of fast interface states at the Si–SiO_2 junction; this gives rise to a high surface recombination velocity and associated higher leakage currents.

The EBS microwave device targets are not passivated with oxide in regions that are directly excited by the beam, and here the secondary influence of generated X-rays on the passivating oxide round the junction edges is avoided by using a thick layer of phosphorus glass coated with a beam shield or a thick metallic overlayer.

In devices such as the photosil, the Digicon, and the ICCD, it is virtually impossible to prevent some fast electrons from reaching the oxide. There is a slow rise in diode leakage with these devices, and they are mainly used in weak light conditions where the rate of irradiation is low. Why can this problem not be solved merely by using an unpassivated silicon target? The rigorous baking and outgassing procedures used for sealed-off vacuum tubes and prolonged electron bombardment of unprotected diodes still result in an increase of reverse-bias leakage current and a fall in breakdown voltage. Further problems associated with alkali metals used in electron tubes containing a photocathode have been mentioned above in Sections A–D. In the present instance some indication will now be given of the influence of the former alone, as would be the case in tubes such as EBS microwave amplifiers and pulse generators.

In one measurement described, an unpassivated mesa silicon Schottky diode had a reverse voltage breakdown of 250 V which fell to 115 V after tube bake-out at 150°C. After bombardment by 9 kV electrons for 1 hour, this voltage fell to about 25 V (Siekanowicz et al. 1972). These figures compare with some mentioned by Silzars et al. (1974) who described how the external electric field at the junction edges in an unpassivated silicon p–n mesa diode caused electrical breakdown in its vicinity when it was reverse biased in a vacuum; this leads to reverse voltage breakdown at lower voltages merely by putting the diode in a vacuum. The effect was found to be reversible upon diode exposure to the atmosphere again, but unpassivated mesas were useless for EBS device applications. Similar measurements on unprotected gallium arsenide mesa Schottky diodes were more favourable. The breakdown voltage for one diode with a Nichrome metal contact about 800 Å thick fell from about 90 V to 80 V after bake-out at 250°C, and then to about 70 V after electron bombardment by 9 keV electrons. Another diode with a chromium contact 400 Å thick did not show any degradation on bombardment by 9 kV electrons. The greater resistance of gallium arsenide to radiation damage was mentioned in Chapter 4, and the resistance of the diodes to deterioration in sealed-off vacuum devices when compared with silicon is explained as follows.

The density of surface states in gallium arsenide is several orders of magnitude greater than that in silicon which is between 10^{10} and 10^{12} cm^{-2}. A very high density of surface states can dominate the space charge just beneath the surface and thus make the surface potential essentially independent of external fields. Therefore GaAs targets are much less sensitive to the tube processing and to electron beam damage than silicon targets. However, as in the case of silicon, if a surface charge tends to cause an accumulation of majority carriers on the high-resistivity side of a p–n junction, the depletion layer will tend to be shortened at the surface, and the breakdown voltage of the junction will be reduced. The presence of a very small surface conductivity can cause a rearrangement of charge, and the resulting field can have a strong perturbing effect on the semiconductor. In the case of silicon, vacuum bake-out and electron bombardment between them can cause a significant increase in the density of surface states and a simultaneous increase in surface conductivity.

As examples taken from electron tubes using passivated p–n junction silicon targets, the EBS microwave tubes (Section F above) were shown to suffer from a decrease in the reverse bias breakdown voltage to about 0·9 of its initial value after the first few hours of operation mainly due to X-rays. However, life tests on sealed-off tubes (Bates *et al.*, 1977) showed more than 21 000 hours, at power levels equal to twice the rated level, without failure.

Tests on a self-scanned Digicon indicated an approximately 15-fold increase in reverse leakage current after a total target exposure to 5×10^{13} els cm^{-2} at 20 kV, and this appeared to be a stabilization value beyond which the reverse leakage did not increase with further electron bombardment (Choisser, 1976). In this tube, as with others of a similar kind in which a layer of SiO$_2$ over the junction edges is unavoidably exposed to electron irradiation, the saturation could be explained in terms of the polarization saturation. This is suggested by the steady-state EBC of SiO$_2$ investigated in the scanning electron microscope at a bombarding voltage of 30 kV by Hezel (1979) whose results are shown in Fig. 7.19 together with similar EBC measurements on Si$_3$N$_4$ for comparison. The conduction current i_s through

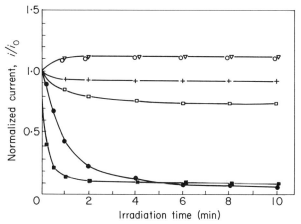

Figure 7.19. EBC in silicon integrated circuit capacitors as a function of irradiation time (g is normalized to its initial value; MOS insulator is SiO$_2$ and MNOS is Si$_3$N$_4$): ■ SiO$_2$, wet, 350 nm; ● SiO$_2$, dry, 630 nm; □ Si$_3$N$_4$, 840°C, 200 nm; ▽ Si$_3$N$_4$, annealed at 1050°C; ○ Si$_3$N$_4$, annealed at 1200°C; + Si$_3$N$_4$, low pressure, 800°C, 64 nm ($V_p = 30$ kV; $i_p = 1·6 \times 10^{-9}$ A). (Hezel, 1979)

the thin insulating films has been normalized to its initial value so that its changes with bombardment time can be seen more clearly. The changes in i_s are indicative of a bulk trapped space charge whose field opposes that initially applied. It can be seen that this effect is very pronounced in SiO_2 on silicon despite the completely penetrating beam, but it is considerably smaller in Si_3N_4, indicating that this material is considerably less susceptible to polarization effects. Hezel suggested that this is because neither sign of carrier is mobile in this insulator, and that equal numbers of electrons and holes are trapped. Thus silicon nitride appears to be a preferred insulator for electron bombarded devices or other silicon devices exposed to any kind of ionizing radiation. No experimental measurements have yet been described for electron bombarded silicon targets passivated with Si_3N_4, or those made using Si_3N_4 photolithography.

In some structures, polarization charges in SiO_2 over a p–n junction diode might increase the EVE gain. This effect is reversible, and it will be explained in connection with computer memory storage tubes in Section III.B.

II. Miscellaneous applications of bombarded semiconductor targets

A. Barrier EVE and the scanning electron microscope

Secondary emission images obtained by scanning the surface of an object by a high velocity electron beam had been employed for several years before a description of the use of the scanning electron microscope (SEM) in the electron bombardment induced current (EBIC) mode was first published (Everhart et al., 1963). In this mode, charges released in the bulk of a semiconductor are collected by a nearby junction and the current due to their motion is detected in an amplifier associated with the specimen. This current is modulated by impurity precipitates, swirl defects, slip dislocations, or almost any structure influencing the efficiency with which the generated charges are collected. If this modulated current is treated as a video signal, these inhomogeneities can be displayed on a viewing cathode ray tube screen much as in television. This method of using the SEM is so powerful that by 1976 no fewer than 113 technical articles on this mode alone had been published (for bibliography on EBIC mode scanning electron microscopy to March 1976, see Leedy, 1977). The technique is now used for qualitative or semiquantitative measurements of defects and dislocations and other features just described. The strong resemblance to the mechanism of charge generation in the electron voltaic effect (Chapter 4) had led to the name barrier EVE for this process in generating a video signal. The picture displayed is representative of the distribution of the electronic barriers over the scanned area to a depth equal to the penetration depth of the electrons in the beam, within the charge collection region.

In most commercially available scanning electron microscopes an electron beam accelerating voltage between 0·5 and 50 kV is used; the focused spot size is typically 0·05–1 μm, and the primary beam current is of the order of 10^{-11}–10^{-6} A. Scanning speeds and amplitudes are adjustable so that the image magnification can be altered by changing the size of the scanned patch while keeping the size of the displayed image constant. Probably the most valuable uses of the SEM in this mode in electronics are for examination of silicon integrated circuits. The materials of interest here are mainly Si, SiO_2, and Al. These all have approximately the same effective atomic number, and for this reason the energy penetration depth data for fast electrons in the range 5–25 kV can be taken as a function of density only. A widely used range law for silicon in SEM studies in the range $5 < V_p < 25$ kV is

$$R = 0 \cdot 017 V_p^{1 \cdot 75} \qquad (7.6)$$

(Everhart and Hoff, 1971) where R is measured in μm and V_p in kV. This law may be compared with others of the kind $R = aV^n$ for materials used in integrated circuits; the parameters and the range of applicability are summarized in Table 7.2.

Table 7.2. Penetration depth–energy parameters for electrons in integrated circuit materials.

Material	a	n	Energy range (kV)	Source of data
Aluminium	0·018	1·9	1–10	Table 2.7
Aluminium	0·019	1·68	2·5–27	Table 2.6
Silicon	0·022	1·65	2–20	Table 2.6
Silicon	0·024	1·7	10–100	Eqns (2.14a) and (2.15a)
Silicon	0·025	1·65	10–50	Gibbons et al. (1975)
SiO_2	0·024	1·68	3–30	Matsukawa et al. (1974)

1. Minority carrier diffusion length

If the scanning beam is slowly passed across an edge marking the limits of a Schottky barrier contact on the surface of a semiconductor, the generated electron–hole pairs will diffuse into the high-field region near the contact, and some will recombine before being collected. The further the generation region from the edge of the electrode, the larger the proportion lost by recombination. A measurement of the EVE current as a function of distance of excitation measured perpendicularly to the edge enables the minority

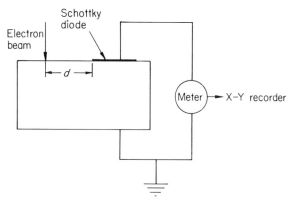

Figure 7.20. Experimental set-up for measuring carrier diffusion length in the scanning electron microscope. (Ioannou and Davidson, 1979)

carrier diffusion length to be determined. The experimental arrangement is shown in Fig. 7.20. The specimen is first prepared with a Schottky barrier contact on the surface; this is preferable to a p–n junction because no high-temperature processing is needed and therefore the SEM measurements apply to the minority carrier diffusion lengths in the bulk material before any possible disturbance or changes which might otherwise be brought about by heating to high temperatures.

The technique for making a good Schottky barrier contact to silicon was described by Leamy et al. (1976). The crystal is ultrasonically cleaned in an organic solvent, and it is further soaked and rinsed in a solution of H_2O–HF (5 : 1) for 1 minute; electrodes about 1000 Å thick are then evaporated immediately in a vacuum better than 10^{-7} torr. For n- and p-type silicon respectively, Ti and Au–Pd (4 : 1) alloy contacts are used; the Schottky barrier heights for each of these metals are about 0·6 and 0·8 eV.

The technique employed for diffusion length determinations is to scan the electron beam slowly away from the Schottky diode, and the EBIC decay with distance (d) from the edge of the electrode is measured parallel to the surface of the specimen (Fig. 7.20). For distances large compared with the diffusion length λ, Ioannou and Davidson (1979) showed that the collected current is given by

$$I(d) = [A \exp(-d/\lambda)]/d^{3/2} \tag{7.7}$$

The value of λ is obtained from the reciprocal slope of a graphical plot of $\ln(Id^{3/2})$ against d. Figure 7.21 shows a series of such plots where this technique has been applied to a number of ion-implanted silicon diodes which were subsequently annealed at various temperatures; the derived values of λ are shown in the figure, and it can be seen how high-temperature annealing reduces λ dramatically. Catalano and Bhushan (1980) measured

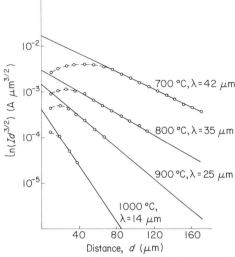

Figure 7.21. Variation of EBIC at $V_p = 15$ kV with distance for several annealing temperatures. (Ioannou and Davidson, 1979)

the electron diffusion length in a sample of Zn_3P_2 in a similar way, but they found that the induced current decreased with distance from the junction according to a simple exponential law, $I(d) = A \exp(-d/\lambda)$, and λ was determined from the slope of a plot of $\ln I$ against d.

This method and its variations were examined mathematically by von Roos (1978, 1979, 1980) in a series of five papers on the analysis of a p–n junction solar cell by the SEM. He showed that the lateral shift method, described above, is the only reliable method of measuring λ, and that determination after all the diffusion steps are complete on a finished cell is almost impossible. However, Possin and Kirkpatrick (1980) showed how it is possible to derive a number of important device and after-processing materials parameters, by reverse biasing p–n junctions that are already part of the structure and using one or more of these as signal collecting diodes. Their method was more elaborate than that already considered, and it involved a knowledge of the depth–dose law [equation (7.6)] and an analysis of the probability of charge collection by a junction when carriers reach it by diffusion [equation (7.10)]. They were able to infer the junction depth, the density of recombination centres in the depletion region, the limits of minority carrier diffusion on either side of it, and the surface recombination velocity.

2. Schottky barrier charge collection microscopy

When carriers are generated by the primary beam of fast electrons in a SEM, the current that arises when these move coherently under the influence of the internal semiconductor fields is modulated by the local recombination rate. This, in turn, is a function of the density of generated carriers, the energy levels and the nature of the defects, and the velocity with which the carriers are drawn out of the generation region. The conduction current may also be modulated by dopant inhomogeneities and charging effects which might occur if the structure is coated with a passivating oxide layer.

In the method described here, a thin metallic Schottky barrier film is prepared by the technique described above in Section 1. The scanning beam penetrates the conducting metallic film and generates carriers within the semiconductor bulk; their motion within the depletion region of the

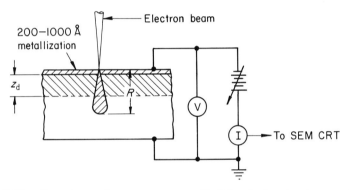

Figure 7.22. Experimental arrangement used in Schottky barrier charge collection microscopy; the shaded pear-shaped volume represents the generation region. (Leamy *et al.*, 1976)

Schottky diode is detected in the circuit containing the specimen. As shown in Fig. 7.22, the diode is back-biased to increase the width of the high-field region. For n-type silicon, the depth of the high-field region (the depletion region) is given by equation (7.5), described above in Section I.F,

$$z_d = 0.5(\rho v)^{1/2}$$

where the depletion depth z_d is measured in μm, the resistivity ρ is in ohm-cm, and the bias voltage $v = v_a + v_i$ is the sum of the applied bias voltage v_a and the in-built barrier potential which is 0·5–0·8 V for most metals.

If we plot the depletion field and the electron beam penetration depth on the same scale, it is possible to obtain a rough idea of the depth z beneath the surface that is being examined. The electric field within the depletion region is given by

$$\mathscr{E}(z) = (2v/z_d)(1 - z/z_d) \tag{7.8}$$

The number of free carriers generated within the generation volume is proportional to the rate of energy loss by the beam. If the total number of generated electron–hole pairs is kept constant for different V_p, and if we neglect surface effects (which might account for energy losses of 2·5–5 keV), the total number of pairs generated will be V_p/E^* where E^* is 3·6 eV for silicon. This number has been used as the normalizing factor in Fig. 7.23 where surface losses have been neglected. In the same figure $\mathscr{E}(z)$ is shown

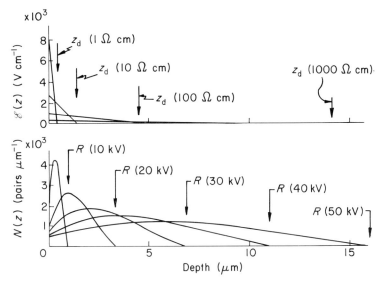

Figure 7.23. Comparison of electric field strength for n-type Si Schottky barriers and (below) depth–dose function for different values of V_p. (de Kock et al., 1977)

for different values of ρ in n-type silicon and for zero applied bias (Leamy et al., 1976). Provided that the depletion depth z_d is greater than the beam penetration depth, it is possible to locate the depth of a defect beneath the surface by changing V_p and hence the depth of carrier generation. Localized defects, such as lattice dislocations, are often centres for impurity precipitates; in either case, a dislocation or a precipitate will affect the charge collection efficiency of the Schottky diode because it will reduce the recombination lifetime. These will therefore show up on the displayed image as a black picture signal on the homogeneous grey background characteristic of the bulk material. A typical SEM photograph showing bulk defects in silicon obtained using the charge collection technique is in Fig. 7.24.

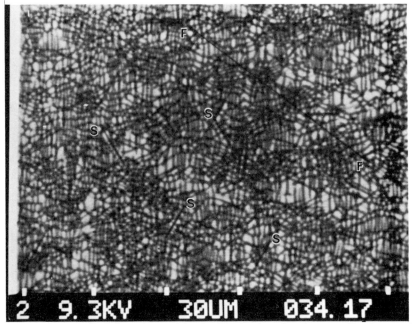

Figure 7.24. Bulk defects imaged in the SEM using the charge collection mode. The dislocation network can be readily seen and individual dislocations readily resolved. A small number of stacking faults (labelled S) are also visible; these give a slightly broader image. The long linear feature (labelled F) is a surface scratch. (Ashburn and Bull, 1979)

3. Analysis of semiconductor devices

By using beam accelerating potentials of about 15 kV, the primary electrons penetrate far enough into silicon to allow imaging of shallow diffused electron devices made by planar technology. The spatial resolution is great enough for separate elements of a buried p–n junction diode or transistor to be scanned, and if it is suitably biased, currents corresponding to the generated electron–hole pairs can be collected by the junction and used as the video signal. No additional junctions are fabricated as in the Schottky barrier technique (see above) because charges may be collected by diffusion or via the drift fields associated with the p- and n-type regions of the device. Since nearly all the microscopic details of an integrated circuit or a planar diffused silicon device can be examined in terms of independent or interacting junction diodes, and the resolution is adequate to image them separately, the principal way in which this is used for inspecting such devices is to measure the behaviour of the separate diodes. Figure 7.25 shows a suitable circuit for connecting the diode to the viewing cathode ray tube. Provision is

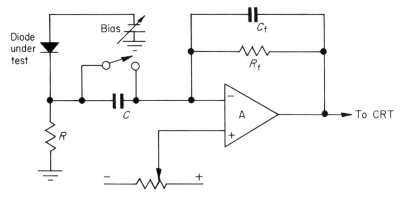

Figure 7.25. Circuit configuration for charge collection microscopy. (Gonzales, 1974)

made for backing-off a zero offset due to a dc component in the signal by supplying a separate voltage input to the differential operational amplifier. If the diode leakage current is too high for this to be done, ac coupling can be used by opening the switch shown in the circuit. For high speed scanning, the video signal is not differentiated by opening the switch, although it will affect images obtained at low speeds unless, of course, C is made very large, and dc restoration circuit techniques are used.

The spatial resolution of the image is determined by the extent of the generation region; it is not determined by the minority carrier diffusion length (Donolato, 1979). The generation region is a pear-shaped volume, cylindrically symmetrical about the primary beam axis. In the energy range used in the SEM, and for materials of atomic number (or compounds of effective atomic number) of the same order as that of silicon, Everhart and Hoff (1971) and Matsukawa et al. (1974) showed that the shape of the energy dissipation versus depth curve is independent of the beam and materials parameters provided that the energy loss dE/dz is normalized appropriately for the atomic number and density, and the depth z is measured in units equal to the extrapolated range. Matsukawa's normalized energy dissipation curve is shown in Fig. 7.26; the experimental data of Cosslett and Thomas (1965) for 20 keV electrons in copper and of Namba et al. (1969) for 25 keV electrons in CdS are included for comparison. It can be seen that the general shape is consistent, and it agrees with those for electron penetration in phosphors described in Chapter 2 (Figs 2.17 and 2.18), with the possible exception that the phosphor curves might appear too broad at low V_p because the glows are near the resolution limit of the optical microscope; fluorescence caused by X-rays would also cause a broadening of the phosphor curves, but X-rays spread the influence of the beam on the

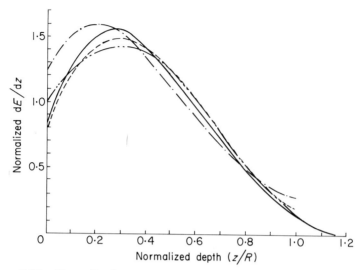

Figure 7.26. Normalized energy dissipation (dE/dz) as a function of normalized depth (z/R) (Matsukawa et al., 1974): ——Matsukawa et al. (18·7 kV); – – – – Everhart and Hoff; – · – · – Cosslett and Thomas (Cu 20 kV); – · · – · · – Namba et al. (CdS 25 kV).

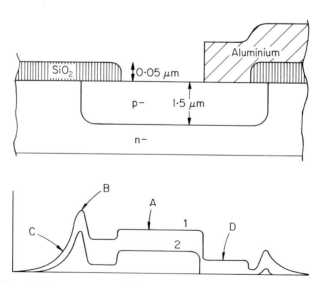

Figure 7.27. Diode structure and charge collection SEM response at beam energies of 15 keV (curve 1) and 10 keV (curve 2). (Gonzales, 1974)

generation volume. Such curves are thus very relevant in the context of carrier generation.

Gonzales (1974) demonstrated the image of a silicon p–n junction planar diffused diode. Figure 7.27 gives a cross-sectional view and dimensions of the structure, and the curves beneath show the SEM signals obtained as the beam at two different energies scanned across it. Curve 1 shows that 15 keV was sufficient to penetrate the aluminium layer covering the junction on the right-hand edge of the diode. Region A is where the carriers are generated mainly in the p-type diffused region and where the electric field is small but constant, region B is where the electric field in the vicinity of the junction is larger, region C can probably be attributed to minority carrier diffusion, and region D is where the electron beam energy is attenuated by the aluminium metallization.

Defect structures in bipolar transistors were investigated by Ashburn and Bull (1979). The output could be obtained by measuring the current due to carriers collected at the base–emitter junction or at the base–collector junction. It was shown that irregularities in the base–emitter junction resulted from localized regions of shallower phosphorus impurity diffusion due to dislocations. Bull *et al.* (1980) showed that, if the emitter region of a silicon planar bipolar transistor is examined in the charge collection mode, using the base–emitter junction to collect the video signal, the normally black-on-grey contrast would change polarity (white-on-grey) if the junction was reverse biased to near breakdown.

Studies of silicon integrated circuit capacitors by the SEM in the charge collection mode provide valuable data relating to the influence of radiation on passivating dielectric layers and, in particular, to the influence of charges that build up when the insulator is bombarded directly by electrons (Hezel, 1979). The structure of such capacitors is shown in Fig. 7.28; in an MOS

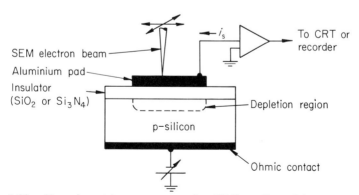

Figure 7.28. Experimental arrangement for EBC studies of integrated circuit capacitors in the scanning electron microscope. (Hezel, 1979)

structure the insulator is SiO_2, and in MNOS it is Si_3N_4. Investigation of SiO_2 under the influence of electron bombardment has been undertaken by Pensak (1949) and by Ehrenberg et al. (1966). These previous measurements showed that fairly large steady EBC gain is attained if the SiO_2 is very pure. However, measurements on electron bombarded silicon targets in devices (see above in Section I) indicated that one of the most important effects associated with an SiO_2 layer on or near a p–n junction is the charging of the insulator because of trapped holes in the bulk. Electrons are mobile in SiO_2, and therefore the oxide charges positively, and the electric fields arising as a result cause changes in p–n junction characteristics adjacent to it. Similar effects in the oxide occur if it is close enough to the electron beam impact area to be irradiated by soft X-rays generated by the fast electrons. One way in which this effect can be prevented was discussed in Section I.F. When the structure shown in Fig. 7.28 is bombarded by the electron beam in the SEM, a current flows in the circuit containing the capacitor mainly as a result of EBC in the dielectric; the Si merely acts as a substrate contact.

Figure 7.29 shows how the induced current depends on the time during which the capacitor is irradiated by a 30 keV electron beam at a current density of $1 \cdot 8 \times 10^{-4}$ A cm^{-2}; the scanned area was 30×30 μm^2 and an average electric field of 1 MV cm^{-1} was applied between the electrodes. Two features are immediately apparent: the EBC gain is smaller for Si_3N_4, and the change in gain with irradiation time is larger in SiO_2. These results were

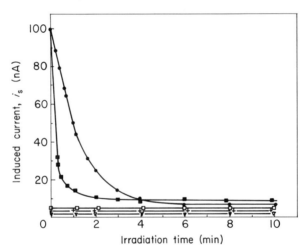

Figure 7.29. EBC of SiO_2 and Si_3N_4 as a function of irradiation time: ■ SiO_2, wet, 350 nm; ● SiO_2, dry, 630 nm; □ Si_3N_4, 840°C, 200 nm; ▽ Si_3N_4, annealed at 1050°C; + Si_3N_1, low pressure, 800°C (V_p = 30 kV, i_p = $1 \cdot 6 \times 10^{-9}$ A). (Hezel, 1979)

discussed in Section I.G but they were interpreted by Hezel as implying that only electrons are mobile in SiO_2 whereas neither electrons nor holes are mobile in Si_3N_4. The results for both insulators from Fig. 7.29 were replotted as seen in Fig. 7.19 where the relative changes in gain with irradiation time are more obvious.

B. Nuclear radiation detectors

The possibility of constructing a solid-state radiation detector, as opposed to the rather bulky and fragile Geiger counter tube or gas-filled ionization chamber, became apparent with the first studies of the crystal counter. The preliminary measurements by van Heerden (1945) with AgCl and by experimentalists such as Ahearn (1948) and Hofstadter (1949, 1950) with diamond were described in Chapter 1. An added advantage was discovered by Voorhies and Street (1949) who showed that the pulse height response of an AgCl counter was proportional to the energy loss for particles throughout the range from almost the most weakly ionizing to the strongly ionizing such as mesons stopping in the crystal. However, as described earlier, the crystal counter failed to fulfil the requirements for a measuring instrument because no convenient method could be found for destroying the polarizing field due to trapped carriers. More than a decade later, when semiconductor nuclear particle counters were arousing interest, Grainger *et al.* (1960) showed that exactly the same law of proportionality between pulse height and particle energy applied in the case of high resistivity diffused silicon p–n junction diodes. This was tested for particles having masses throughout the range from that of the electron to that of carbon ions and fission fragments.

In nuclear spectroscopy the distribution of pulse heights (and hence the energy spectrum of the radiation due to radioactive emissions) is used to identify an isotope; the term "channel number" is often used in referring to the pulse height when measured by a pulse height analyser, and usually the channel number is proportional to the particle energy. Since individual particles can be detected, an instrument of this kind can be used to count the number within predetermined energy limits. As already shown above, in the description of a similar system for photon counting (Section I.A), this method can be used for pulse rates up to about 10^6 s^{-1}.

The basic principle of the semiconductor counter is similar to that of transient EBC time of flight measurements. The crystal is provided with conducting electrodes on opposite faces, and an electric field is maintained by applying a fixed potential between them. Usually the potential is made sufficiently high to empty the useful volume of the detector completely so that any carrier released in the bulk due to the passage of a nuclear particle will be separated and be collected by the electrodes. The passage of the released electrons and holes gives rise to a conduction pulse whose total charge is measured by the amplifier associated with the circuit containing the crystal. Unlike the method used for drift mobility measurements, the specimen need not be rectangular in cross-section and may be hemispherical (Canali et al., 1976) or coaxial (Bradshaw, 1978). Since the total charge collected is important for this application, little significance is attached to the shape of the pulse.

The size, and sometimes the construction, of a detector depends on the penetrating power of the particle. For example, a γ-ray is far more penetrating than an α-particle or a β-ray of the same energy, so this will demand a detector of greater stopping power and nearly always of greater size.

For nuclear spectroscopy, one of the most important parameters is energy resolution; this is the width of the output pulse measured between its half-height points, when the pulse height is converted to units of particle energy. The pulse width (FWHM) is a function of the energy gap of the semiconductor and tends to favour low-gap solids for a high resolution. When the entire energy of an ionizing particle is not deposited inside the sensitive area of the detector, the variance in the useful number of generated electron–hole pairs (N) is given by Poissonian statistics as var$N = \bar{N}$, the mean value of N. The pulse height/noise ratio is thus equal to $(\bar{N})^{1/2}$ where $\bar{N}$ is dependent on the geometrical factors (shape, size, etc.) of the detector and on the number of pairs generated per unit energy absorbed. However, if all the energy is deposited inside the useful volume, var$N = F\bar{N}$ where F is the Fano factor; the pulse height/noise ratio is then equal to $(\bar{N}/F)^{1/2}$ where F has a value of about 0·2 for Ge and Si. Qualitatively this is seen as fluctuations in pulse height arising as a result of partly coherent noise sources rather than as the uncorrelated fluctuations that give rise to purely Poissonian statistical variations. For complete charge collection, a depletion width extending right across the sensitive volume between the electrodes is desirable; this requires a high resistivity semiconductor. A semiconductor with a long minority carrier lifetime will also have a lower reverse bias leakage current when in a diode configuration. For γ-ray spectroscopy a high atomic number (Z) is also desirable because the stopping power to this radiation is strongly dependent on Z. For charged particle detectors (α-particles, protons, etc.) this requirement is not so important since their range is generally smaller.

Bradshaw (1978) gave the following list summarizing the materials requirements just described: (a) low energy gap; (b) high resistivity; (c) long minority carrier lifetime; (d) high carrier mobility; (e) good stopping power; (f) ease of forming p–n junctions. The last requirement is listed mainly for low gap materials where a p–n junction is necessary to keep the leakage current small. The technological processes for making both p-type and n-type material by doping are by no means universal and it cannot even be

Table 7.3. Semiconductor materials for solid-state nuclear radiation detectors.

Material	Energy gap (eV)	Carrier drift length (mm)	Absorption coefficient for 100 keV γ-rays (cm^{-1})
Ge	0·76 at 77 K	10^3	2·7
Si	1·13	10^3	0·4
GaAs	1·42	1·0	2·7
CdTe	1·5	1·0	9·7
HgI_2	2·1	1·0	21
PbI_2	2·5	10^{-3}	high

assumed *a priori* that such techniques will ever be possible with all semiconductors.

Table 7.3 lists possible materials that satisfy at least some of the requirements given above. Many of the high mobility semiconductors described in Chapter 6 which were evaluated by transient EBC methods were investigated in the first place because they were candidates for semiconductor nuclear particle detectors. For γ-ray spectroscopy, silicon carbide, cadmium telluride, gallium arsenide, and mercuric iodide are promising, although ultrapure germanium cooled to liquid-nitrogen temperatures is used at present. For charged particles, silicon Schottky barrier detectors are usually employed owing to the advanced knowledge of the techniques for making them, and because silicon with a wide range of resistivities and minority carrier lifetimes is consistently available.

The depletion layer thickness in silicon or germanium when a barrier is back-biased is given by

$$w = B(\rho v)^{1/2} \tag{7.9}$$

where w is in μm, the resistivity ρ is in ohm-cm, and the bias v is in volts. The constant B for a wide range of parameters has the following values: n-Si 0·5; p-Si 0·3; n-Ge 1·0; p-Ge 0·65.

Figure 7.30 shows the principle of an ultrapure germanium detector. The starting material is typically germanium with a residual impurity concentration of a few parts per 10^{10} cm^3. A low-temperature process for making a p–n junction is to use shallow diffused n-type lithium doping in very weakly p-type Ge and then Au for the ohmic back contact. The energies of interest in γ-ray spectroscopy are typically in the region between 100 keV and 10 MeV. Such rays have a 95% range in Ge of about 1·5 cm and 11 cm respectively. A germanium detector for high energy γ-rays might weigh as much as 600 g.

The γ-ray interacts with the solid by a number of mechanisms; their relative importance varies with energy and thus alters as the γ-ray nears the

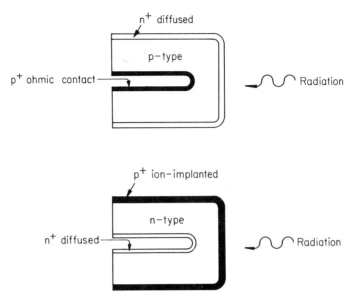

Figure 7.30. Configuration for ultrapure germanium γ-detectors. (Bradshaw, 1978)

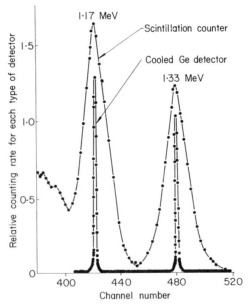

Figure 7.31. Pulse height spectrum for a germanium detector at 145 K (FWHM = 2·8 keV) and a scintillation counter at room temperature (FWHM = 60 keV); the γ-ray source was Co-60.

7. APPLICATIONS IN ELECTRON DEVICES 309

end of its range. Initially, a high energy γ-ray interacts with inner and outer shell electrons of the atoms from which the solid is composed, the prime processes being Compton scatter and absorption. In this energy range, ionization well away from the primary γ-ray can occur by fluorescence transfer and Auger processes; then, as the γ-ray loses more energy, Rayleigh scattering takes place and also photoelectrons are produced internally. Ultimately about a third of the band gap energy is expended in generating electron–hole pairs. Accurate pair production energies (radiant ionization energies, E^*) were given in Fig. 2.24. An important factor in detector design is the almost complete loss of sense of direction by the γ-ray as it nears the end of its path; the geometrical shape of the detector takes this into account.

The difference in resolution capability between a germanium detector and a scintillation detector is shown in Fig. 7.31. The FWHM for cooled Ge is about 20 times less than that of a scintillation counter, but a scintillation counter is more portable, does not need cooling, and is less expensive.

Charged particle detectors are less bulky than γ-ray detectors since the particles are less penetrating. Range–energy curves for a number of nuclear particles in silicon are shown in Fig. 7.32. Very often silicon surface barrier detectors are made from high purity silicon using a gold Schottky barrier contact on the face through which the particles enter. Methods of fabrication are similar to those of the Schottky barriers described above in Section A.

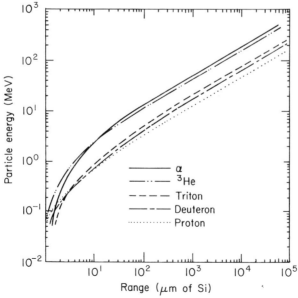

Figure 7.32. Range–energy curves for charged particles in silicon. (Bradshaw, 1979)

310 ELECTRON BOMBARDMENT INDUCED CONDUCTIVITY

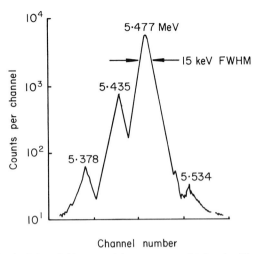

Figure 7.33. Typical Am-241 α-particle spectrum obtained with a partly depleted silicon Schottky barrier detector. (Bradshaw, 1979)

A typical pulse height (energy) spectrum of α-particles emitted from an Am-241 radioactive source is shown in Fig. 7.33. The energy resolution or FWHM of 15 keV at 5·477 MeV is comparable to the best resolution obtainable from a gas-filled ionization chamber (an argon–methane Frisch grid chamber) which is about 30 keV.

Although diamond has a fairly small atomic number and also a wide energy gap, it has been used for α and β spectroscopy at high temperatures. Kozlov et al. (1975) demonstrated this in class II natural diamonds (diamonds containing naturally incorporated nitrogen and aluminium impurities). An extremely important feature was their ability to produce injecting contacts, since this enabled them to overcome polarization phenomena. An injecting contact for holes could be produced by silver paste followed by heating in air for 2–3 h at 600°C. An injecting contact for electrons could be made by ion-implantation of P, Li, or C ions. In use, the direction of applied bias was so chosen that the top (bombarded) electrode was non-injecting and the bottom contact was injecting for the other sign of carrier. A non-injecting contact could be made by an evaporated thin film of Au, Ag, or Pt. As might have been expected, the energy resolution was not as great as that of Ge or Si detectors; the FWHM resolution was 120 keV for 5·5 MeV α-particles and 85 keV for 482 keV β-particles.

The most comprehensive study of CdTe γ and β particle radiation counters was published by Taguchi et al. (1978). A newly made CdTe detector would initially exhibit a FWHM resolution of 8 keV for 59·8 keV γ-rays and 12 keV for 122 keV γ-rays. However, the drift mobilities both of electrons and of holes were reduced after continuous exposure of the detector to Co-60 γ-rays (energies 1·17 and 1·33 MeV) at a dose rate of 10^5 rads h^{-1}. This was accompanied by the appearance of an electron trapping level at 0·5 eV and a hole trap at 0·14 eV; they caused a marked loss in pulse height resolution, but it could be recovered after annealing to 600°C to reduce the electron trap density. This shows that prolonged γ-irradiation above the displacement threshold causes defects which can subsequently be annealed. Shorter-term polarization phenomena in CdTe could be overcome by using thin metal insulator semiconductor structures (Siffert *et al.*, 1976). The performance of CdTe as an α-particle detector was reported by Canali *et al.* (1971b); a FWHM resolution of 29 keV was observed when counting 5 MeV α-particles.

III. EBC of insulating films in electron devices

A. Ebicon (Ebitron, Uvicon) television pickup tube

The first television pickup tube to employ an amplifying target using the EBC gain of an insulator was named Ebicon (Westinghouse Electric Corporation, 1957). An ultraviolet sensitive version (Uvicon) developed for space astronomy was described by Schneeberger *et al.* (1961); a diagram of a modern version (Ebitron) of such a tube is shown in Fig. 7.34. The tube consists of an evacuated glass envelope containing an image section and a scanning section; the image section has a semitransparent photoemissive cathode deposited on the wall at one end, and a lightly supported thin continuous EBC insulating target at the other end. The side of the EBC target that faces the photocathode is covered with a thin electron-permeable

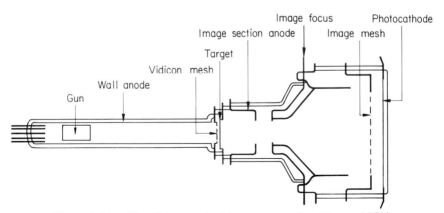

Figure 7.34. The Ebitron television camera tube. (Dawe, 1971)

conducting film; the scanning section is similar to a 12 mm vidicon, and its electron gun faces the other surface of the EBC target.

In operation, the free surface of the target is stabilized at the potential of the gun cathode by scanning with low-velocity electrons. The image of the scene under observation is focused on the photoemissive cathode, and photoelectrons are emitted in proportion to the local brightness of the scene. Photoelectrons corresponding to the image are accelerated to an energy of about 12 keV and focused on the EBC target by means of the electrostatic field provided by the internal electrodes in the tube. The photocathode is operated at a high negative potential so that the target and its associated circuits can be operated near ground potential. The signal contact consists of an aluminium film on the image section side supported on a 30 nm thick membrane of Al_2O_3 produced by an anodizing process. The operating technique is very similar to that of the SIT television pickup tube described above in Section I.D.

A potential of about 30 V is maintained across the target which consists of evaporated ZnS about 0·1 μm thick. When this is bombarded by the 12 keV electrons, conduction currents flow through the ZnS thin film insulator in direct proportion to the incident current. Thus a charge pattern is built up on the surface of the ZnS dielectric corresponding to the scene under observation, and this remains without degradation by virtue of the high surface and bulk resistivities of the target. When the target is scanned by the electron beam, the surface is restabilized at cathode potential. A current is induced in the signal contact corresponding to the negative charge deposited, and this is equal in magnitude to the positive charge instantaneously being scanned. The EBC gain provides a pre-scanning charge amplification of between 100 and 300.

With photocathode and target areas of 1 cm^2 and 0·25 cm^2 respectively the tube will have a luminous sensitivity of between 2 and 60 mA per lumen, depending on target bias; the limiting resolution is about 32 lines mm^{-1}, and the particular design of image section provides a uniform resolution over the sensitive area (Mayo and Bennett, 1972).

The Ebitron was employed in a satellite experiment described by Gricourt (1971) which involved a study of cosmic rays in space using a spark chamber. The high sensitivity camera tube just described was used to track the trajectories of the particles by using a television raster scan having only 10 lines. This information over a period of about three months in orbit was transmitted back to an earth station where the results were analysed by computer.

Further applications of this tube have been for low light level surveillance in security areas illuminated by no more than starlight (10^{-5} lux), using a television 625-line scanning system with a picture repetition rate of 25 s^{-1}

but without the normal interlace on the camera tube occurring midway between successive scans. This effectively makes full use of the entire 40 ms between scans to allow efficient integration of the amplified emitted photo-electrons, which never normally occurs fully with an interlaced scan on any camera tube. In a camera designed for intensification of 1 MeV electron microscope images, the tube is arranged beneath the phosphor screen outside the microscope vacuum and the tube coupling is achieved through an optical system. Another use in a low light level television camera is for the examination of unopened packages in the security area of aircraft loading bays or in customs examination; a typical search times is 4 s when a camera tube of the kind described is used in conjunction with an X-ray tube and a fluoroscopic screen.

It is interesting to notice from the reported results that much higher EBC gains are achieved with an electron beam contact of the kind used in this tube than are obtained with sandwich layers made in the same laboratory; the latter yielded results substantially in agreement with those described in Chapter 3. The highest gain measured at room temperature on an experimental As_2S_3 Ebicon was 1400 at 10^{-11} A bombarding current, a bias of 60 V, and a bombarding voltage of 29 kV. Bowman (1972) reported a similar effect of a low velocity electron beam contact to thin evaporated layers of ZnS (see Fig. 3.11).

B. Computer mass memory (Beamos, Ebam)

Although polarization effects in SiO_2 passivated silicon p–n junction diodes cause a significant change in diode characteristics when they are subjected to electron bombardment (see Section I.G), in the first instance this effect is reversible. The polarization due to trapped holes can be eliminated by further bombarding the oxide with reverse bias applied across it, or even under zero bias conditions. This effect has been employed to provide a memory target in a number of electron beam addressed computer storage tubes.

MacDonald and Everhart (1968) made a storage tube containing a target consisting of a large number of discrete MOSFETs (metal–SiO_2–silicon field effect transistors) and used the storage effect caused when the SiO_2 gate insulator was polarized by electron bombardment. The bulk trapping of holes in SiO_2 is a fairly long-term phenomenon at room temperature, and charges "written in" by the electron beam remained for several months without substantial change. The information stored as a polarization pattern in the oxide was read out by row–column access to sense the channel conductance of the semiconductor, which in turn was a function of the oxide charge.

An oxidized silicon plate covered with a thin conducting layer of aluminium was used as a storage tube target by Cohen and Moore (1974). The base material was p-type silicon coated with a p-type epitaxial layer of Si, and writing could be achieved by scanning the target with a modulated electron beam while applying a potential between the silicon and the aluminium. Reading was achieved by scanning the target with a fixed beam current under zero bias conditions. Within charged areas of the oxide, the positively charged insulator caused inversion of the semiconductor, and lateral conduction of electrons in this inversion layer resulted in efficient separation of electron–hole pairs which were generated by the electron beam. Thus a large signal due to pairs generated within the space charge region of the silicon could be generated across the Si–Al contacts. The uncharged oxide regions, corresponding to unwritten areas or a computer "zero", on the other hand gave rise to a very small output signal since the inversion layer was not formed and such processes did not occur. This tube was given the name Ebam (Speliotis, 1975).

A modified version of this last mentioned tube was extensively developed, and its target structure and mode of operation were described by Ellis *et al.* (1974); part of the target in cross-section is sketched in Fig. 7.35 which also shows the external circuit connections and the electron beam. The useful area of the target is about 30 mm square and consists of a planar slice of 400 ohm-cm p-type silicon on which an epitaxial layer of n-type material with a resistivity of 1 ohm-cm is grown. One connection to the target is made to the p-type substrate and another to the n-type layer which is covered by a thermally grown layer of SiO_2 about 0·4 μm thick; a top connection is made to this by a thin aluminium film 800 Å thick. Unlike the target described by

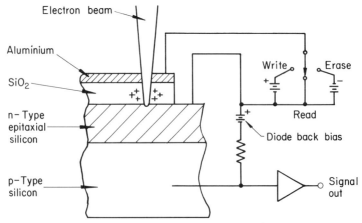

Figure 7.35. Cross-section of MOS memory chip. (Hughes *et al.*, 1975)

MacDonald and Everhart, the storage surface is not structured, and the size of a storage element is determined almost entirely by the size of the focused electron beam.

The writing process depends on the establishment of a polarization pattern corresponding to the computer "ones" and "zeros" within the SiO_2 layer. This is achieved by applying a positive potential to the aluminium metallization with respect to the Si substrate while scanning a modulated 10 kV beam current over the target surface. The beam current is typically 0·15 μA in a focused spot of about 4 μm diameter. The charge pattern, which is relatively stable at room temperature, is sensed by scanning the target again with a steady beam current, but this time with zero bias applied across the oxide. The beam range is sufficient to penetrate the overlying Al and SiO_2 layers and to generate carriers within the n-type epitaxial layer. Since the minority carrier (hole) diffusion length in the epitaxial layer is greater than its thickness, surface recombination at the Si–SiO_2 interface dominates over electron–hole recombination in the bulk. The generated carriers leave the generation region mainly by diffusion to the p–n junction at the oxide interface. Holes generated in the epilayer and which diffuse to the junction constitute a signal which can be detected and amplified by the circuit associated with the p–n junction.

The probability (P) that a hole generated at a distance x from the Si–SiO_2 interface will reach the edges of the p–n junction at a distance d from the interface is given by

$$P = [1 + (S/D)x]/[1 + (S/D)d] \qquad (7.10)$$

where D is the hole diffusion coefficient in Si (see Chapter 6) and S is the surface recombination velocity at the Si–SiO_2 interface. When the oxide is polarized by positive charge trapping, the silicon surface is in accumulation and S can be very small (less than 10^3 cm s^{-1}), whereas in the flat-band condition or in depletion it can be very large (more than 10^6 cm s^{-1}). The stored positive charge in the oxide modulates S, and thus changes the magnitude of the electron bombarded silicon diode current according to the magnitude of the charge originally written in. The EVE diode gain g is shown in Fig. 7.36 as a function of the total flux of electrons used for polarizing the SiO_2 at a bias of 45 V; also shown is g during the depolarizing process at a bias of zero. It can be seen that g changes from about 10 after depolarizing to about 400 after polarizing the oxide.

This target was used in a computer mass memory device described by Hughes *et al.* (1975) and given the name Beamos (Fig. 7.37). The process of writing corresponded to oxide polarization by modulating the amplitude of the current according to the pattern of ones and zeros being stored. Reading was accomplished as shown above by scanning a fixed beam current across

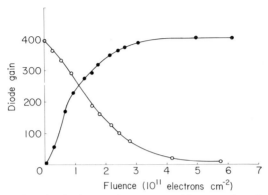

Figure 7.36. Diode gain during readout at zero bias as a function of irradiation dose at a bias of 45 V; the discharge characteristic after writing to saturation is also shown. (Ellis et al., 1974)

the target but at 5% amplitude so that the oxide charge was not all neutralized in a single scan. An "erase" process could be employed by scanning a fixed beam over the target but with reverse bias applied to the oxide; the same procedure could be used for destructive reading. These bias conditions are indicated diagrammatically in Fig. 7.35. The extremely high resolution in the focused electron beam was achieved by an unusual double-step deflection and focusing system. As indicated in Fig. 7.37, the electron beam was deflected on to a matrix lens in front of the target. This lens provided post-deflection focusing by using a triple element einzel lens grid consisting

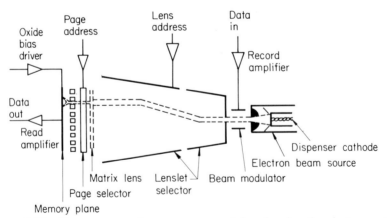

Figure 7.37. Beamos computer mass memory tube, showing the electron optical system; for simplicity, the evacuated glass envelope is not shown. (Hughes et al., 1975)

of 289 square apertures in a plane about 30 mm square, which significantly reduced deflection defocusing. The tube as described had a capacity of 32×10^6 bits, a random position access time of 30 μs, and a serial data transfer rate of 10 Mbit s^{-1}. The latent storage time was about 1 month without rewriting, and about 20 consecutive readouts were possible before significant loss of data occurred.

C. Graphechon scan converter

As a direct requirement for bright distributed radar displays, a number of different types of storage tube for scan conversion or signal processing applications have been developed. Charge storage tubes are electron devices where an electrical signal is written on the face of an insulating target (typically evaporated ZnS) in the form of a modulated two-dimensional charge pattern which is subsequently read out as a television signal by means of a scanned electron beam. Owing to the high volume and surface resistivities, when there is no excitation, of the insulators used as targets in tubes of this kind, the charge pattern remains without appreciable deterioration in the absence of readout for times varying from several seconds to many weeks depending on the particular insulator. One tube built specifically for scan conversion is the Graphechon which uses the EBC of an insulating target for writing and storage. This tube, first described by Pensak (Chapter 3, Figs 3.4 and 3.5), originally used a thin film of SiO$_2$ as the storage target. The tube may be constructed in more than one version; Pensak described experimental tubes in which the writing gun and the reading gun are on the same side of the target at a slight angle to each other, a similar tube in which both guns are electrostatically focused and deflected, and another in which the two guns are on opposite sides of the target which is normal to the axis of both of them.

The last mentioned tube (Fig. 7.38) consists of an evacuated envelope containing an electron gun at each end, facing opposite sides of the storage target which is approximately in the centre of the envelope. The target consists of a thin insulating layer supported on a fine electroformed silver, copper, or nickel mesh having about 40 threads mm^{-1}. The mesh may be coated with a thin aluminium skin by vacuum evaporating a layer of aluminium 0·1 μm thick on to a nitrocellulose layer which is subsequently removed by baking. A thin layer of insulator is then deposited over the aluminium skin which acts as one of the contacts when the insulating film is used as a charge storage target with EBC writing. In some tubes no aluminium is used over the nitrocellulose. The velocity of the writing electrons is sufficient for them to penetrate the aluminium without appreciable energy loss in those versions where this is applicable.

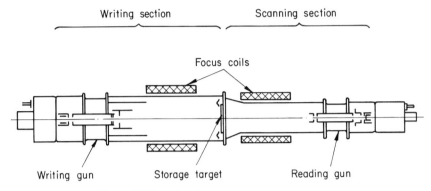

Figure 7.38. Graphechon scan converter tube.

In operation, a target bias field is established by holding the mesh supporting it at about 200 V negative with respect to the tube collector electrode. The collector electrode (not shown in the figure) is a conducting ring close to the envelope walls and fixed on the free side of the target in a plane parallel to it at a distance of about 40 mm. The target is prepared for the writing process by scanning a modulated electron beam from the other gun over the target. The about 1·5 keV. Since the secondary emission coefficient of the insulator exceeds unity at this bombarding energy, its free surface stabilizes by secondary emission at collector potential. The writing process is now established by scanning a modulated electron beam from the other gun over the target. The writing beam voltage is between 8 and 12 kV which is sufficient for complete penetration of the insulator and to induce conductivity in proportion to the instantaneous value of the beam current. The induced currents, which are much larger than the beam currents, thus tend to discharge the elementary areas of the target corresponding to those immediately beneath the scanned spot. In this way, when the entire target has been scanned by the writing beam, the free surface contains a stored surface charge pattern corresponding directly to the signal applied to the writing gun.

The charge pattern may be read out simultaneously or later. The reading beam is scanned over the target in a similar way to that used for preparing the target for writing; variations to the scanning pattern are also possible as described below. The process of reading thus restabilizes the target surface at collector potential by depositing a charge equal but opposite to that which flowed during the writing process. This charge is detected as a video signal in the collector electrode circuit after removing the dc component corresponding to the unused portion of the beam. Since the reading beam current may be insufficient to discharge the target completely in a single scan, a number of successive reading operations may be performed.

This tube has several variations depending on the particular application. In the case of bright distributed radar displays it is normal to show the radar picture in the form of radar echoes from a rotating aerial. The echoes are typically completely bright spots on the display, and a radial scan synchronized with the radar aerial is used for the writing process; this might take approximately 1–10 s for a complete cycle. Scan conversion is now achieved by using a standard TV rectangular raster for readout, which may be e.g. 625 lines at 25 frames s^{-1}. Since the process of reading also partly erases the stored charge pattern, the signal output gradually fades in the absence of a new writing signal. The rate at which it fades may be controlled to a degree by reducing the reading beam current and compensating for the smaller output signal by a higher video amplifier gain. This cannot be extended indefinitely, however, since eventually amplifier noise becomes significant compared with the signal.

Another method may be employed to increase the reading time, by making use of the secondary EBC effect. If the target is treated so that it contains class II centres, secondary currents flow which increase the time delay between writing and cessation of induced conduction currents. A typical tube may be adjusted to have a storage time from about 6 s to 60 s with continuous readout. Another version makes use of a double-layer dielectric. The insulating target, which is chosen for its EBC properties in this application, is coated with a thin evaporated layer of an insulator such as MgF_2 which is selected for its secondary electron emission properties. A theoretical analysis of the double-layer dielectric will be given in the Appendix; in this application, the basic principle is for the charge pattern which is initially stored on the two component capacities in series to be transferred by the reading beam scanning process to the two capacities in parallel. Since the CaF_2 layer is so thin, most of the charge will thus be transferred to this overlayer. The storage time may be made longer by this process.

The final sequence in the operation cycle of a scan converter storage tube of this type is known as erasing which is achieved by scanning the target with the reading beam in the absence of a writing signal and with a sufficiently large current for the target to stabilize all points on its surface at collector potential. In most cases, vestiges of earlier inputs can be completely erased faster if this operation is preceded by writing a uniform peak white signal over the entire target.

The most important application is for conversion of radar radial displays into a television format having standard parameters. In this way it is also possible to maintain accurate angular information on the displays in different locations without complicated interconnections. It is possible to display the video information on standard bright TV monitors which are quite

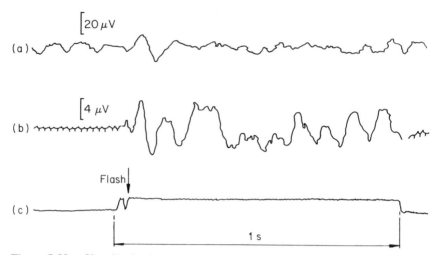

Figure 7.39. Signal/noise improvement by integration of 25 evoked electroencephalograph signals in a storage tube: (a) output in the presence of noise; (b) integration of 24 such outputs; (c) trace showing position of stimulating light flash. (Gibbons, 1961)

suitable for viewing in a well lit control room. It is further possible to mix video signals from the radar and another source so that a map of the area being scanned is superimposed on it.

Another application is for integration of repetitive signals buried in noise to improve the signal/noise ratio. If such signals are superimposed in successive writing scans, the signals will add coherently as the charge pattern is built up, but the noise components will add in quadrature since there is no correlation between them. Thus, if there are N successive superimposed noisy signals, the signal/noise ratio of the output will be increased by a factor of $N^{1/2}$. Storage tubes have been used in medical electronics to improve the signal/noise ratio in the case of a weak electromyograph signal from a stimulated muscle, and also in the case of an evoked electroencephalograph signal. In the latter case the noise components are not due to the amplifier but are disturbances due to random brain activity. The result of such a signal processing experiment is shown in Fig. 7.39.

The basic EBC target electron beam readout system just described has also been applied experimentally to a tube for television standards conversion by reducing the image retention time to about 40 ms by lowering the target capacity and increasing σ (Guillard, 1966). This might be used for converting an 819-line picture to a standard 625-line format.

Appendix

I. Two-layer dielectric

The problem of the double-layer dielectric is relevant to the application of EBC in drift velocity measurements where carrier injection is inhibited by the use of a thin blocking glass film (an example was given in Chapter 5), and also in connection with a scan-converter storage tube described in Chapter 7.

The physical picture of the two-layer dielectric is shown in Fig. A.1. The first part of the dielectric film is assumed to be built up from a material having leakage resistance R_1 and capacitance C_1. The product R_1C_1 is the dielectric relaxation time-constant of the material from which the first layer is made. This layer is now coated with another dielectric for which the corresponding values of capacitance and leakage resistance are C_2 and R_2.

We now wish to answer a question that might arise if a step function voltage pulse is applied across the two capacitors in series. How does the voltage across either capacitor change with time? Since the voltage step is known, it is sufficient to investigate how the potential varies at the point where the two capacitors are joined.

This problem is readily solved through use of Thévenin's theorem which states that, if a network is viewed from two points, it can be replaced by an equivalent voltage source V_{oc} and an equivalent impedance Z_{th} in series.

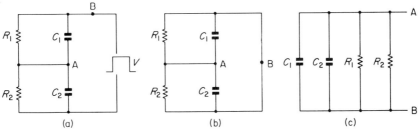

Figure A.1. Two-layer dielectric: (a)–(c) the transformations needed to solve the problem as a network using Thévenin's theorem.

Here V_{oc} is the open-circuit voltage appearing at the terminals, and the series impedance is what would be seen by looking into the pair of terminals with all the energy sources turned off and shorted. These transformations are shown in Fig. A.1(a)–(c).

The voltage V_{oc} is determined by the initial conditions. It can now be seen that, if a voltage step of magnitude V is applied as shown in (a), the condenser C_1 charges up at the same rate as a capacitance $(C_1 + C_2)$ through a resistance $R_1 R_2/(R_1 + R_2)$ from a potential $V' = VC_2/(C_1 + C_2)$ to a potential $V'' = VR_1/(R_1 + R_2)$. This immediately yields

$$V_1 = [VC_2/(C_1 + C_2)]\exp(-t/\tau) + [VR_1/(R_1 + R_2)][1 - \exp(-t/\tau)] \quad (A.1)$$

where $\tau = [R_1 R_2/(R_1 + R_2)](C_1 + C_2)$. In the special case where $R_1 C_1 = R_2 C_2$, $V' = V''$ for all t, and the potential of the junction between the two dielectrics after applying a voltage step remains constant.

II. Secondary emission of insulators (Gibbons, 1966)

A. Surface layers and evaporated films

Material	Form	σ_{max}	$V_{p\,max}$	V_{p1}	V_{p2}
LiF	film	5·6	—	21 V	—
NaF	film	5·7	—	21 V	—
NaCl	film	6–6·8	600 V	15 V	1·4 kV
KCl	film	7·5–8·0	1500 V	15 V	>10 kV
RbCl	film	5·8	—	—	—
CsCl	film	6·5	—	—	—
NaBr	film	6·25	—	—	—
KI	film	5·6	—	22 V	—
NaI	film	5·5	—	—	—
CaF_2	film	3·2	—	34–48 V	—
BaF_2	film	4·5	—	—	—
MgF_2	film	4·1	410 V	50 V	>4 kV
BeO	layer	3·4–8	200–400 V	120 V	900 V
MgO	layer	2·4–17·5	400 V–1·6 kV	<100 V	—
Al_2O_3	layer	1·5–3·2	350 V–1·3 kV	37–40 V	—
Cu_2O	layer	1·19–1·25	440 V	—	—
PbS	layer	1·2	500 V	—	—
MoS_2	layer	1·10	—	—	—
WS_2	layer	0·96–1·04	—	—	—
ZnS	layer	1·8	350 V	—	—
MoO_2	layer	1·09–1·33	—	—	—
Ag_2O	layer	0·90–1·18	—	—	—
SiO_2	film	2·2	300 V	48 V	1·4 kV
Cs_2O	layer	2·3–11	800 V	—	—

B. Single crystals

Material	σ_{max}	$V_{p\,max}$	V_{p1}	V_{p2}
Ge	1·2–1·4	400 V	—	—
Si	1·1	250 V	—	—
Se	1·35–1·40	400 V	—	—
C (Diamond)	2·8	750 V	—	—
NaCl	14	1·2 kV	10 V	—
NaBr	24	1·8 kV	—	—
KCl	12	—	10 V	—
KI	10·5	1·6 kV	12 V	—
KBr	12–14·7	1·8 kV	10 V	—
MgO	23	1·2 kV	—	—
SiO$_2$ (Quartz)	2·1–2·9	400–440 V	50 V	2·3 kV
Mica	2·4–2·75	325 V	25–40 V	3·0 kV
Al$_2$O$_3$–MgO (Spinel)	3·6	450 V	<300 V	>3·5 kV
Al$_2$O$_3$–Cr$_2$O$_3$ (Ruby)	3·8	900 V	<300 V	>3·5 kV
GaAs	1·1	350 V	—	—
InSb	1·15	700 V	—	—

C. Glasses

Material	σ_{max}	$V_{p\,max}$	V_{p1}	V_{p2}
Pyrex	2·3	240–400 V	40 V	2·3–2·4 kV
Nonex	—	—	150 V	1·2–5·0 kV
Soda	2·1	300 V	—	900 V
Lime	—	—	70 V	>2 kV
Cover	1·9	330 V	<60 V	1·7 kV
Ground	3·1	420 V	—	3·8 kV
Jena filter glasses				
GG12	2·7	400 V	—	1·5 kV
BG2	3·2	500 V	—	2·2 kV
UG6	3·8	500 V	—	2·8 kV

III. Optimum scanning speed for constant charge imaging device

If a preamplifier is used in conjunction with a constant charge imaging device, the signal/noise ratio of the output is dependent on the rate at which the stored charge is discharged. If the system bandwidth is adjusted so that it is only just adequate to resolve the individual image points spatially (the Nyquist criterion), there is an optimum scan rate for the largest signal/noise

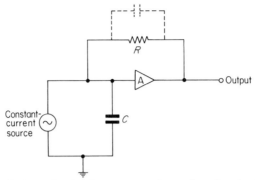

Figure A.2. Equivalent circuit of a constant charge imaging device: A, the head amplifier; R, feedback resistor; C, stray capacitance in shunt with the source (i.e. the imaging device).

ratio or pulse-height resolution. This optimum applies to all electron devices that depend on charge accumulation, storage, and then sequential discharge. The analysis can therefore be applied to the self-scanned Digicon and low light level television camera tubes.

We shall assume that the device accumulates a fixed charge independent of the scanning rate. We are free to select the discharge speed, and there is no restriction except to choose the system bandwidth (f_0) to equal the Nyquist sampling frequency whenever we change the scanning rate. A sketch of the circuit to be analysed is shown basically in Fig. A.2. In general, the amplifier A will be a low-noise one- or two-fet input opamp, possibly with a single or dual low-noise jfet at the input.

According to an analysis by James (1952), the noise current referred to the input is given by

$$I_n^2 = 4k\theta f_0(1/R + 4\pi^2 C^2 R_n/3) \tag{A.2}$$

where R is the feedback resistance, R_n is the equivalent (series) noise resistance of A, and C is the stray capacitance in shunt with the source. In practice it is possible to make R very large indeed (e.g. 10^{10} ohms), and if we now put $R = \infty$ in equation (A.2), and use the Nyquist criterion to set the signal current $I_s = af_0$, this yields I_s/I_n proportional to $f_0^{-1/2}$.

Thus the scanning should be made as slow as possible consistent with remaining above the frequency at which fet ($1/f$) noise starts to make a significant contribution. Then, since the output voltage will be $V_0 = 2qf_0R$, where q is the stored charge per image element, R is chosen to make V_0 as large as possible consistent with staying within the output voltage capabilities of the amplifier.

The feedback resistance R should be increased until its noise contribution is negligible in comparison with the signal. If this makes the amplifier saturate on large signals, either R must be reduced, with a corresponding noise penalty, or a separate feedback path through diodes can be provided for large amplitude signals. In the latter case the amplifier transfer function is non-linear, and subsequent compensation is required.

Supplying the fet bias current through R should not be a problem for the following reason (Green, 1980). If the bias current is I_b, the resulting shot noise current will be $I_n = (2qI_b f)^{1/2}$. The Johnson noise contributed from R at temperature θ is $I_r = (4k\theta f/R)^{1/2}$ where k is Boltzmann's constant. For equal noise contributions from the feedback resistor and the bias current, $2qI_b = 4k\theta/R$. Since the voltage developed across R by the fet bias current is $V_b = I_b R$, for equal noise contributions, $V_b = 2k\theta/q = 52$ mV at 300 K.

This shows that, if the value of R can be optimized according to these criteria, the total voltage drop due to the bias current is very small, and in practice, if it is not possible to make R as large as $2k\theta/qI_b$, the voltage drop will be less than 52 mV.

IV. Ramo's theorem

When charges move within the gap between a pair or a series of conducting electrodes, a displacement current can be detected in an external circuit even though the charges might not have moved far enough to reach any of them. This theorem shows how to calculate the externally measured current or charge.

In common with many other rules and laws in science, the theorem attributed to Ramo (1939) was not really discovered by the man whose name is usually associated with it. A perfectly correct statement, although unexplained, was made by Becker (1904). A simple derivation was published in a paper on photoconductivity by Gudden and Pohl (1923). The rule was used, again without explanation, by Hecht (1932). The theorem was generalized and formally derived by Shockley (1938) without reference to these earlier papers. A particular example was explained, in connection with the passage of electrons in a thermionic triode, by Ramo (1939) who was apparently quite unaware at the time that anything had been done before. An elegant proof given by Moore (1949) is reproduced here.

Consider the motion of a charge packet q from one plate towards another inside a plane parallel capacitor (Fig. A.3). The separation of the plates is L and the charge is situated at a distance x from the left-hand plate. A battery is included in the circuit to provide a bias v between the electrodes, and also a charge-sensitive meter to measure the charge dQ as seen externally.

When the charge q moves a distance dx between the plates, it expends energy $(v/L)q\,dx$ since the internal electric field between the plates is equal to

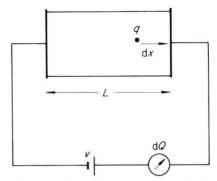

Figure A.3. Ramo's theorem. The charge q moves a distance dx inside the gap between two plane parallel condenser plates separated by a distance L; the battery provides a bias potential v between them, and the externally measured induced charge dQ is indicated on the meter.

v/L. This energy must have been supplied by the battery, so it is equal to vdQ where dQ is the externally measured charge. Thus, by the first law of thermodynamics,

$$(v/L)qdx = vdQ \qquad (A.3)$$

Equation (A.3) gives Ramo's theorem directly. In words, the externally measured charge displacement dQ due to the motion dx of a charge q between plane parallel electrodes at a separation distance L is $dQ = q(dx/L)$.

References

Abraham, J. M., Wolfgang, L. G. and Inskeep, C. N. (1966), *Adv. El. El. Phys.* **22B**, 671.
Adams, A. R. and Spear, W. E. (1964), *J. Phys. and Chem. Solids* **25**, 1113.
Adams, A. R., Gibbons, D. J. and Spear, W. E. (1964), *Solid State Commun.* **2**, 387.
Ahearn, A. J. (1948), *Phys. Rev.* **73**, 524.
Ahmad, C. N., Adams, A. R. and Pitt, G. D. (1979), *J. Phys. (C) Solid State Phys.* **12**, L379.
Aitken, D. W., Marcum, A. I. and Zulliger, H. R. (1965), *IEEE Trans. Nucl. Sci.* **NS-13**, 287.
Alberigi Quaranta, A., Casadei, G., Martini, M., Ottaviani, G and Zanarini, G. (1965), *Nucl. Instr. Meth.* **35**, 93.
Alberigi Quaranta, A., Canali, C. and Ottaviani, G. (1970), *Rev. Sci, Instr.* **41**, 1205.
Alberigi Quaranta, A., Jacoboni, C. and Ottaviani, G. (1971), *Revista del Nuovo Cimento* **1**, 445.
Alcock, R. N. (1962), thesis, University of London.
Allan, D. (1978), *Phil. Mag.* **B38**, 381.
Allan, D., Spear, W. E. and Le Comber, P. G. (1977), in *Proc. 7th Internat. Conf. on Amorphous and Liquid Semiconductors* (ed. W. E. Spear), p. 323, CICL, University of Edinburgh.
Allen, A. O. (1976), *Drift Mobilities and Conduction Band Energies of Excess Electrons in Dielectric Liquids*, U.S. Nat. Bur. Stand., National Standard Reference Data Series NSRDS–NBS 58.
Anderson, P. W. (1958), *Phys. Rev.* **109**, 1492.
Ansbacher, F. (1948), in *Opening of Biomolecular Research Laboratory, Birkbeck College, July 1st*, p. 6, Birkbeck College, London.
Ansbacher, F. (1950), thesis, University of London.
Ansbacher, F. and Ehrenberg, W. (1949), *Nature* (London) **164**, 144.
Ansbacher, F. and Ehrenberg, W. (1951), *Proc. Phys. Soc.* **A64**, 362.
Archangelskaya, V. A. and Bonch-Bruevich, A. M. (1951), *Dokl. Akad. Nauk SSSR* **77**, 229 (in Russian).
Archard, G. D. (1961), *J. Appl. Phys.* **32**, 1505.
Argue, A. N., Craven, P. G., Fruin, D. J., Jelley, J. V. and Smith B. J. (1977), *Astron. and Astrophys.* **58**, 27.
Aris, F. C., Davies, P. M. and Lewis, T. J. (1976), *J. Phys. (C) Solid State Phys.* **9**, 797.
Arrhenius, S. A. (1887), *Sber. Akad. Wiss. Wien* **96**, 831.
Ashburn, P. and Bull, C. J. (1979), *Solid State Electr.* **22**, 105.
Atkins, K. R. (1959), *Phys. Rev.* **116**, 1339.
Bakker, G. J. and Segrè, E. (1951), *Phys. Rev.* **81**, 489.
Bardeen, J. and Shockley, W. (1950), *Phys. Rev.* **80**, 72.
Bates, D. J., Knight, I. and Spinella, S. (1977), *Adv. El. El. Phys.* **44**, 221.
Bäuerlein, R. (1962), in *Radiation Damage in Solids* (ed. R. S. Billington), p. 358, Academic Press, New York, London, (Proc. Internat. School of Physics "Enrico Fermi", Ispra, 1960; course 18).
Baynham, A. C. (1969), *IBM J. Res. Dev.* **13**, 568.
Beaver, E. A., McIlwain, C. E., Choisser, J. P. and Wysoczanski, W. (1972), *Adv. El. El. Phys.* **33B**, 863.
Beaver, E. A., Harms, R. J. and Schmidt, G. W. (1976), *Adv. El. El. Phys.* **40B**, 745.

Becker, A. (1904), *Ann. Physik* **13**, 394.
Becker, A. and Kruppke, E. (1937), *Z. Physik* **107**, 476.
Becquerel, E. (1839), *Compt. Rend.* **9**, 561.
Becquerel, H. (1903). *Compt. Rend.* **136**, 1173.
Bell, R. O., Wald, F. V., Canali, C., Nava, F. and Ottaviani, G. (1974), *IEEE Trans. Nucl. Sci.* **NS-21**, 331.
Benda, H. (1951), *Ann. Physik* **6**, 413.
Benoit, J., Benalloul, P. and Mattler, J. (1969), *Z. Naturforsch.* **24a**, 1569.
Berger, M. J. (1963), in *Methods of Computational Physics,* Vol. 1 (ed. A. Alder, S. Fernbach and M. Rotenberg), Academic Press, New York.
Berger, M. J. and Seltzer, S. M. (1964), *Tables of Energy Losses and Ranges of Electrons and Positrons*, Report SP-3012, NASA, Washington.
Bethe, H. A. (1930), *Ann. Physik* **5**, 325.
Bethe, H. A. (1933), in *Handbuch der Physik*, Vol. 24, p. 519, Springer, Berlin.
Bethe, H. A. and Ashkin, J. (1953), *Experimental Nuclear Physics*, p. 276, Wiley, New York.
Billington, E. W. (1960), thesis, University of London.
Billington, E. W. and Ehrenberg, W. (1960), *Proc. Internat. Conf. on Semiconductor Physics* (Prague), p. 473.
Billington, E. W. and Ehrenberg, W. (1961), *Proc. Phys. Soc.* **78**, 845.
Birkhoff, R. D. (1958), in *Handbuch der Physik* (ed. S. Flügge), Vol. 34, p. 53, Springer, Berlin.
Bixby, W. E. and Ullrich, A. U. (1951), USA Pat. 2,753,278 (publ. 1956).
Blanchard, C. H. (1954), Circular 527, Nat. Bur. Stand., Washington, D.C.
Blatt, F. J. (1957), *Phys. Rev.* **105**, 1203.
Bleil, C. E. (1962), *J. Phys. and Chem. Solids* **23**, 1729.
Bloch, F. (1933), *Z. Physik* **81**, 363.
Bloembergen, N. (1945), *Physica* (Eindhoven) **11**, 343.
Bohr, N. (1915), *Phil. Mag.* **30**, 581.
Borsari, V. and Jacoboni, C. (1972), *Phys. Stat. Sol.* (b) **54**, 649.
Bothe, W. (1933), *Handbuch der Physik*, Vol. 22(II), 1.
Bowlt, C. (1967), *Brit. J. Appl. Phys.* **18**, 1585.
Bowlt, C. and Ehrenberg, W. (1969), *J. Phys. (C). Solid State Phys.* **2**, 159.
Bowman, A. (1972), M.Sc. report, Thames Polytechnic, London.
Bradberry, G. and Spear, W. E. (1964), *Brit. J. Appl. Phys.* **15**, 1127.
Bradshaw, A. (1978), *New Electronics* **11**, (May 16) 35.
Bradshaw, A. (1979), *New Electronics* **12** (April 17), 38.
Bril, A. and Gelling, W. G. (1962), *Philips Res. Rep.* **17**, 329.
Brivati, J. A., Gross, J. M., Symons, M. C. R. and Tinling, D. J. A. (1965), *J. Chem. Soc.* 1199.
Brodsky, M. H. (ed.) (1979), *Amorphous Semiconductors*, Springer, Berlin, Heidelberg, New York.
Broerse, P. H. (1966), *Adv. El. El. Phys.* **22A**, 305.
Brooks, H. (1955), *Adv. El. El. Phys.* **7**, 85.
Bube, R. H. (1960), *Photoconductivity of Solids*, Wiley, New York.
Bull, C. J., Ashburn, P. and Gowers, J. P. (1980), *Solid State Electr.* **23**, 953.
Bullard, E. C. and Massey, H. S. W. (1930), *Proc. Cambridge Phil. Soc.* **26**, 556.
Butkevich, V. G. and Butslov, M. M. (1958), *Radiotekhnika i Elektronika* **3**, 355.
Canali, C., Jacoboni, C., Ottaviani, G. and Alberigi Quaranta, A. (1971a), *J. Phys. and Chem. Solids* **32**, 1707.

Canali, C., Martini, M., Ottaviani, G., Alberigi Quaranta, A. and Zanio, K. R. (1971b), *Nucl. Instr. Meth.* **96**, 561.
Canali, C., Ottaviani, G., Taroni, A. and Zanarini, G. (1971c), *Solid State Electr.* **14**, 661.
Canali, C., Nava, F., Ottaviani, G. and Paorici, C. (1972), *Solid State Commun.* **11**, 105.
Canali, C., Loria, A., Nava, F. and Ottaviani, G. (1973), *Solid State Commun.* **12**, 1017.
Canali, C., Ottaviani, G., Bell, R. O. and Wald, F. V. (1974), *J. Phys. and Chem. Solids* **35**, 1405.
Canali, C., Jacoboni, C., Nava, F., Ottaviani, G. and Alberigi Quaranta, A. (1975a), *Phys. Rev.* **B12**, 2265.
Canali, C., Jacoboni, C., Ottaviani, G. and Alberigi Quaranta, A. (1975b), *Appl. Phys. Lett.* **27**, 278.
Canali, C., Gavioli, G., Losi, A. and Ottaviani, G. (1976), *Solid State Commun.* **20**, 57.
Carlson, D. E. (1980), *IEEE Spectrum* **17**, 39.
Carlson, D. E. and Wronski, C. R. (1979), in *Topics in Applied Physics*, Vol. 36, *Amorphous Semiconductors*, Ch. 10, Springer, Berlin.
Catalano, A. and Bhushan, M. (1980), *Appl. Phys. Lett.* **37**, 567.
Chang, D. M. and Ruch, J. G. (1968), *Appl. Phys. Lett.* **12**, 111.
Chen, I. (1970), *Phys. Rev.* **B2**, 1053.
Chittick, R. C., Alexander, J. H. and Sterling, H. F. (1969), *J. Electrochem. Soc.* **116**, 77.
Choisser, J. P. (1976), *Adv. El. El. Phys.* **40B**, 735.
Choisser, J. P. (1977), *Opt. Eng.* **16**, 262.
Choisser, J. P., Nather, R. E. and Tull, R. G. (1974), *Proc. SPIE* **44** (Astronomy), p. 83.
Clark, H. A. M. and Vanderlyn, P. B. (1949), *J. Inst. Elect. Eng.* (Part 3), **96**, 189.
Coblentz, W. W. and Emerson, W. B. (1917), *J. Wash. Acad. Sci.* **7**, 525.
Cohen, M. H. and Lekner, J. (1967), *Phys. Rev.* **158**, 305.
Cohen, M. S. and Moore, J. S. (1974), *J. Appl. Phys.* **45**, 5335.
Conwell, E. M. and Weiskopf, V. F. (1950), *Phys. Rev.* **77**, 388.
Cook, B. E. and Spear, W. E. (1969), *J. Phys. and Chem. Solids* **30**, 1125.
Corsini-Mena, A., Elli, M., Paorici, C. and Pelosini, L. (1971), *J. Crystal Growth* **8**, 297.
Cosslett, V. E. and Thomas, R. N. (1965), *Brit. J. Appl. Phys.* **16**, 779.
Crowell, M. H. and Gordon, E. I. (1967), IEEE International Meeting on Electron Devices, Washington (unpublished).
Crowell, M. H. and Labuda, E. F. (1969), *Bell Syst. Tech. J.* **48**, 1481.
Curie, D. (1963), *Luminescence in Crystals* (transl. G. F. J. Garlick), Methuen, London.
Currie, D. C. and Choisser, J. P. (1976). *Proc. SPIE* **78**, 83.
Curtis, S. B. (1976), *IEEE Trans. Nucl. Sci,* **NS-23**, 1355.
Dalrymple, R. J. F. and Spear, W. E. (1972), *J. Phys. and Chem. Solids* **33**, 1071.
Davies, D. E., Lorenzo, J. P., Ryan, T. G. and Fitzgerald, J. J. (1979), *Appl. Phys. Lett.* **35**, 631.
Dawe, A. C. (1971), in *Applications Spatiales des Tubes de Prises de Vues*, p. 221, CNES Paris.
Dean, P. J., Lightowlers, E. C. and Wright, D. R. (1965), *Phys. Rev.* **A140**, 352.

Decker, R. W. and Schneeberger, R. J. (1957), *I.R.E. Nat. Conv. Rec.* (March 1957), Part III, p. 156.
Dexter, D. L. and Seitz, F. (1952), *Phys. Rev.* **86**, 964.
Didenko, A. A., Nemilov, Yu. A. and Fomina, V. I. (1959), *Fiz. Tverd. Tela* **2**, 1434 [Eng. transl. (1961) *Soviet Phys., Solid State* **2**, 1304].
Diemer, G. and Hoogenstraaten, W. (1957), *J. Phys. and Chem. Solids* **2**, 119.
Diemer, G., van Gurp, F. J. and Hoogenstraaten, W. (1958), *Philips Res. Rep.* **13**, 458; (1959) **14**, 11.
Distad, M. F. (1938), thesis, University of Minnesota; see *Phys. Rev.* (1939) **55**, 1146.
Doggett, J. A. and Spencer, L. V. (1956), *Phys. Rev.* **103**, 1597.
Dolezalek, F. K. and Spear, W. E. (1970), *J. Non-cryst. Solids* **4**, 97.
Dolezalek, F. K. and Spear, W. E. (1975), *J. Phys. and Chem. Solids* **36**, 819.
Donolato, C. (1977), *J. Phys. (D) Appl. Phys.* **10**, 1781.
Donolato, C. (1979), *Solid State Electr.* **22**, 797.
Drude, P. (1900), *Ann. Physik* **1**, 566; **3**, 369.
Duh, C. Y. and Moll, J. L. (1967), *Solid State Electr.* **11**, 917.
Dumke, W. P. (1956), *Phys. Rev.* **101**, 531.
Ehrenberg, W. (1940), *J. Sci. Instr.* **17**, 41.
Ehrenberg, W. and Franks, J. (1953), *Proc. Phys. Soc.* **B66**, 1057.
Ehrenberg, W. and Ghosh, B. (1969), *J. Phys. (C) Solid State Phys.* **2**, 152.
Ehrenberg, W. and Henderson, S. T. (1937/39), Brit. Pat. 500, 805.
Ehrenberg, W. and Hidden, N. J. (1962), *J. Phys. and Chem. Solids* **23**, 1135.
Ehrenberg, W. and King, D. E. N. (1963), *Proc. Phys. Soc.* **81**, 751.
Ehrenberg, W. and Lang, C. S. (1954), *Physica* **20**, 1137.
Ehrenberg, W. and Shrivastava, R. P. (1973), *J. Phys. (D) Appl. Phys.* **6**, 2079.
Ehrenberg, W., Lang, C. S. and West, R. (1951), *Proc. Phys. Soc.* **64**, 424.
Ehrenberg, W., Gutan, V. B. and Vodopyanov, L. K. (1966), *Brit. J. Appl. Phys.* **17**, 63.
Ehrenreich, H. and Overhauser, A. W. (1956), *Phys. Rev.* **104**, 649.
Elliott, B. J., Gunn, J. B. and McGoddy, J. C. (1967), *Appl. Phys. Lett.* **11**, 253.
Ellis, G. W., Possin, G. E. and Wilson, R. H. (1974), *Appl. Phys. Lett.* **24**, 419.
Emin, D. (1975), *Adv. Phys.* **24**, 305.
Emin, D. and Holstein, T. (1969), *Ann. Phys. (N.Y.)* **53**, 439.
Engstrom, R. W. and Rodgers, R. L. (1971), *Optical Spectra* **5**, 26.
Epstein, D. W. and Pensak, L. (1946), *RCA Rev.* **7**, 5.
Evans, A. G. R. and Robson, P. N. (1974), *Solid State Electr.* **17**, 805.
Evans, A. G. R., Robson, P. N. and Stubbs, M. G. (1972), *Electr. Lett.* **8**, 195.
Everhart, T. E. and Hoff, P. H. (1971), *J. Appl. Phys.* **42**, 5837.
Everhart, T. E., Wells, O. C. and Matta, R. K. (1963), extended abstracts Electr. Div. **12**, 2 (Electrochem. Soc., New York).
Fang, F. F. and Fowler, A. B. (1970), *J. Appl. Phys.* **41**, 1825.
Fano, U. (1953), *Phys. Rev.* **92**, 328.
Fawcett, W. (1973), in *Electrons in Crystalline Solids*, p. 531, IAEA, Vienna.
Fawcett, W. and Paige, E. G. S. (1971), *J. Phys. (C) Solid State Phys.* **4**, 1801.
Fawcett, W. and Ruch, J. G. (1969), *Appl. Phys. Lett.* **15**, 368.
Fegan, D. J. and Craven, P. G. (1977), *J. Phys. (E)* **10**, 510.
Feldman, C. (1960), *Phys. Rev.* **117**, 455.
Firmin, J. C. and Oatley, C. W. (1955), *Proc. Phys. Soc.* **B68**, 620.
Fleeman, J. (1954), Circular 527, p. 91, Nat. Bur. Stand., Washington.
Förster, T. (1959), *Discuss. Faraday Soc.* **27**, 7.

Frerichs, R. (1947), *Phys. Rev.* **72**, 594.
Frerichs, R. (1949), *Phys. Rev.* **76**, 1869.
Garlick, G. F. J. (1949), *Luminescent Materials,* Clarendon Press, Oxford.
Garlick, G. F. J. and Gibson, A. F. (1949), *J. Opt. Soc. Amer.* **39**, 935.
Ghosh, B. (1967), thesis, University of London.
Ghosh, P. K. and Spear, W. E. (1968), *J. Phys. (C) Solid State Phys.* **1**, 1347.
Gibbons, D. J. (1960), *Adv. El. El. Phys.* **12**, 203.
Gibbons, D. J. (1961), *Electronic Eng.* **33**, 404.
Gibbons, D. J. (1966), in *Handbook of Vacuum Physics*, Ch. 2, Part 3 (ed. A. H. Beck), Pergamon Press, Oxford.
Gibbons, D. J. (1970), *Mol. Cryst. Liq. Cryst.* **10**, 137.
Gibbons, D. J. (1974a), *J. Phys. (D) Appl. Phys.* **7**, 433.
Gibbons, D. J. (1974b), *J. Phys. (D) Appl. Phys.* **7**, 439.
Gibbons, D. J. and Papadakis, A. C. (1968), *J. Phys. and Chem. Solids* **29**, 115.
Gibbons, D. J. and Spear, W. E. (1966), *J. Phys. and Chem. Solids* **27**, 1917.
Gibbons, D. J., Hogarth, C. A. and Waters, D. G. (1975), *J. Phys. (D) Appl. Phys.* **8**, 262.
Gill, W. D. (1972), *J. Appl. Phys.* **43**, 5033.
Gill, W. D. and Street, G. B. (1973), *J. Non-cryst. Solids* **13**, 120.
Gill, W. D., Street, G. B. and Macdonald, R. E. (1967), *J. Phys. and Chem. Solids* **28**, 1517.
Glazebrook, R. (ed.) (1922), *A Dictionary of Applied Physics*, Vol. 2, p. 596, Macmillan, London.
Gobrecht, H. and Bartschat, A. (1953), *Z. Physik* **136**, 224.
Gonzales, A. J. (1974), in *Scanning Electron Microscopy 1974*, Part IV, p. 941, ITT Research Institute, Chicago.
Gordon, E. I. and Crowell, M. H. (1968), *Bell Syst. Tech. J.* **47**, 1855.
Grainger, R. J., Mayer, J. W., Wiggins, J. S. and Friedland, S. S. (1960), *Bull. Amer. Phys. Soc.* **5**, 265.
Green, I. M. (1980), personal communication.
Gricourt, M. (1971). in *Applications Spatiales des Tubes de Prises de Vues,* p. 135, CNES, Paris.
Gross, B. (1957), *Phys. Rev.* **107**, 368.
Gudden, B. and Pohl, R. (1923), *Z. Physik* **16**, 170.
Guillard, C. (1966), *Ann. Radioel.* **21**, 195.
Guillard, C. and Charles, D. R. (1966), *Adv. El. El. Phys.* **22A**, 315.
Guldberg, J. and Schroder, D. K. (1971), *IEEE Trans. Elec. Devices* **ED-18**, 1029.
Gunn, J. B. (1963), *Solid State Commun.* **1**, 88.
Harris, L. (1955), *J. Opt. Soc. Amer.* **45**, 26.
Hartke, J. L. (1968), *J. Appl. Phys.* **39**, 4871.
Hayes, R. (1975), *IEEE Trans. Elec. Devices* **ED-22**, 931.
Haynes, J. R. (1948), see Shockley, W. (1950).
Haynes, J. R. and Shockley, W. (1949), *Phys. Rev.* **75**, 691.
Haynes, J. R. and Shockley, W. (1951), *Phys. Rev.* **81**, 835.
Hecht, K. H. (1932), *Z. Physik* **77**, 235.
van Heerden, P. J. (1945), *The Crystal Counter,* thesis, University of Utrecht; see *Physica* (Eindhoven) (1950) **16**, 505, 517.
Heiland, G. (1952), *Z. Physik* **132**, 367.
Helfrich, W. and Schneider, W. G. (1966), *J. Chem. Phys.* **44**, 2902.
Herring, C. (1955), *Bell Syst. Tech. J.* **34**, 237.

Herzog, R. F., Greenich, J. S., Everhart, T. E. and van Duzer, T. (1972), *IEEE Trans. Elect. Devices* **ED-19**, 635.
Hezel, R. (1979). *Solid State Electr.* **22**, 735.
Hidden, N. J. (1960), thesis, University of London.
Hirsch, J. (1966), *J. Phys. and Chem. Solids* **27**, 1385.
Hittorf, J. W. (1869), *Pogg. Ann.* **136**, 77.
Hofstadter, R. (1949), *Nucleonics* **4**, 2, 29.
Hofstadter, R. (1950), *Proc. IRE* **38**, 726.
Holliday, J. E. and Sternglass, E. J. (1959), *J. Appl. Phys.* **30**, 1428.
Holstein, T. (1959), *Ann. Phys.* (N.Y.), **8**, 343.
Holt, D. B., Chase, B. D. and Censlive, M. (1973), *Phys. Stat. Sol.* (a) **20**, 459.
Holt, D. B., Muir, M. D., Grant, M. D. and Boswara, I. B. (ed.) (1974), *Quantitative Scanning Electron Microscopy*, Ch. 8, p. 2241, Academic Press, London, New York.
Houston, P. A. and Evans A. G. R. (1977), *Solid State Electr.* **20**, 197.
Hughes, W. C., Lemmond, C. O., Parks, H. G., Ellis, G. W., Possin, G. E. and Wilson, R. H. (1975), *Proc. IEEE* **63**, 1230.
Ing. S. W., Neyhart, J. H. and Schmidlin, F. (1971), *J. Appl. Phys.* **42**, 696.
Ioannou, D. E. and Davidson, S. M. (1979), *J. Phys. (D), Appl. Phys.* **12**, 1339.
Jacoboni, C. and Reggiani, L. (1979), *Adv. Phys.* **28**, 493.
Jacoboni, C., Canali, C., Ottaviani, G. and Alberigi Quaranta, A. (1977), *Solid State Electr.* **20**, 77.
Jaffé, G. (1913), *Radium (Paris)* **10**, 126; (1913) *Ann Physik* **42**, 303; (1914) *Phys. Z.* **15**, 353.
Jakubowski, W. and Whitmore, D. H. (1971), *J. Amer. Ceram. Soc.* **54**, 161.
James, I. J. P. (1952), *Proc. IEEE* **99**, Part IIIA, 796.
Jelley, J. V. (1973), *Observatory* **93**, 9.
Jelley, J. V. (1980), *Quart. J. Roy. Astr. Soc.* **21**, 14.
Kanter, H. (1961), *Phys. Rev.* **121**, 677.
Katz, L. and Penfold, A. S. (1952), *Rev. Mod. Phys.* **24**, 28.
Keating, P. N. and Papadakis, A. C. (1964), *Proc. 7th Internat. Conf. on Semiconductor Physics* (Paris), p. 519.
Kent, S. M. (1979), *Pub. Astron. Soc. Pacific* **91**, 394.
King, D. E. N. (1960), thesis, University of London.
Kino, G. S. and Neukermans, A. (1973), *Phys. Rev.* **7**, 2693.
Klasens, H. A. (1953), *J. Electrochem. Soc.* **100**, 72.
Klein, C. A. (1968), *J. Appl. Phys.* **39**, 2029.
Klemperer, O. and Barnett, M. E. (1971), *Electron Optics*, 3rd edn, Cambridge University Press.
de Kock, A. J. R., Ferris, S. D., Kimerling, L. C. and Leamy, H. J. (1977) *J. Appl. Phys.* **48**, 301.
Koller, L. R. and Alden, E. D. (1951), *Phys. Rev.* **83**, 684.
Konopleva, R. F., Novikov, S. R. and Rubinova, E. E. (1966), *Soviet Phys., Solid State* **8**, 264.
Konorova, E. A. and Schevchno, S. A. (1977), *Soviet Phys., Semicond.* **1**, 299.
Kosmata, J. and Huber, K. (1941), see article by Stetter.
Kot, M. V. and Simashkevich, A. V. (1962), *Radio Eng. Elec. Phys.* **7**, 1556.
Kozlov, F. F., Stuck, R., Hage-Ali, M. and Siffert, P. (1975), *IEEE Trans. Nucl. Sci.* **NS-22**, 160.
Kröger, F. A. (1954), *Physica* **20**, 1149.
Krönig, R. de Laer (1924), *Phys. Rev.* **24**, 377.

Kubo, R. (1952), *Phys. Rev.* **86**, 929.
Kurtin, S., McGill, T. C. and Mead, C. A. (1969), *Phys. Rev. Lett.* **22**, 1433.
de Laet, L. H., Guislain, H. J., Meens, H. J. and Schoenmaekers, W. K. (1971), Second Symposium on Semiconductor Detectors for Nuclear Radiation (Munich), unpublished.
Lambe, J. and Klick, C. C. (1955), *J. Phys. Radium (Paris)* **17**, 663.
Lane, R. O. and Zaffarano, D. J. (1954), *Phys. Rev.* **94**, 960.
Leamy, H. J., Kimerling, L. C. and Ferris, S. D. (1976), in *Scanning Electron Microscopy 1974*, ITT Research Institute, Chicago. Part IV, p. 529.
Leamy, H. J., Kimerling, L. C. and Ferris, S. D. (1978), in *Scanning Electron Microscopy 1978*, ITT Research Institute, Chicago, Part I, p. 717.
Le Comber, P. G. and Spear, W. E. (1979), in *Amorphous Semiconductors* (ed. M. H. Brodsky), Ch. 9, Springer, Berlin.
Le Comber, P. G., Spear, W. E. and Weinmann, A. (1966), *Brit. J. Appl. Phys.* **17**, 467.
Le Comber, P. G., Madan, A. and Spear, W. E. (1972), *J. Non-cryst. Solids* **11**, 19.
Le Comber, P. G., Loveland, R. J. and Spear, W. E. (1975), *Phys. Rev.* **B11**, 3124.
Le Comber, P. G., Wilson, J. B. and Loveland, R. J. (1976), *Solid State Commun.* **18**, 377.
Le Comber, P. G., Spear, W. E. and Ghaith, A. (1979), *Electronics Lett.* **15**, 179.
Leedy, K. O. (1977), *Solid State Tech.* **20**, 45.
Lemke, H. and Müller, G. O. (1970), *Phys. Stat. Sol.* (a) **17**, 287.
Lenz, H. (1925), *Ann. Physik* **77**, 449; *Phys. Z.* **10**, 365.
Lewis, H. W. (1950), *Phys. Rev.* **78**, 526.
Lorentz, H. A. (1905), *Proc. Amst. Acad.* **7**, 438.
Loveland, R. J., Le Comber, P. G. and Spear, W. E. (1972), *Phys. Rev.* **B6**, 3121.
Lowrance, J. L., Zucchino, P., Renda, G. and Long, D. C. (1979), *Adv. El. El. Phys.* **52**, 441.
Lubszynski, H. G. (1971), personal communication.
MacDonald, N. C. and Everhart, T. E. (1968), *Proc. IEEE* **56**, 158.
McKay, K. G. (1948a), *Bull. Amer. Phys. Soc.* **23**, 31.
McKay, K. G. (1948b), *Phys. Rev.* **74**, 47.
McKay, K. G. (1949), *Phys. Rev.* **76**, 1537.
McKay, K. G. (1951), *Phys. Rev.* **84**, 829.
McKay, K. G. and McAfee, K. B. (1953), *Phys. Rev.* **91**, 1079.
Madan, A. and Le Comber, P. G. (1977), in *Proc. 7th Internat. Conf. on Amorphous and Liquid Semiconductors* (ed. W. E. Spear), p. 377, CICL, Edinburgh.
Madan, A., Le Comber, P. G. and Spear, W. E. (1976), *J. Non-cryst. Solids* **20**, 239.
Makhov, A. F. (1960), *Fiz. Tverd. Tela* **2**, 2161 [Eng. transl. *Soviet Phys., Solid State* **2**, 1934].
Many, A. and Rakavy, G. (1962), *Phys. Rev.* **126**, 1980.
Many, A., Simhony, M. and Grushkevicz, Y. (1965), *J. Phys. and Chem. Solids* **26**, 1925.
Marshall, J. M. (1977a), in *Proc. 7th Internat. Conf. on Amorphous and Liquid Semiconductors* (Edinburgh) (ed. W. E. Spear), p. 541, CICL, Edinburgh.
Marshall, J. M. (1977b), *Phil. Mag.* **36**, 959.
Marshall, J. M. (1978), *Phil. Mag.* **B38**, 335, 407.
Marshall, J. M. and Allan, D. (1979), *Phil. Mag.* **B40**, 381.
Marshall, J. M. and Owen, A. E. (1971), *Phil. Mag.* **24**, 1281.
Martini, M., Mayer, J. W. and Zanio, K. R. (1972), in *Advances in Solid State Science*, Vol. 3, p. 181, Academic Press, London.

Matsukawa, T., Shimizu, R., Harada, K. and Kato, T. (1974), *J. Appl. Phys.* **45**, 733.
Mayer, J. W. (1968), in *Semiconductor Detectors* (ed. G. Bertolini and A. Coche), Ch. 5, North Holland, Amsterdam.
Mayo, B. J. and Bennett, A. W. (1972), *Adv. El. El. Phys.* **33**, 571 (Proc. 5th Symposium on Photoelectronic Imaging Devices, London, 1971).
Mead, C. A. (1966), *Solid State Electr.* **9**, 1023.
Mead, C. A. and McGill, T. C. (1976), *Physics Lett.* **58A**, 249.
Milevskii, L. S. and Garnyk, V. S. (1979), *Soviet Phys., Semicond.* **13**, 801.
Miller, L. S. (1967), thesis, University of Leicester.
Miller, L. S., Howe, S. and Spear, W. E. (1968), *Phys. Rev.* **166**, 112.
Milnes, A. G. (1973), *Deep Impurities in Semiconductors*, Wiley, New York.
Mitchell, J. P. (1967), *IEEE Trans. Elect. Devices* **ED-14**, 764.
Miyashiro, S. and Shirouzu, S. (1971), *IEEE Trans. Elect. Devices* **ED-18**, 1023.
Miyashiro, S. and Shirouzu, S. (1972), *Adv. El. El. Phys.* **33A**, 207.
Miyazaki, E., Maeda, H. and Miyaji, K. (1966), *Adv. El. El. Phys.* **22A**, 331.
Molière, G. (1947), *Z. Naturforsch.* **211**, 133.
Mollwo, E. (1948), *Ann. Physik* **63**, 230.
Mollwo, E. and Stöckman, F. (1948), *Ann. Physik* **63**, 223, 240.
Moore, A. R. (1949), thesis, Cornell University; see *Semiconducting Materials* (ed. H. K. Henisch), p. 116, Butterworths Scientific Publications, 1951.
Moore, A. R. and Hermann, F. (1951), *Phys. Rev.* **81**, 472.
Morin, F. J. and Maita, J. P. (1954), *Phys. Rev.* **96**, 28.
Mort, J. and Lakatos, A. I. (1970), *J. Non-cryst. Solids* **4**, 117.
Mort, J. and Pai, D. M. (ed.) (1976), *Photoconductivity and Related Phenomena*, Elsevier, Amsterdam.
Mort, J., Pfister, G. and Grammatica, S. (1976), *Solid State Commun.* **18**, 693.
Moss, S. C. and Graczyk, J. F. (1970), in *Proc. 10th Internat. Conf. on Semiconductor Physics* (Cambridge, Mass.) (ed. S. P. Keller, J. C. Hensel and F. Stern), p. 658, U.S. Atomic Energy Commission.
Mott, N. F. (1929), *Proc. Roy. Soc.* **A124**, 425.
Mott, N. F. (1932), *Proc. Roy. Soc.* **A135**, 429.
Mott, N. F. (1969), *Phil. Mag.* **20**, 835.
Mott, N. F. and Davis, E. A. (1971), *Electronic Processes in Non-crystalline Materials*, 1st edition, p. 371, Clarendon Press, Oxford.
Mott, N. F. and Davis, E. A. (1979), *Electronic Processes in Non-crystalline Materials*, 2nd edition, Clarendon Press, Oxford.
Mott, N. F. and Gurney, R. W. (1940), *Electronic Processes in Ionic Crystals*, p. 185, Clarendon Press, Oxford.
Mott, N. F. and Twose, W. D. (1961), *Adv. Phys.* **10**, 107.
Moulin, M. (1908), *Radium* (Paris) **5**, 136.
Nag. B. R. (1972), *Theory of Electrical Transport in Semiconductors*, Pergamon Press, Oxford.
Namba, S., Artiome, H. and Masuda, K. (1969), *Record of 10th Symposium on Electron, Ion and Laser Beam Technology*, p. 197, San Francisco Press.
Nava, F., Canali, C., Catellini, F., Gavioli, G. and Ottaviani, G. (1976), *J. Phys. (C) Solid State Phys.* **9**, 1685.
Nava, F., Canali, C., Artuso, M., Gatti, E. and Manfredi, P. F. (1979a), *IEEE Trans. Nucl. Sci.* **NS-26**, 308.
Nava, F., Canali, C. and Reggiani, L. (1979b), *J. Appl. Phys.* **50**, 922.

Nava, F., Canali, C., Reggiani, L., Gasquet, D., Vaissiere, J. C. and Nougier, J. P. (1979c), *J. Appl. Phys.* **50**, 922.
Nava, F., Canali, C., Jacoboni, C., Reggiani, L. and Kozlov, S. F. (1980), *Solid State Commun.* **33**, 475.
Nelms, A. T. (1956), Circular 577, National Bureau of Standards, Washington.
Neukermans, A. and Kino, G. S. (1970), *Appl. Phys. Lett.* **17**, 102.
Newton, R. R. (1949), *Phys. Rev.* **75**, 234.
Norris, C. B. (1972), *J. Appl. Phys.* **43**, 4060.
Norris, C. B. and EerNisse, E. P. (1974), *J. Appl. Phys.* **45**, 3876.
Norris, C. B. and Gibbons, J. F. (1967), *IEEE Trans. Elect. Devices* **ED-14**, 38.
Oelgart, G. (1979), *Phys. Stat. Sol.* (a) **52**, K205.
Oksman, Ya.A. and Tikhomirov, G. P. (1959), *Radiotekh. i Elektr.* **4**, 344 (in Russian; Engl. summary, PB141106-T-13, Office of Tech. Services, U.S. Dept. Commerce, Washington, D.C.).
Onsager, L. (1938), *Phys. Rev.* **54**, 554.
Ottaviani, G., Canali, C., Jacoboni, C. and Alberigi Quaranta, A. (1972), *Solid State Commun.* **10**, 745.
Ottaviani, G., Canali, C., Jacoboni, C., Alberigi Quaranta, A. and Zanio, K. (1973a), *J. Appl. Phys.* **44**, 360.
Ottaviani, G., Canali, C., Nava, F. and Mayer, J. W. (1973b), *J. Appl. Phys.* **44**, 2917.
Ottaviani, G., Reggiani, L., Canali, C., Nava, F. and Alberigi Quaranta, A. (1975), *Phys. Rev.* **B12**, 3315.
Owen, A. E. and Robertson, J. M. (1970), *J. Non-cryst. Solids* **2**, 40.
Owen, A. E. and Spear, W. E. (1976), *Phys. and Chem. Glasses* **17**, 174.
Papadakis, A. C. (1967), *J. Phys. and Chem. Solids* **28**, 641.
Partin, D. L., Chen, J. W., Milnes, A. G. and Vassamillet, L. F. (1979), *Solid State Electr.* **22**, 455.
Pauling, L. (1960), *The Nature of the Chemical Bond*, 3rd edition, Cornell University Press, pp. 93–96.
Pearson, G. L. and Bardeen, J. (1949), *Phys. Rev.* **75**, 865.
Pensak, L. (1948), *Bull. Amer. Phys. Soc.* **23**, 472.
Pensak, L. (1949), *Phys. Rev.* **75**, 472.
Pensak, L. (1950), *Phys. Rev.* **79**, 171.
Perron, J. C. (1967), *Adv. Phys.* **16**, 657.
Persky, G. and Bartelink, D. J. (1971), *J. Appl. Phys.* **42**, 4414.
Persky, G. and Bartelink, D. J. (1969), *IBM J. Res. Develop.* **13**, 607.
Petravicius, A., Juska, G. and Smilga, A. (1977), *Soviet Phys. Collection* **17**, 54, Allerton Press.
Pfister, G. (1974), *Phys. Rev. Lett.* **33**, 1474.
Pfister, G. (1976), *Phys. Rev. Lett.* **36**, 271.
Pfister, G. and Scher, H. (1978), *Adv. Phys.* **27**, 747.
Pfister, H. Z. (1957), *Z. Naturforsch.* **12a**, 217.
Plücker, T. (1859), *Pogg. Ann.* **107**, 77.
Pooley, D. (1966), *Proc. Phys. Soc.* **87**, 245.
Possin, G. E. and Kirkpatrick, C. G. (1980), *J. Microscopy* **118**, Part 3, 291.
Prenner, J. S. and Williams, F. E. (1956), *J. Phys. Radium* (Paris) **17**, 667.
Prince, M. B. (1955), *J. Appl. Phys.* **26**, 534.
Pruett, H. D. and Broida, H. P. (1967), *Phys. Rev.* **164**, 1138.
Ramo, S. (1939), *Proc. IRE* **27**, 584.

Randall, J. T. and Wilkins, M. H. F. (1945), *Proc. Roy. Soc.* **A184**, 347, 365, 390.
Rappaport, P., Loferski, J. J. and Linder, E. G. (1956), *RCA Rev.* **17**, 100.
Reggiani, L. (1979), lecture for the NATO-ASI on Non-linear Transport in Semiconductors (Urbino, Italy), Plenum Press (1980).
Reggiani, L., Majni, G. and Minder, R. (1975), *Solid State Commun.* **16**, 151.
Reggiani, L., Canali, C., Nava, F. and Ottaviani, G. (1977), *Phys. Rev.* **B16**, 2718.
Reggiani, L., Canali, C., Nava, F. and Alberigi Quaranta, A. (1978), *J. Appl. Phys.* **49**, 4446.
Reggiani, L., Bosi, S., Canali, C. and Nava, F. (1979), *Solid State Commun.* **30**, 333.
Rhoderick, E. W. (1978), *Metal–Semiconductor Contacts*, Clarendon Press, Oxford.
Rittner, E. S. (1948), *Phys. Rev.* **74**, 1212.
Roberts, G. G. (1967), *Brit. J. Appl. Phys.* **18**, 749.
Rode, D. L. (1972), *Phys. Stat. Sol.* (b) **53**, 245.
Rodriguez, V. and Nicolet, M. A., (1969), *J. Appl. Phys.* **40**, 496.
Röntgen, W. C. (1895), preliminary announcement at December meeting of Würzburg Physico-medical Society; see Dam, H. J. W. (1896), *McClure's Mag.* **6**, 403.
von Roos, O. (1978a), *Solid State Electr.* **21**, 1063.
von Roos, O. (1978b), *Solid State Electr.* **21**, 1069.
von Roos, O. (1979a), *Appl. Phys. Lett.* **35**, 408.
von Roos, O. (1979b), *Solid State Electr.* **22**, 773.
von Roos, O. (1980), *Solid State Electr.* **22**, 177.
van Roosbroeck, W. (1953), *Phys. Rev.* **91**, 282.
Rose, A. (1951), *RCA Rev.* **12**, 362.
Rosenzweig, W. (1962), *Bell Syst. Tech. J.* **41**, 1573.
Ruch, J. G. and Kino, G. S. (1967), *Appl. Phys. Lett.* **10**, 40.
Ruch, J. G. and Kino, G. S. (1968), *Phys. Rev.* **174**, 921.
Ruthemann, G. (1948), *Ann. Physik* **62**, 113.
Rutherford, E. (1911), *Phil. Mag.* **21**, 669.
Ryvkin, S. M., Konovalenko, B. M. and Smetannikova, U.S. (1954), *Zh. Tech. Fiz.* **24**, 961.
Sacton, J. (1956), *Bull. Acad. Roy. Belg., Cl. Sci.* **42**, 1118.
Saelee, H. T., Lucas, J. and Limbeek, J. W. (1977). *IEE J. Solid State and Electr. Devices* **1**, 111.
Safratova-Eskertova, L. (1955), *Czech. J. Phys.* **5**, 551.
Saidoh, M. and Townsend, P. D. (1975), *Radiation Effects* **27**, 1.
Saleh, M. (1980), *Phys. Scr. (Sweden)* **21**, 220.
Sato, T., Takeishi, Y., Hara, H., Ohnuma, H. and Okamoto, Y. (1971), *J. Phys. Soc. Japan* **31**, 1846.
Scharfe, M. E. (1973), *Bull. Amer. Phys. Soc.* **18**, 454.
Scher, H. and Montroll, E. W. (1975), *Phys. Rev.* **B12**, 2455.
Scher, H. and Pfister, G. (1975), *Bull. Amer. Phys. Soc.* **20**, 322.
Schmidt, W. F., Bakalc, G. and Tauchert, W. (1973), *Ann. Rep. Conf. Elect. Insul. Dielect. Phenom.* (Varennes, Canada), Nat. Acad. Sci., Washington, D.C., 1974.
Schmidt, W. F., Bakale, G. and Sowada, U. (1974), *J. Chem. Phys.* **61**, 5275.
Schneeberger, R. J., Skorinko, G., Doughty, D. D. and Fiebelman, W. A. (1961), *Proc. 2nd Image Intensifier Symposium*, October 24–26, p. 27.
Schön, M. (1951), *Z. Naturforsch.* **6a**, 287.
Schonland, B. F. J. (1923), *Proc. Roy. Soc.* **A104**, 235; (1925) **A108**, 187.
Schwartz, L. M. and Hornig, J. F. (1965), *J. Phys. and Chem. Solids* **26**, 1821.
Sclar, N. and Kim, Y. C. (1957), Electron Devices Conference (Washington); see *IEEE Trans. Elect. Devices* **5**, 110 (1958).

Scott, D. W., McCullough, J. P. and Kruse, F. H. (1964), *J. Mol. Spectr.* **13**, 313.
Seitz, F. (1946), *Rev. Mod. Phys.* **18**, 408.
Seitz, F. (1948), *Phys. Rev.* **73**, 549.
Seitz, M. A. and Whitmore, D. H. (1968), *J. Phys. and Chem. Solids* **29**, 1033.
Seki, H. (1974), in *Proc. 5th Internat. Conf. on Amorphous and Liquid Semiconductors* (ed. J. Stuke and W. Brenig), p. 105, Taylor and Francis, London.
Seliger, H. H. (1955), *Phys. Rev.* **100**, 1029.
Sessler, G. M. and West, J. E. (1973), *J. Acoust. Soc. Amer.* **55**, 345.
Sevchenko, A. N., Tkachev, V. D. and Lugakov, P. F. (1967), *Soviet Phys. Dokl.* **11**, 610.
Shapiro, M. M. and Siberberg, R. (1974), *Phil. Trans.* **A277**, 319.
Shockley, W. (1938), *J. Appl. Phys.* **9**, 635.
Shockley, W. (1950), *Electrons and Holes in Semiconductors*, Van Nostrand, New York.
Shockley, W. (1951), *Bell Syst. Tech. J.* **30**, 990.
Shockley, W. (1961a), *Solid State Electr.* **2**, 35.
Shockley, W. (1961b), *Czech. J. Phys.* **B11**, 81.
Shukla, M. M. and Salzberg, J. B. (1973), *J. Phys. Soc. Japan* **35**, 996.
Siebrand, W. (1964), *J. Chem. Phys.* **41**, 3574.
Siekanowicz, W. W., Huang, H. C., Engstrom, R. E., Martinelli, R. U, Ponczak, S. and Olmstead, J. A. (1972), Final Report AD-757 749, NTIS, Springfield, Vancouver.
Siekanowicz, W. W., Ho-Chung Huang, Engstrom, R. E., Martinelli, R. U., Ponczak, S. and Olmstead, J. (1974), *IEEE Trans. Elect. Devices* **ED-21**, 691.
Siffert, P., Berger, J., Cornet, A., Bell, R. O., Serreze, H. B. and Wald, F. V., (1976), *IEEE Trans. Nucl. Sci.* **NS-23**, 159.
Silver, M. and Cohen, L. (1977), *Phys. Rev.* **B15**, 3276.
Silver, M., Dy, K. S. and Huang, D. L. (1971), *Phys. Rev. Lett.* **27**, 21.
Silzars, A., Bates, D. J. and Ballonoff, A. (1974), *Proc. IEEE* **62**, 1119.
Simhony, M. and Gorelik, J. (1965), *J. Phys. and Chem. Solids* **26**, 1133.
Smith, W. (1873), *Soc. Tel. Eng. J.* **2**, 31.
Smith, J. E., Nathan, M. I., McGroddy, J. C., Proowski, S. A. and Paul, W. (1969), *Appl. Phys. Lett.* **15**, 242.
Snow, E. H., Grove, A. S. and Fitzgerald, D. J. (1967), *Proc. IEEE* **55**, 1168.
Somerford, D. J. and Spear, W. E. (1971), *J. Phys. (C) Solid State Phys.* **4**, 795.
Spear, W. E. (1955), *Proc. Phys. Soc.* **B68**, 991.
Spear, W. E. (1956), *Proc. Phys. Soc.* **B69**, 1139.
Spear, W. E. (1957), *Proc. Phys. Soc.* **B70**, 669.
Spear, W. E. (1958), *Phys. Rev.* **112**, 362.
Spear, W. E. (1961), *J. Phys. and Chem. Solids* **21**, 110.
Spear, W. E. (1969), *J. Non-cryst. Solids* **1**, 197.
Spear, W. E. (1974), *Adv. Phys.* **26**, 523.
Spear, W. E. (1977), *Adv. Phys.* **26**, 811.
Spear, W. E. and Le Comber, P. G. (1969), *Phys. Rev.* **178**, 1454.
Spear, W. E. and Le Comber, P. G. (1972), *J. Non-cryst. Solids* **8–10**, 727.
Spear, W. E. and Le Comber, P. G. (1975), *Solid State Commun.* **17**, 1193.
Spear, W. E. and Le Comber, P. G. (1976), *Phil. Mag.* **33**, 935.
Spear, W. E. and Le Comber, P. G. (1977), in *Rare Gas Solids* (ed. M. L. Klein and J. A. Venables), Ch. 18, Academic Press, London.
Spear, W. E. and Mort, J. (1962), *Phys. Rev. Lett.* **8**, 314.
Spear, W. E. and Mort, J. (1963), *Proc. Phys. Soc.* **81**, 130.

Spear, W. E., Adams, A. R. and Henderson, G. A. (1963), *J. Sci. Instr.* **40**, 332.
Spear, W. E., Allan, D., Le Comber, P. G. and Ghaith, A. (1980a), *J. Non-cryst. Solids* **35/36**, 357.
Spear, W. E., Allan, D., Le Comber, P. G. and Ghaith, A. (1980b), *Phil. Mag.* **B41**, 419.
Speliotis, D. E. (1975), *Proc. 1975 Nat. Computer Conf.*, p. 501.
Spencer, L. V. (1955), *Phys. Rev.* **98**, 1597.
Spencer, L. V. (1959), Monograph 1, National Bureau of Standards, Washington, D.C.
Springett, B. E., Jortner, J. and Cohen, M. H. (1968), *J. Chem. Phys.* **48**, 2720.
Sternglass, E. J. (1955), *Rev. Sci. Instr.* **26**, 1202.
Stetter, G. (1941), *Verh. Deutsch. Phys. Ges.* **22**, 13.
Stinchfield, J. M. (1940), quoted in G. F. J. Garlick, *Luminescent Materials*. Clarendon Press, Oxford, 1949, p. 194.
Sturner, H. W. and Bleil, C. E. (1963), *J. Phys. and Chem. Solids* **24**, 735.
Su, J. L., Nishi, Y., Moll, J. L. and Neukermans, A. (1970), *Solid State Electr.* **13**, 115.
Sverdlova, A. M. and Rokakh, A. G. (1964), *Izv. Akad, Nauk SSSR, Ser. Fiz.* **28**, 1514 [Eng. transl., *Bull. Acad. Sci. USSR, Phys. Ser.* (U.S.A.) **28**, 1413 (1965) (11th All-Union Conf. Cathode Electronics, Kiev, 1963)].
Symons, M. C. R. (1963), *J. Chem. Soc.* 570.
Sze, S. M. and Irvin, J. C. (1968), *Solid State Electr.* **11**, 599.
Taguchi, T., Shirafuji, J. and Inuishi, Y. (1978), *Nucl. Instr. Meth.* **150**, 43.
Tahmasbi, A. R. and Hirsch, J. (1980), *Solid State Commun.* **34**, 75.
Takahashi, T. (1979), *J. Non-cryst. Solids* **34**, 307.
Takeya, K. and Nakamura, K. (1958), *J. Phys. Soc. Japan* **2**, 223.
Tandon, J. L., Goleki, I., Nicolet, M.-A., Sadana, D. K. and Washburn, J. (1979), *Appl. Phys. Lett.* **35**, 867.
Taroni, A. and Zanarini, G. (1969), *J. Phys. and Chem. Solids* **30**, 1861.
Terrill, H. M. (1923), *Phys. Rev.* **22**, 101.
Terrill, H. M. (1924), *Phys. Rev.* **24**, 616.
Thomson, J. J. (1906), *Conduction of Electricity through Gases*, p. 84 *et seq.*, Cambridge University Press.
Townsend, P. D., Browning, R., Garlant, D. J., Kelley, J. C., Mahjoobi, A., Michael, A. J. and Saidoh, M. (1976), *Radiation Effects* **30**, 55.
Trodden, W. G. and Jenkins, R. O. (1965), *GEC Journal* **32**, 85.
Tsein, S. T., Marty, C. and Dreyfus, B. (1947), *J. Phys. Radium* **9**, 269.
Tull, R. G., Choisser, J. P. and Snow, E. H. (1975), *Appl. Optics* **14**, 1182.
Tyndall, A. M. and Grindley, G. C. (1926), *Proc. Roy. Soc.* **110**, 342.
Urgell, J. and Leguerre, J. R. (1974), *Solid State Electr.* **17**, 239.
Urli, B. and Corbett, J. W. (ed.) (1977), *Radiation Effects in Semiconductors*, Institute of Physics, Bristol and London (Internat. Conf. on Radiation Effects in Semiconductors, Dubrovnik, 1976).
Valentine, J. M. and Curran, S. C. (1958), *Rep. Progr. Phys.* **21**, 1.
Varol, H. S. (1978), thesis, University of Surrey.
Voorhies, H. G. V. and Street, J. C. (1949), *Phys. Rev.* **76**, 1100.
Vossen, J. L. and Kern, W. (ed.) (1978), *Thin Film Processes*, Academic Press, New York, London.
Vyatskin, A. Ya, and Khramov, V. Yu. (1975), *Soviet Phys., Solid State* **17**, 1023.
Vyatskin, A. Ya. and Makhov. A. F. (1958), *Soviet Phys., Tech. Phys.* **3**, 690.
Warfield, G. (1950), Ph.D. thesis, Cornell University.

Weimer, P. K. (1950), *Phys. Rev.* **79**, 171.
Weimer, P. K., Forgue, S. V. and Goodrich, R. R. (1950), *Electronics* **23**, 70.
Wertheim, G. K. and Augustyniak, W. M. (1956), *Phys. Rev.* **27**, 1062.
White, D. L. (1962), *J. Appl. Phys.* **33**, 2547.
Williams, E. J. (1932), *Proc. Roy. Soc.* **A135**, 108.
Wilson, J. B. and Le Comber, P. G. (1977), *Europhysics Conf. Abstr.,* Vol. 2D, p. 19 (3rd EPS Conf. on Electron Transport in Molecular Solids, Leeds, England).
Wolfang, L. G., Abraham, J. M. and Inskeep, C. N. (1966), *IEEE Trans. Nucl. Sci.* **NS-13**, 46.
Young, J. R. (1956), *Phys. Rev.* **103**, 292.
Young, R. H., May, W. and Marchetti, P. (1977), *Appl. Phys. Lett.* **30**, 38.
Young, R. H., Walker, E. I. P. and Marchetti, A. P. (1979), *J. Chem. Phys.* **70**, 443.
Zaininger, K. H. and Holmes-Siedle, A. G. (1967), *RCA Rev.* **28**, 208.
Zanio, K. R., Akutagawa, W. M. and Kikuchi, R. (1968), *J. Appl. Phys.* **39**, 2818.
Zulliger, H. R., Norris, C. B., Sigmon, T. W. and Pehl, R. H. (1969), *Nucl. Instr. Meth.* **70**, 125.

Index

Acoustoelectric saturation
  in CdS, 233–234
  in ZnS, 242–243
Activation centres, 103, 116
Al_2O_3
  EBC, 101, 110
  preparation by anodizing, 83
Alpha particles
  radioactive source, 78
  range in Si, 309
  range–energy table, 78
  spectrum, 310
Alpha-particle excitation
  comparison with transient EBC method, 158–160
  of point contact rectifier, 125
Aluminized screen, 3, 264
Ambipolar velocity and mobility, 162
Amorphous solids, 9–11 (see also specific properties)
  carrier transport, 24–30
Anderson localization, 31
Annealing, effect on diffusion length in Si, 296–297
Apparatus
  microwave time-of-flight, 193–194
  solid gases, 187–188
  steady-state EBC of thin films, 34–42
  transient EBC, 183–192
Ar (liquid)
  carrier transport, 247
  transient EBC pulse shape, 244
Ar (liquid and solid), band gap and radiation ionization energy, 66
Ar (solid), carrier transport, 246, 247, 248, 249
As_2S_3 (thin film)
  β-ray induced currents, 111
  Ebicon target, 313
  gain as function of bias, 106
  gain as function of thickness, 104–105
  primary gain, 93–94
  purification of starting material, 113
  range–energy for fast electrons, 64
  rise and decay of induced currents, 111–113
  secondary gain, 103
  space charge, 96–97, 101
As_2Se_3, 171
Astronomy
  star spectra, 275–277
  telescope autoguider, 269
Attempt-to-escape frequency, 11–12
Avalanche effect, 137–140
  Digicon target diode, 274
  GaP diode, 139–140
  Si diode, 137–139

Back face irradiated Si diode, 136
Band bending, 20
Band gap, table, 16
Band structure of solids, 6–9
Band theory in carrier transport, 20–23
  limits of applicability, 24–25
Barrier EVE, *see* Scanning electron microscope
Barrier layer cell, 127
Barrier potential, 298
Beamos, 313–317
Beta particles
  apparatus for measuring EBC, 35
  sources, 34
Beta-ray induced currents
  in As_2S_3 thin film, 111
  in SiO_2, 113
Bias supply for specimens in transient EBC measurements, 192

INDEX   341

Bimolecular recombination, rate equations, 114
Bloch solutions of wave equation, 6
amorphous solids, 30
Blocking contacts, preparation, 82
Bubble state, 29

Carrier lifetime
  experimental determination, 168–170
  measurements of transit time when lifetime is short, 160–161
Cathode ray tube, 264
Cathodoconductivity, see Thin films
Cathodoluminescence, 2, 16, 54, 62, 67
CdS (crystalline), 228–235
  electron drift mobility, 229
  electron traps, 230–233
  excitons, 91
  gain dependence on contacts, 120–122
  high field electron transport, 234
  hole drift velocity, 234
  lateral EBC, 116
  transfer of excitation, 90
CdS (thin film), lateral EBC, 116–117, 118, 120
CdSe (crystalline)
  carrier drift velocities, 239–240
  lateral EBC, 116
  transfer of excitation, 90
CdSe (layer), lateral EBC, 116, 120
CdTe, 235–241
  band gap, 16
  carrier drift mobilities, 235–237, 240
  high field electron drift velocity, 238–239
  lateral EBC, 116
  nuclear particle detector, 310–311
  transient EBC current waveform, 238
CH_4 (solid), electron mobility, 252, 253–254
Charge collection microscopy circuit, 301
Charge sensitive preamplifier circuit, 267
Charge sheet, see Limited space charge transits

Chord range, see Range–energy
Cluster formation, 29, 30
CO (solid), electron mobility, 252, 253
Collection efficiency, 129
Colour centres, see Point defects in alkali halides
Conditioning effects, 109, 129
Conductivity glow curves, 67
Contacts
  electron beam contacts, 86–88
  electron beam stopping power, table, 290
  influence on EBC gain, 120–122
  injecting, 68
  solid, 86
Continuous time random walk, 170–173
Contrast in SEM, 303
Copper oxide rectifier, 125
Corona discharge, 37
Cosmic rays, 75–76
Crystal counter, 3
Crystalline solids, 5–9 (see also specific materials)

Damage
  by diffusion of impurities, 136
  by radiation, 71–75, 130
  to EVE devices by prolonged electron bombardment, 291–294
Dark current, effect of contacts, 121–122
Defect imaging in SEM, 298–300
Deflected beam microwave power tube, 287
Deflection system for microwave time-of-flight, 193
Degradation phenomena (see also Virgin state), in devices, 291–294
Delayed generation, 91, 154
Depletion layer
  field, 163, 299
  thickness, 163
  thickness in Si and Ge, 307
Detector for charged particles, 309
Detrapping, 149 (see also Traps)
Diamond
  carrier drift mobility, 208
  crystal counter, 3

342 INDEX

Diamond (cont.)
　EBC gain, 4
　high-field drift velocity, 206–207
　injecting contact, 310
　nuclear particle detector, 311
Dielectric relaxation time
　measuring transit time when dielectric relaxation time is short, 161–164
Diffusion of minority carriers, 315
Diffusion length of minority carriers
　in solar cell, 132–133
　measurements using SEM, 295–297
Digicon, 269–274
Dislocations
　carrier scattering by, 23
　imaging in SEM, 279–300
Dispersive carrier transport by trapping, 221–222
Dispersive transits, 171–175
Displacement threshold energy, 72
Donor band hopping, 213, 215, 216
Donors and acceptors, 16
　energy levels in Ge, Si, GaAs, and GaP, 14–15
Doping, inhomogeneities, 298
Double-layer dielectric, 319, 321–322
Drift mobility from transient EBC, 178–181 (see also specific materials)

Ebam, 313–314
Ebicon (Ebitron), 311–313
EBS tube, output current rise time, 288
Einstein relation, 196
Electrets, 79
Electric field
　in depletion region, 163, 298–299
　in electron bombarded insulator, 20, 94–95
　in targets containing junctions, 124
Electric field profiles, technique for measuring, 166
Electrodes
　for thin film EBC specimens, 86 (see also Contacts)
　for transient EBC specimens, 186
Electrons
　accelerator, 191–192

beams, 34–75
　passage through matter, see Range–energy
　sources, 34–44
Electron beam annealing, 75
Electron beam induced current (EBIC), see Scanning electron microscope
Electron bombarded semiconductor devices (EBS), 285–291
Electron bombardment conductivity (EBC), basic measuring circuit, 2 (see also Thin films)
Electron devices, 264–320
Electron gun, 34–42
　for EBS tube, 287
　for picosecond rise-time pulses, 190
　for transient EBC, 185
Electron microscope gun for measuring EBC, 36, 40
Electron voltaic effect (EVE), 123–140 (see also specific materials)
　theory, 129–131
Energy bands
　diagram for depleted diode structure, 124
　selenium and sulphur, 260–262
　structure of solids, 5–8
Energy dissipation of fast electrons, 302–303 (see also Range–energy; Penetrating power)
Energy dissipation function, 51–53
Energy levels in solids, 5–16
Escape time, 175
Evoscope, 284–285
Excitation energy, 47
Excitons, 90–91

Fano factor, 306
Fatigue in selenium cells, 129
F-centres, see Point defects in alkali halides
Fluorescence as means for transfer of excitation, 91–92 (see also Luminescence)
Focus coil design, 36, 41
Focusing electron beam
　in Beamos tube, 316
　in EBC measuring apparatus, 41

INDEX 343

Frenkel defect, 68
Full width at half maximum (FWHM)
  in ICCD, 277
  in nuclear radiation detector, 306
  (see also specific detector materials)

GaAs
  donor and acceptor energies, 14
  electron velocity, 223–226
  electron voltaic effect (EVE), 126, 135
  radiation ionization energy, 127
  Schottky barrier, 129, 224
  transient EBC pulse shape, 224
$GaAs_{0.7}P_{0.3}$
  electron voltaic effect, 126–127
  radiation ionization energy, 127
Gain, 103–122 (see also Electron voltaic effect; specific materials)
  definition, 104
  effect of current density, 108
  function of electrode thickness, 285
    (see also X-rays)
  function of field applied, 105–108
  high temperature effects, 108–110
  lateral EBC of surface layers, 116
  maximum, 100–101
  secondary EBC, 70
  thickness dependence, 104–105
GaP
  avalanche diode, 139–140
  band gap, 16
  donor and acceptor energies, 15
Gap states, 11
Gases, conduction of electricity, 17–19
  (see also Solid gases)
Ge
  band gap, 16, 307
  carrier diffusivity, 201–202
  carrier drift mobility and high field drift velocity, 203–206
  depletion layer thickness, 307
  donor and acceptor energies, 14
  nuclear radiation detector, 307, 308
Geminate recombination, 158
Generation region, 54–59, 302
Glasses, see Se; As_2Se_3; Amorphous solids
Graphechon, 317
Gun, see Electron gun

H_2 (solid), electron mobility, 252, 253
Haynes–Shockley technique, 33, 161–162
Hecht curve, 146–148
High field mobility, 23 (see also specific materials)
High mobility solids, apparatus for measuring transient EBC, 188–192
Historical introduction, 1–5
Holes, 6 (see also specific materials)
Hopping
  adiabatic, 29, 249
  non-adiabatic, 28, 31, 213, 257
  variable range, 32, 172, 173
Hot electrons, 23

Impurities
  diffusion damage in silicon, 131–132, 136, 300, 303
  effect on EBC gain in SiO_2, 113 (see also Donors and acceptors; Scattering)
  energy levels, 7, 14–16
  precipitates, 299
  scattering by, 22
Induced conductivity in thin films, 93–114
Induced current, definition, 2, 104
Inert gases, see Noble gases
Initial recombination, 107
InSb
  band gap, 16
  electron velocity, 228
  transient EBC pulse shape, 227
Insensitive region
  in CdS crystal, 117
  in p–n junction cells, 128
Insulators, conduction of electricity, 17–23
Integrated circuit
  capacitance, 293, 303, 304
  examination by EBIC mode of SEM, 294–305
Integrated path length, see Range–energy
Integration of repetitive signals in storage tube, 320
Intensified charge coupled devices (ICCD), 274–277

Intermetallic (III–V) compounds, 223–228
  transport properties, table, 222
Ion implantation damage, 297
Ionic transport, 30
Ionization by holes, 140
Ionization chamber
  construction, 35
  Thomson's model, 17–19

Junctions in semiconductors, 123 (see also Contacts; Electron voltaic effect; Depletion layer)

Kr (liquid), electron drift velocity, 246
Kr (liquid and solid), band gap and radiation ionization energy, 66
Kr (solid)
  carrier transport, 245, 247–249
  electron drift velocity, 245

Lateral induced conductivity, 114–122 (see also Electron voltaic effect; specific materials)
Lattice scattering, 21–22
Leakage
  allowance in definition of gain, 104
  between diodes in SIT tubes, 281
  effect of contacts in thin films, 122
  in targets of electron devices, see Degradation phenomena
  in thin films, 104, 113
  in ZnO films, 117, 119
Lenard's law, 65
Lifetime of hot electrons in CdTe, 238, 239 (see also Carrier lifetime; Degradation phenomena)
Light flash excitation, comparison with transient EBC technique, 158–160
Limited space charge transits (LSC), 157
Localization radius, 172–173
Low-S surface, 280
Luminescence, 67
  physical models, 13, 16
  technique for measuring range–energy, 55–59

Mean free path
  carriers in low mobility solids, 32
  electrons in ZnS film, 102
Metallization, see Contacts
Microelectronics devices, see Scanning electron microscope
Microwave power tube, see Electron bombarded semiconductor
Microwave time-of-flight technique, 157–158, 192–194
Miller's law, 138
Mobility, 21–32 (see also specific materials)
  thickness-dependent, 172, 173
Molecular gas solids and liquids, 250–254
Motile trap model, 182–183
Multichannel array circuit, 272

N_2 (liquid and solid), transient EBC pulse shape, 251
N_2 (solid), electron and hole mobilities, 252
Ne (liquid), drift mobility of carriers, 247
Ne (solid), hole transport, 249
Negative differential mobility (NDM), 222, 223 (see also specific materials)
Neutralization of space charge, 177–178
Noble gases
  purifying methods, 245
  solids, hole–polaron interaction parameters, 247–250
  solids and liquids, 244–250
Noise sources
  in astronomy, 276
  in preamplifier, 323–325 (see also Signal/noise ratio)
Nuclear radiation, general properties, 77–78
Nuclear radiation detector, 305–311
  noise, 306
  pulse height, 305

O_2 (solid), electron and hole mobilities, 252, 253
Optical link, 42, 184

INDEX 345

Pair production energy for gases, 65 (*see also* Radiation ionization energy)
Path length of fast electrons, tables, 61
PbO, 100
Penetrating power of fast electrons, approximate, 48 (*see also* Range–energy)
Persistence characteristics of SIT tube, 282
Phosphor screen in CRT, 264
Phosphorescence, *see* Luminescence
Photocell using EVE, *see* Photosil
Photoconduction (Photoconductivity), 67–70
Photocurrent, saturation value, 69
Photoemissive layer, lateral EBC, 119
Photon counting, 267–268, 276–277
  with ICCD, 275
Photosil, 265–269
  pulse height distribution, 268
  target, 266
Photovoltaic effect, 70–71
Physical principles, 1–33
Pile-up, 268
p–n junction, 128–129
  grown, 137
  protection from degradation, 289
Point contacts, 125
Point defects
  in alkali halides (colour centres), 71–73
  in thermal equilibrium, 182–183
Polarization, 78–79 (*see also* Neutralization of space charge)
  in CdTe, 240
  integrated circuit capacitance, 304
Poole–Frenkel effect in CdTe, 239–240
Potential distribution
  in ionization chamber, 18–19 (*see also* Electric field)
  in solid with metal contacts, 19–20
Power supplies, 36–41
Primary current, definition, 2
Primary photocurrent, 69
Projected range, *see* Range–energy
Prompt component, 110–113 (*see also* Transient EBC)
Pulse generation by scanned electron beam, 190, 191–192

Pulse height analysis
  in nuclear spectroscopy, 305
  in photon counting, 267–268, 276–277
Pulse unit, 184
Pulsed EBC, 110–113 (*see also* Transient EBC)

Radar display, 319
Radial distribution function, 9–10
Radiation belts, 75–76
Radiation damage, 71–75 (*see also* Degradation phenomena)
  to Si junction cells by β-rays, 130
Radiation ionization energy, 66–67
  (see also specific materials)
Ramo's theorem, 325–326
Random phase model, 30
Range–energy
  alpha particles, 78
  charged particles in Si, 309
Range–energy (electrons), 43–65 (see also specific materials)
  chord range, 45, 79
  energy below 10 keV, 61–65
  extrapolated range, 53
  integrated circuit materials, 295
  penetrating power (practical range), 49–64
  scaling of variously defined ranges, 53, 56, 59–60
  tables, 53, 56, 61, 62, 295
  ultimate range, 54, 60
Rare gases, *see* Noble gases
Recombination
  bimolecular, 114
  bulk, experimental technique, 166–168
  centres, 13–16
  columnar, 107–108
  Langevin, 263
  Shockley–Read, 106
  surface, experimental technique, 168
  surface, in GaP diode, 140
Reduced mobility, *see* Trap controlled mobility
Resistive sea, 281
Resonance transfer, 92
Rise and decay of induced currents, *see* Pulsed EBC; Transient EBC

346 INDEX

S_8 (ions), structure, 259–260
S_8 (liquid), carrier transport, 259–260
S_8 (orthorhombic), 256–263
  band structure, 260–262
  carrier recombination, 262–263
  electron transport, 257–260
  hole transport, 182, 257
  interrupted transit, 170
  transient EBC pulse shape, 258
Saturation drift velocity, 23 (see also specific materials)
Scaling
  graph, 60
  tables, 53, 56
  variously defined electron ranges, 59
Scan converter tube
  using EBC target, 317–320
  using SIT target, 283
Scanning electron beam, Beamos tube, 316
Scanning electron microscope (SEM), 295–305
  analysis of semiconductor devices, 300–305
  contrast, 303
  for measuring EBC, 43
Scanning speed, 282, 323–324
Scattering of carriers, 21–23
Schottky barrier
  charge collection microscopy, 297–300
  contact to Si, technique, 296
  detector, 307, 309, 310
  EVE in, 125–128
Schottky defect, 68
Schubweg, 106, 147
  in low mobility solid, 32
Se (amorphous)
  carrier drift mobilities, 255, 256
  method of producing films, 254
  shallow electron trapping, 256
  thin film, 99–100
  transient induced conduction pulse shape, 174
  tube for measurements, 89
Se (grey)
  diffusion length of carriers, 133–134
  EVE gain in photoelement, 134
  fatigue, 129
  lateral EBC, 119
  photocell manufacture, 115

Se (monoclinic) (Se$_\alpha$)
  band structure, 261–262
  electron drift mobility, 255
Secondary EBC, 12, 103 (see also specific materials)
Secondary emission coefficient, 87
  tables, 322–323
Self compensation, 228–229, 235, 237
Semiconductor junction devices, 265–294
Si (amorphous), 208–222
  density of states distribution, 31, 209, 220
  electron transport, 211–218
  hole transport, 218–220
  preparation, 82–83
  transient EBC pulse shape, 210, 214, 218
Si (crystalline), 196–200 (see also Target)
  band gap, 16
  barrier potential, 298
  carrier diffusion length, 132–133, 297
  carrier diffusivity, 196–197
  carrier drift mobility, 198–199
  damage by impurity diffusion, 136, 299–300
  depletion layer thickness, 307
  donor and acceptor energies, 14
  EVE for back face electron bombardment, 136 (see also SIT camera tube)
  EVE of grown junction, 137
  EVE in p–n junction diodes, 131–132
  EVE in Schottky barriers, 126, 127
  EVE gain as function of oxide charge, 316
  high field carrier mobility, 199–200
  radiation ionization energy, 127
  range–energy for charged particles, 308
  range–energy for fast electrons, 61, 295
  target for storage tubes, 278, 283, 314–316
  transient EBC pulse shape, 189
Signal/noise ratio, 323–324
  Digicon, 271
  enhancement in transient EBC measurements, 164

INDEX 347

Signal/noise ratio (*cont.*)
  nuclear radiation detector, 306
  photon counting, 286
  photosil, 267
SiO_2
  fused silica, 113
  Graphechon target, 317
  integrated circuit capacitor, 293, 304
  radiation damage, 74–75
  thin film, EBC, 3, 88, 95, 101
  vacuum tube for measuring EBC, 89
SIT camera tube, 278–282
Slow component, 110–113 (*see also* Continuous time random walk; Pulsed EBC; ZnO; Delayed generation)
Slow scan television, 282
Small polaron, 26–27
Solar cell
  diffusion length of carriers, 132–133 (*see also* Photovoltaic effect)
  measurements in SEM, 297
Solid gases
  apparatus for measuring transient EBC, 186–188
  carrier transport, 244–254
Space charge
  neutralization, 177–178
  static, 166, 177
Space-charge-free transits (SCF), 144–151
Space-charge-limited current (SCL), 151, 152, 153
Space-charge-perturbed current (SCP), 151, 152, 153
Specimen
  thin film, 85
  transient EBC, 186
Specimen holder
  β-ray measurements, 35
  microwave time-of-flight, 193
  steady-state EBC, 37, 38, 40
  transient EBC, 185–186
Spencer theory of energy loss, 50–53
Stabilization potentials of floating surface, 87–88
Straggling of ranges, fast electrons, 48–49
Storage tube
  computer memory, 313–317
  scan conversion, 283, 317–320

Stored charge, 95
Substepping, 273
Surface layers, lateral EBC in, 115–119 (*see also* specific materials)
Surface recombination velocity, modulation by charges, 315
Switching effect in thin-film EBC, 109–110

Tail of transient induced current, 170–173 (*see also* Pulsed EBC; Delayed generation)
Tangents to determine transit time, 145, 165
Tantalum oxide, anodic films, 83
Target (*see also* Specimen)
  Beamos tube, 314
  Digicon, 270, 271, 272, 273
  Ebitron, 312
  EBS tubes, 286–291
  Evoscope, 284–285
  leakage, 281
  photosil, 266
  SIT camera tube, 278–279, 280–281
Television camera tube, *see* SIT camera tube; Ebicon
Temperature measurements, 186
Thin films, 80–122
  preparation, 80–84
  specimens for measurement, 84–85
Thomson–Whiddington–Bethe law, 44–65 (*see also* specific materials)
Threshold bombarding voltage, 93
Threshold field for NDM, 205, 222
Time dependence of EBC effect, *see* Pulsed EBC
Time-of-flight technique (ToF), *see* Transient EBC
Tracks of electrons, 45 (*see also* Range–energy)
Transient charge technique (TCT), *see* Transient EBC
Transient EBC, 141–194 (*see also* specific materials)
  history of technique, 142–143
  pulse shape, 145–157
  pulse shape, depleted diode, 154–155
  pulse shape, dispersive transport, 171–175

348 INDEX

Transit time, 149 (see also Transient EBC)
Transverse induced conductivity, see Lateral induced conductivity
Traps, 8, 144 (see also specific materials)
   carrier release time, experimental determination, 175–176
   density, experimental determination, 181
   depth, from transient EBC measurements, 179, 180–181
   distribution, effect on transient EBC pulse shape, 173–175
   thermal release rate, 11–12, 176
Trap-controlled mobility, 175, 180–181, 212
Two-layer dielectric, 319, 321–322

Uvicon, see Ebicon

Variable range hopping, see Hopping
Virgin state, measurements of properties, 143–144

Walden's rule, 30
Writing speed of SIT scan converter, 283

Xe (liquid), electron drift velocity, 246
Xe (liquid and solid), band gap and radiation ionization energy, 66
Xe (solid), carrier transport, 245, 247
X-rays
   efficiency of generation, 77
   target degradation by, 289
   transfer of excitation by, 89–92

ZnO
   decomposition, 73
   crystalline, drift mobility of electrons, 243–244
   surface layer, 117–118, 119
ZnS (crystalline)
   band gap, 16
   drift velocity of electrons, 242–243
ZnS (thin film)
   bulk charge storage, 102–103
   deposition, 98
   Ebitron target, 312
   electron beam contacts, 98
   gain as function of bias, 106
   Graphechon target, 317
   mobilization energy, 95
   purity of starting material, 113
   schubweg, 102
ZnSe, lateral EBC, 116